Subtelomeres

Edward J. Louis · Marion M. Becker
Editors

# Subtelomeres

Springer

*Editors*
Edward J. Louis
Marion M. Becker
Centre for Genetic Architecture of
Complex Traits
Department of Genetics
University of Leicester
Leicester
UK

ISBN 978-3-642-41565-4 ISBN 978-3-642-41566-1 (eBook)
DOI 10.1007/978-3-642-41566-1
Springer Heidelberg New York Dordrecht London

Library of Congress Control Number: 2013956423



Printed on acid-free paper

Springer is part of Springer Science+Business Media (www.springer.com)

*To Charlie*

# Preface

We have had an interest in subtelomeric regions for many years, first in yeast, then in parasites and pathogens and finally in all organisms with linear chromosomes. Over the years we have met many other people with an interest in subtelomeres in their favourite organism but it has taken a lot of time to bring such people together. After a few false starts at getting together at one conference or another, this book first was discussed at the 24th International Conference on Yeast Genetics and Molecular Biology 2009 in Manchester with Sabine Schwarz, a Life Sciences Editor from Springer. A few months later, discussions started in earnest about how a book on Subtelomeres would be structured and what would be in it. Then things slowed down due to other commitments, however we succeeded in obtaining funding from the Royal Society to hold a 2-day special interest meeting on Subtelomeres at their new conference centre near Milton Keynes, the Kavli Royal Society Centre at Chicheley Hall in January of 2011. This meeting with 16 attendees of diverse backgrounds, brought together people working on subtelomeres in many different organisms and from different perspectives. Many of those in attendance are contributors to the book, which really got its start after this meeting. Although we never did come to a definition of 'Subtelomere' that fit all cases, we did come to a better appreciation of the global nature of the issues, problems and genuine exciting nature of the regions near the ends of chromosomes. Like all academics, time is always in short supply and it has taken quite a long time to finally put the book together. Like the dynamic changing subtelomeres, players involved have also undergone dynamic change. Many people have moved institutions, some people have retired since the start, many originally keen contributors had to drop out due to their own commitments and even the editor has changed as we now have Anette Lindqvist at Springer handling the book. Like the technical challenges faced by the genomic analysis of subtelomeres, it must be frustrating to all involved and we can only thank everyone, especially the editors, for their patience. As the subtelomeric regions are ever changing, such a collection of articles on subtelomeres will evolve over time and we're sure there will be new additions and advances in future efforts.

# Contents

**1 Introduction** . . . . . . . . . . 1
Edward J. Louis

**2 Subnuclear Architecture of Telomeres and Subtelomeres in Yeast** . . . 13
Emmanuelle Fabre and Maya Spichal

**3 Subtelomeric Regions Promote Evolutionary Innovation of Gene Families in Yeast** . . . . . . . . . . 39
Tim Snoek, Karin Voordeckers and Kevin J. Verstrepen

**4 Subtelomere Organization, Evolution, and Dynamics in the Rice Blast Fungus *Magnaporthe oryzae*** . . . . . . . . . . 71
Mark Farman, Olga Novikova, John Starnes and David Thornbury

**5 *Pneumocystis carinii* Subtelomeres** . . . . . . . . . . 101
James R. Stringer

**6 Subtelomeres of *Aspergillus* Species** . . . . . . . . . . 117
Elaine M. Bignell

**7 *Trypanosoma brucei* Subtelomeres: Monoallelic Expression and Antigenic Variation** . . . . . . . . . . 137
Luisa M. Figueiredo and David Horn

**8 Human and Primate Subtelomeres** . . . . . . . . . . 153
M. Katharine Rudd

**9 FSHD: A Subtelomere-Associated Disease** . . . . . . . . . . 165
Andreas Leidenroth and Jane E. Hewitt

**10 Characterization of Chromosomal Ends on the Basis of Chromosome-Specific Telomere Variants and Subtelomeric Repeats in Rice (*Oryza sativa* L.)** . . . . . . . . . . 187
Hiroshi Mizuno, Jianzhong Wu and Takashi Matsumoto

**11 What is the Specificity of Plant Subtelomeres?** . . . . . . . . . . . . . . . . . . 195
A. V. Vershinin and E. V. Evtushenko

**12 Subtelomeres in *Drosophila* and Other Diptera** . . . . . . . . . . . . . . . . . . 211
James M. Mason and Alfredo Villasante

**13 Accumulation of Telomeric-Repeat-Specific Retrotransposons in Subtelomeres of *Bombyx mori* and *Tribolium castaneum*** . . . . . . . . 227
Haruhiko Fujiwara

**14 Subtelomere Plasticity in the Bacterium *Streptomyces*** . . . . . . . . . . . . 243
Annabelle Thibessard and Pierre Leblond

**15 Genomics of Subtelomeres: Technical Problems, Solutions and the Future** . . . . . . . . . . . . . . . . . . . . . . . . . . . . . . . . . . . 259
Marion M. Becker and Edward J. Louis

# Abbreviations

| | |
|---|---|
| 2L | Left arm of chromosome 2 |
| 2R | Right arm of chromosome 2 |
| 3L | Left arm of chromosome 3 |
| 3R | Right arm of chromosome 3 |
| 4L | Left arm of chromosome 4 |
| 4R | Right arm of chromosome 4 |
| ACP | Acyl carrier protein |
| Alu | Member of the SINE repeat family |
| ANT1 | Adenine nucleotide translocator |
| ApoL1 | Apolipoprotein L1 |
| ASF1A | Anti-silencing factor 1A |
| AUD | Amplifiable units of DNA |
| BAC | Bacterial artificial chromosome |
| BAH | Bromo-associated homology |
| Base J | Glucosylated version of thymidine |
| BES | Bloodstream expression site |
| BFB | Breakage-fusion bridge |
| BIR | Break-induced replication |
| bp | Base pair |
| C1TFA | Class I transcription factor A |
| CAF1-b | Chromatin assembly factor 1-b |
| cDNA | Complementary DNA |
| CDS | Coding sequence |
| ChIP | Chromatin immunoprecipitation |
| CNV | Copy number variation |
| CO-FISH | Chromosome orientation FISH |
| Crje | Conserved recombination junction element |
| DAC1/DAC3 | Histone deacetylase 1/3 |
| DHN | Dihydroxynapthalene |
| DMAT | Dimethylallyl tryptophan synthases |
| DNA | Deoxyribonucleic acid |
| DNMT3b | DNA methyltransferase 3b |

| | |
|---|---|
| DOT1B | Disruptor of telomeric silencing 1B |
| DRC | D4Z4 repressor complex |
| DSB | Double-strand break |
| DUX4 | Double homeobox 4 |
| EAA | Extrinsic allergic alveolitis |
| EHMT1 | Euchromatic histone methyltransferase 1 |
| ELP3b | Elongator protein 3b |
| ES | Expression site |
| ESAG | Expression site associated gene |
| ESB | Expression site body |
| EST | Expressed sequence tag |
| FACS | Fluorescence-activated cell sorting |
| FACT | Facilitates chromatin transcription |
| FISH | Fluorescence in situ hybridization |
| FRG1 | FSHD region gene 1 |
| FRG2 | FSHD region gene 2 |
| FSHD | Facioscapulohumeral dystrophy |
| GC | Guanine cytosine base pair |
| GFP | Green fluorescent protein |
| GOC | Gene order conservation |
| H3K9me3 | Histone 3 lysine 9 tri-methylation |
| HAT1 | Histone acetyltransferase 1 |
| HDAC | Histone deacetylase |
| HMG-CoA | 3-Hydroxy 3-methylglutaryl CoA |
| HMGB2 | High-mobility group B2 |
| HR | Homologous recombination |
| HTT | Retrotransposons HeT-A, TART and TAHRE |
| ICF | Immunodeficiency-centromeric instability facial anomalies |
| IPA | Invasive pulmonary aspergillosis |
| IS | Insertion sequence |
| ISWI | Imitation switch |
| ITS | Interstitial telomere sequences |
| kb | Kilobase |
| kbp | Kilobase pairs |
| L | Long chromosome arm |
| LINE | Long interspersed nuclear element |
| LTR | Long terminal repeat |
| Mbp | Megabase pairs |
| MES | Metacyclic VSG expression site |
| MSG | Major surface glycoprotein |
| MSR | Major-surface-glycoprotein related |
| Mya | Millions of years ago |
| NHEJ | Non-homologous end joining |
| NLP | Nucleoplasmin-like protein |

| | |
|---|---|
| NLR | Non-LTR retrotransposons |
| NMP | Nuclear matrix protein |
| NOR | Nucleus organizer region or nucleolar organizer |
| NPC | Nuclear pore complex |
| NRPS | Non-ribosomal peptide synthases |
| OR | Olfactory receptor |
| Orf | Open reading frame |
| QTL | Quantitative trait locus |
| PAC | P1-derived artificial chromosome |
| PcG | Polycomb group |
| PCR | Polymerase chain reaction |
| PDLIM3 | PDZ and LIM domain protein 3 |
| PEV | Position-effect variegation |
| PFGE | Pulse-field gel electrophoresis |
| PICh | Proteomics of isolated chromatin fragments |
| PKS | Polyketide synthases |
| Pol I | RNA polymerase I |
| Pol II | RNA polymerase II |
| Poly(G) | DNA sequence comprised of a run of guanosine residues |
| Prt1 | Protease |
| qPCR | Quantitative real-time PCR |
| RAD51 | Recombinase |
| RAP1 | Repressor/activator protein 1 |
| rDNA | Ribosomal DNA |
| REL | Restriction enzyme-like |
| RFLP | Restriction fragment length polymorphism |
| RNA | Ribonucleic acid |
| RPA | RNA polymerase A (Pol I) |
| RPB | RNA polymerase B (Pol II) |
| rRNA | Ribosomal RNA |
| RT-PCR | Reverse-transcription PCR |
| S | Short chromosome arm |
| SCE | Sister chromatid exchange |
| SD | Segmental duplication |
| SHANK3 | SH3 and multiple ankyrin repeat domain 3 |
| SINE | Short interspersed nuclear element |
| SIR2rp1 | Silent information regulator related protein 1 |
| SNP | Single nucleotide polymorphism |
| SPB | Spindle pole body |
| SRA | Serum resistance-associated gene |
| STR | Subtelomeric Repeats |
| T-loops | Telomere loops |
| TAR | Transformation associated recombination |
| TAS | Telomere-associated sequence |

| | |
|---|---|
| TATR | Telomere-associated tandem repeat |
| TE | Transposable element |
| TERRA | Telomeric repeat-containing RNA |
| TIR | Terminal inverted repeats |
| TLH | Telomere linked helicase |
| TP | Terminal proteins |
| TPE | Telomeric position effect |
| TR | Tandem repeat, or telomere repeat |
| TRAP | Telomeric repeat amplification protocol |
| TRF | Telomeric restriction fragment |
| TUBB4Q | Tubulin, beta polypeptide member 4 |
| UCS | Upstream conserved sequence |
| UTR | Untranslated region |
| VNTR | Variable number of tandem repeats |
| VSG | Variant surface glycoprotein |
| XCR | X element combinatorial repeats |
| XL | Left arm of chromosome X |
| XR | Right arm of chromosome X |
| YAC | Yeast artificial chromosome |
| YY1 | Ying Yang 1 |
| $\gamma$-H2AX | Gamma histone H2A family, member X |

# Chapter 1
# Introduction

**Edward J. Louis**

**Abstract** Dynamic, polymorphic, problematic yet intriguing are the general view of the genomic region near the ends of chromosomes. Unlike the generally conserved caps of chromosome ends, the telomeres, for which our understanding of their biology has advanced greatly in recent years, the adjacent telomere-associated sequences (TAS) or subtelomeres remain an enigma. This is in large part due to the technical difficulties in working with repeated sequence regions of the genome both experimentally and in genome projects. The repetitive nature makes it difficult to observe a signal from a specific chromosome end among the noise of all the other ends that look and behave similarly. It also has precluded complete assembly of the regions in genome projects. In virtually all eukaryotes and some prokaryotes, linear chromosomes have dynamic and polymorphic subtelomeric regions. In many cases, a great deal of important biology of the organism is encoded in and regulated by the subtelomeric regions. One generality is that the region tends to encode for genes involved in interacting with the extracellular environment though this is not universal. Recombination, chromatin, gene density, and other properties of the region differ from those of the core of the genome in many organisms, though the specific differences vary between organisms. Perhaps the most well-understood subtelomeres are in the budding yeast *Saccharomyces cerevisiae*, while the epitome of adaptive use of the properties of the region is found in parasites, such as *Plasmodium falciparum* and *Trypanosoma brucei*, causing malaria and sleeping sickness. These parasites utilize the region to escape their hosts' immune systems through generation of diversity and exquisite control of surface antigen expression. A great deal has been learned from comparison between subtelomeres in different organisms, and the interest in subtelomeres is growing. This book does not cover every aspect of subtelomeres in every organism where they are studied, but

E. J. Louis (✉)
Department of Genetics, Centre for Genetic Architecture of Complex Traits,
University of Leicester, Leicester LE1 7RH, UK
e-mail: ejl21@le.ac.uk

E. J. Louis and M. M. Becker (eds.), *Subtelomeres*, DOI: 10.1007/978-3-642-41566-1_1, 

provides a broad coverage of the field of subtelomeres in diverse organisms from bacteria to yeast and fungi through plants, insects, parasites, and humans. It should serve as an entry point into the field, hopefully generating an interest in this fascinating region of genomes.

## 1.1 Introduction

### *1.1.1 History of Subtelomeres*

Perhaps the earliest recognition of interesting repeated sequences at chromosome termini came from observations in yeast (Carlson et al. 1985; Chan and Tye 1983a, b; Horowitz and Haber 1984; Horowitz et al. 1984) and insects in the mid-1980s (Saiga and Edstrom 1985; Young et al. 1983). Soon thereafter, the regions near the ends of chromosomes were being described in *Plasmodium falciparum* (Corcoran et al. 1988; Dore et al. 1990; Pace et al. 1990, 1995) and humans (Brown et al. 1990; Cross et al. 1990; Riethman et al. 1989), other yeasts (De Las Penas et al. 2003; Fairhead and Dujon 2006), fungi (Farman and Leong 1995; Galagan et al. 2005; Underwood et al. 1996), other parasites such as Giardia (Adam et al. 1991; Le et al. 1991; Prabhu et al. 2007; Upcroft et al. 1997) and trypanosomes (Becker et al. 2004; Crozatier et al. 1990; Hertz-Fowler et al. 2008), other insects (Biessmann et al. 1998; Okazaki et al. 1995; Roth et al. 1997; Takahashi et al. 1997), plants (Ganal et al. 1991, 1992; Kuo et al. 2006; Sykorova et al. 2003, 2006; Vershinin et al. 1995), and even linear mitochondria (Fukuhara et al. 1993; Nosek et al. 1995; Nosek and Tomaska 2003; Tomaska et al. 2004) and bacteria (Fischer et al. 1997; Hinnebusch et al. 1990; Kitten and Barbour 1990; Leblond et al. 1996; Restrepo et al. 1992). Even in specific cases where the telomeres were not the canonical telomerase-maintained G-rich repeats (Biessmann et al. 1990, 1998; Roth et al. 1997; Saiga and Edstrom 1985; Sykorova et al. 2003, 2006), there are subtelomeric repeat structures similar to other organisms. A comparison of what was known by the mid-1990s leads to the realization that there was a general phenomenon of a mosaic of repeated elements in the subtelomeric regions of most eukaryotes and that this region allowed the evolution of use for adaptive purposes as well as for non-telomerase maintenance of the chromosome ends (Barry et al. 2003; Mefford and Trask 2002; Pryde et al. 1997; Scherf et al. 2001). In the years since, there have been advances in the structural and functional analysis of the subtelomeric regions in many organisms, but the big questions and the main framework were in place by the mid-1990s. Figure 1.1 displays the basic structures of subtelomeric regions in many organisms and sets of organisms.

The big questions still are: How can a genome maintain a stable core over many generations yet tolerate a dynamic plastic region that undergoes frequent changes? Is the nature of the subtelomeric region due to the vicinity of the end of

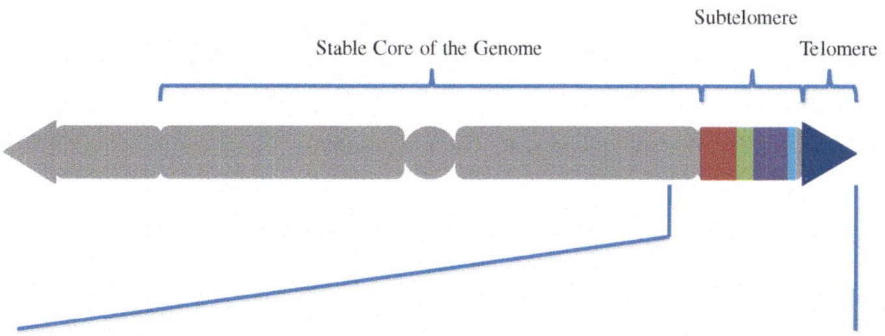


| Organism | long homology blocks | extracellular genes | VNTRs | helicase | TEs | rDNA | TRs |
|---|---|---|---|---|---|---|---|
| Saccharomyces | Yes | Yes | Yes | Yes | Yes | No | Yes |
| Other yeasts | Yes | Yes | Yes | Yes | Yes | Yes | Yes |
| Human | Yes | Yes | Yes | No | Yes | Yes | Yes |
| Pneumocyctis | Yes | Yes | Yes | ? | ? | No | Yes |
| Fungi | Yes | Yes | Yes | Yes | Yes | Yes | Yes |
| Drosophila | No | ? | Yes | No | Yes | No | No |
| Bombyx | No | ? | No | No | Yes | No | Yes |
| Chironomus | No | ? | Yes | ? | ? | ? | No |
| Anopheles | No | ? | Yes | ? | ? | No | No |
| Giardia | Yes | Yes | Yes | Yes | No | Yes | Yes |
| Plasmodium | Yes | Yes | Yes | No | Yes | No | Yes |
| Trypanosoma | Yes | Yes | Yes | No | Yes | No | Yes |
| Alliaceae | Yes | ? | Yes | ? | ? | No | No |
| Plants | Yes | ? | Yes | ? | Yes | No | Yes |

**Fig. 1.1** The generic structure of eukaryotic chromosomes with subtelomeric details of several organisms (see text for references). The core of genomes is generally quite stable over long periods of time. This is flanked by the chromosome ends, which are composed of the subtelomeric region followed by the actual telomere, usually telomere repeats (*TRs*) maintained by telomerase. The subtelomeres are generally large blocks of homology shared by more than one chromosome end, which can vary between individuals. Within these large blocks of homology can be genes encoding proteins utilized for interacting with the environment. In yeast, these are enzymes and transporters for carbon source utilization. In humans, these include olfactory receptor genes. Many fungi have secreted proteins that are involved in virulence, and *Pneumocystis carinii* has a major surface glycoprotein analogous to the surface antigens of parasites. The parasites using antigenic variation to evade their host immune systems encode many if not all of the variable surface antigens in the subtelomeres. In insects and plants, it is not known in general whether there are such genes concentrated in the region. Virtually, all subtelomeres have variable number of tandem repeats (*VNTRs*), and in some cases, they function as the telomere (*Chironomus*, *Anopheles*, and *Alliaceae*). Most subtelomeres have transposable elements (*TEs*), some subtelomere specific, such as Ty5 in yeast, and in some cases, they function as the telomere (*Drosophila*). Many yeasts and fungi have subtelomeric-specific helicases of the RecQ family, though the Y'-helicase of yeast is of a different family. The functions of these helicases are unknown for the most part. In many organisms, the rDNA arrays are found in the subtelomeres

the chromosome? Are adaptive uses of the region such as generation of diversity and control of gene expression a consequence of the dynamics of the region? How does the subtelomeric region facilitate generation of diversity? How is the expression of these diverse gene families controlled?

Progress was, and continues to be, hampered by the technical difficulties encountered. The most well-characterized eukaryotic genome is that of the budding yeast *Saccharomyces cerevisiae*, which was the first eukaryotic genome completely sequenced (Goffeau et al. 1996). It stands to this day as one of the few sequenced from end to end, in large part due to the efforts to uniquely tag and clone each chromosome end for sequence, assembly, and mapping onto the core of the genome (Louis and Borts 1995). No other yeast strain has been 'completely' assembled despite the large efforts in their sequencing (Liti et al. 2009; Novo et al. 2009). There are a few other examples of 'complete' genomes, and in every case, special effort has been required to complete the subtelomeric regions.

### *1.1.2 What is a Subtelomere?*

The definition of subtelomeres is not straightforward and in some respects is organism specific. In some cases, the subtelomeres have lower gene density than the core of the genome, while in others, it is higher; some organisms have a different chromatin structure in the regions compared the rest of their genomes, some have less recombination in the region while others have more, and most organisms have more multi-gene families embedded in the region than the rest of the genome, but not all. There are no universal generalities, but there is agreement on the existence of a genomic domain near the ends of chromosomes that differs from the rest of the genome in some fundamental ways. A basic definition can be derived from the comparisons of chromosome structure in different organisms as shown in Fig. 1.1. The subtelomeric region can be considered the sequences between the core of the genome and the telomere where the core is conserved in structure between individuals within a species and even between close species relatives in many cases. Beyond this core, polymorphisms are observed between individuals. The regions can vary in size from a few kilobases to many hundreds of thousands of bases. In most cases, different subtelomeres from different chromosome ends will share homologies. On top of this observed property of polymorphism, subtelomeric regions generally display any of a number of additional properties such as copy number variation, dynamic exchange between different subtelomeres, diversity in repeated sequence families, chromatin differences over the whole region and/or adjacent to the telomere, recombination differences over the whole region and/or adjacent to the telomere, gene density differences, replication differences, and strand biases in base composition.

## 1.2 Dynamics

### *1.2.1 Short-Term Changes and Generation of Diversity*

The polymorphism observed in general between individuals at specific chromosome ends indicated that the region must undergo frequent change through recombination and other processes. From the earliest observations in the regions, it became clear that subtelomeres were more dynamic than the rest of the genome with frequent exchanges between homologous but non-allelic locations. In yeast, this was observed in meiosis despite the reduced recombination in the region (Horowitz et al. 1984) as well as in mitosis (Louis and Haber 1990a; Louis et al. 1994). In Plasmodium chromosome, length polymorphisms were attributed to subtelomeric recombination (Corcoran et al. 1988; Dore et al. 1994; Pace et al. 1990; Ponzi et al. 1992). Structural analysis of human subtelomeres was consistent with exchanges between different chromosome ends (Flint et al. 1997; Mefford and Trask 2002; Riethman et al. 2005; Trask et al. 1998). At the gene level, diversity within copies of a multi-gene family can be formed by recombination between members as has been observed in yeast (Charron et al. 1989), fungi (Wada and Nakamura 1996), and parasites (Barry et al. 2003; Freitas-Junior et al. 2000; Horn and Barry 2005; Scherf et al. 2001) as well as in bacteria where antigenic variation analogous to eukaryotic parasites occurs (Donelson 1995; Restrepo et al. 1992; Saint and Barbour 1991).

This dynamic exchange does not seem compatible with overall genome stability. The hypothesis that perhaps the subtelomeric regions are sequestered from the rest of the genome, allowing a plastic dynamic domain while retaining genome integrity, was developed (Pryde et al. 1997; Pryde and Louis 1997) to explain this paradox. One possibility was that in yeast at least, the tethering of telomeres to the nuclear periphery prevented interactions between homologous sequences residing in both the core of the genome and the subtelomeric region while allowing such interactions between subtelomeric copies (Pryde and Louis 1997). This model is supported by mutations that increase levels of recombination between internal and subtelomeric repeats (Marvin et al. 2009a, b), which also results in the loss of tethering to the nuclear periphery (Laroche et al. 1998). How generalizable this is to other organisms remains to be determined; however, there are structural similarities consistent with subtelomeric domains being in different recombinational compartments between yeast and humans (Flint et al. 1997), and it is likely that the recombination generating antigenic diversity in parasites is sequestered (Barry et al. 2003; Scherf et al. 2001).

### *1.2.2 Long-Term Dynamics and Evolution*

The long-term consequences of the short-term dynamics result in a rapid evolution of the region such that closely related species may not share much similarity in their subtelomeres yet be very similar in the core genome. This can be seen in the *Trypanosomatids* where there is conservation of the core genomes between the three originally sequenced species, but no homology between the species in the subtelomeres (El-Sayed et al. 2005).

Perhaps a more interesting long-term effect is the continued generation of diversity among members of a gene family allowing the generation and testing of new variants, which may confer novel functionality, without the loss of the original members and their function. This is likely to be the process behind the generation of families of genes related to but with different functions from the surface antigens used for antigenic variation in both *Trypanosoma brucei* (Barry et al. 2003; Horn and Barry 2005) and *P. falciparum* (Freitas-Junior et al. 2000; Scherf et al. 2001), the long-term dynamics of olfactory receptor gene diversity in humans (Mefford et al. 2001; Trask et al. 1998) and of carbon source utilization in *S. cerevisiae* (Brown et al. 2010; Charron et al. 1989).

## 1.3 Epigenetics

The study of the epigenetics of subtelomeres exploded in the 1990s after the rediscovery of position-effect variegation in yeast (Gottschling et al. 1990). A marker inserted adjacent to a telomere exhibited variegated expression with some cells in a clone expressing the gene and others not. The expression state was metastable and would switch to the opposite state after several mitotic generations. Ironically, this telomere position effect was found after deleting all the subtelomeric sequences from a specific chromosome end. This was done in order to solve the problem of observing effects at a specific telomere without the complications of noise from the shared repeats at other chromosome ends. Studies of the natural subtelomeres of yeast at the same time using the same marker did not observe variegated expression when the marker was embedded in various subtelomeric sequences (Louis and Haber 1990a, b). Eventually, this difference was reconciled with the determination of the domains of transcriptional repression in yeast subtelomeres, which were limited in extent (Pryde and Louis 1999). Such repression of gene expression near telomeres has now been observed in many organisms from other fungi (Castano et al. 2005; De Las Penas et al. 2003) and parasites (Horn and Barry 2005) to humans (Baur et al. 2001). It is clear though that the exquisite control of gene expression used in some parasites is more than just TPE (Alsford et al. 2007), and so there is still more to learn about the biology of the region.

## 1.4 Genomics

In the early days of genomics and whole-genome shotgun sequencing, the telomere regions and subtelomeres were underrepresented in the genomic libraries constructed (Becker et al. 2004), leaving gaps in the genome assemblies. Various efforts have gone into filling those gaps, sometimes with a great deal of time and effort beyond that of the rest of the genome. In some cases such as yeast, each end can be uniquely marked facilitating the cloning and sequencing of individual subtelomeres (Louis and Borts 1995). Other efforts have gone into telomere enrichment protocols that also enrich the adjacent subtelomeres. Even with clones of the regions, there can be problems with assembly and with mapping onto the existing contigs of individual chromosomes due to the repetitive nature of the regions [in *T. brucei* for example (Hertz-Fowler et al. 2008)]. This is still a problem today with second-generation sequencing even though read depth for the regions has improved.

A more problematic genomic issue is attributing phenotype to genetic variation in complex traits. Depending on the organism and the deficiencies in knowledge of the subtelomeric regions and diversity in populations and the trait of interest, the mapping of causal genetic variants in linkage studies or genome-wide association studies will be hampered to a greater or lesser extent. In the example of yeast, where a population genomic survey of genetic diversity has been correlated with a large number of varying phenotypes, the mapping of quantitative trait loci (QTL) for a given phenotype is incomplete. Approximately 8 % of any given yeast strain's genome is unassembled subtelomeric regions, which contain approximately 25 % of QTLs for a given trait as the QTLs map beyond the last-known assembled sequences (Cubillos et al. 2011; Liti and Louis 2012). This is likely to be a big unrecognized problem in human disease studies as well as breeding programs for agriculture.

## 1.5 Conclusions and Outlook

The subtelomeric regions of linear chromosomes, wherever they are found, are fascinating subjects for study, and over the past 30 years, their importance has become increasingly brought to the attention of most biologists. They represent the last frontier of individual genome projects and will become even more important in the study of complex traits through their analysis in populations of individuals. The dynamics, evolution, and adaptive use of the regions will continue to be the focus of study in many organisms. This collection provides an introduction to the field, and some of the state-of-the-art studies currently being undertaken. It is not all encompassing nor comprehensive; for example, there is only one parasite chapter; however, all the major groups are covered and the obvious omissions, such as *P. falciparum* and other parasites, are covered well in recent reviews (Barry et al. 2012; Guizetti and Scherf 2013; Hayashida et al. 2012; Moraes Barros et al. 2012; Witmer et al. 2012). Despite the technical difficulties in studying the region, the future is very promising for our continued understanding and appreciation of subtelomeres and their importance to biology.

We start with the yeast *S. cerevisiae*, where much of the field got its start with chapters on the subnuclear architecture of subtelomeres (and telomeres), which addresses the functional consequences of specific localization of chromosome ends (Chap. 2), and on the evolutionary consequences of generation of diversity and recombination in subtelomeric gene families (Chap. 3). From here, we move onto fungi starting with a plant pathogen, *Magnaporthe oryzae* (Chap. 4), a human pathogen, *Pneumocystis carinii* (Chap. 5), and the *Aspergillus* species, which comprise plant, human, and non-pathogens (Chap. 6). From here, we have one example of a parasite, *T. brucei*, and how antigenic variation and monoallelic expression are under subtelomeric influence (Chap. 7). Human subtelomeres are dealt first in comparison with other primates (Chap. 8) and in association with a particular disease, FSHD (Chap. 9). Plant subtelomeres are covered in rice (Chap. 10) and in rye (Chap. 11). Insects are covered in the next two chapters, first those without canonical telomeres, *Drosophila* and other diptera (Chap. 12), and second those with telomerase-maintained telomeres, *Bombyx mori* and *Tribolium castaneum* (Chap. 13). In the next chapter, the subtelomere dynamics of linear chromosomes in the bacterium *Streptomyces* are discussed (Chap. 14). All of these organisms have interesting subtelomere properties with species-specific interest but also general properties that can inform subtelomere biology in other organisms. The last chapter deals with the genomics of subtelomeres and the various problems encountered with possible solutions (Chap. 15).

## References


Adam, R. D., Nash, T. E., & Wellems, T. E. (1991). Telomeric location of Giardia rDNA genes. *Molecular and Cellular Biology, 11*, 3326–3330.

Alsford, S., Kawahara, T., Isamah, C., & Horn, D. (2007). A sirtuin in the African trypanosome is involved in both DNA repair and telomeric gene silencing but is not required for antigenic variation. *Molecular Microbiology, 63*, 724–736.

Barry, J. D., Ginger, M. L., Burton, P., & McCulloch, R. (2003). Why are parasite contingency genes often associated with telomeres? *International Journal for Parasitology, 33*, 29–45.

Barry, J. D., Hall, J. P., & Plenderleith, L. (2012). Genome hyperevolution and the success of a parasite. *Annals of the New York Academy of Sciences, 1267*, 11–17.

Baur, J. A., Zou, Y., Shay, J. W., & Wright, W. E. (2001). Telomere position effect in human cells. *Science, 292*, 2075–2077.

Becker, M., Aitcheson, N., Byles, E., Wickstead, B., Louis, E., & Rudenko, G. (2004). Isolation of the repertoire of VSG expression site containing telomeres of *Trypanosoma brucei* 427 using transformation-associated recombination in yeast. *Genome Research, 14*, 2319–2329.

Biessmann, H., Carter, S. B., & Mason, J. M. (1990). Chromosome ends in *Drosophila* without telomeric DNA sequences. *Proceedings of the National Academy of Sciences of the United States of America, 87*, 1758–1761.

Biessmann, H., Kobeski, F., Walter, M. F., Kasravi, A., & Roth, C. W. (1998). DNA organization and length polymorphism at the 2L telomeric region of *Anopheles gambiae*. *Insect Molecular Biology, 7*, 83–93.

Brown, C. A., Murray, A. W., & Verstrepen, K. J. (2010). Rapid expansion and functional divergence of subtelomeric gene families in yeasts. *Current Biology, 20*, 895–903.

Brown, W. R., Mac, K. P., Villasante, A., Spurr, N., Buckle, V. J., & Dobson, M. J. (1990). Structure and polymorphism of human telomere-associated DNA. *Cell, 63*, 119–132.

Carlson, M., Celenza, J. L., & Eng, F. J. (1985). Evolution of the dispersed *SUC* gene family of *Saccharomyces* by rearrangements of chromosome telomeres. *Molecular and Cellular Biology, 5*, 2894–2902.

Castano, I., Pan, S. J., Zupancic, M., Hennequin, C., Dujon, B., & Cormack, B. P. (2005). Telomere length control and transcriptional regulation of subtelomeric adhesins in *Candida glabrata*. *Molecular Microbiology, 55*, 1246–1258.

Chan, C. S. M., & Tye, B.-K. (1983a). A family of *Saccharomyces cerevisiae* repetitive autonomously replicating sequences that have very similar genomic environments. *Journal of Molecular Biology, 168*, 505–523.

Chan, C. S. M., & Tye, B.-K. (1983b). Organization of DNA sequences and replication origins at yeast telomeres. *Cell, 33*, 563–573.

Charron, M. J., Read, E., Haut, S. R., & Michels, C. A. (1989). Molecular evolution of the telomere-associated *MAL* loci of *Saccharomyces*. *Genetics, 122*, 307–316.
Corcoran, L. M., Thompson, J. K., Walliker, D., & Kemp, D. J. (1988). Homologous recombination within subtelomeric repeat sequences generates chromosome size polymorphisms in *P. falciparum*. *Cell, 53*, 807–813.
Cross, S., Lindsey, J., Fantes, J., McKay, S., McGill, N., & Cooke, H. (1990). The structure of a subterminal repeated sequence present on many human chromosomes. *Nucleic Acids Research, 18*, 6649–6657.
Crozatier, M., Van, D. P. L., Johnson, P. J., Gommers, A. J., & Borst, P. (1990). Structure of a telomeric expression site for variant specific surface antigens in *Trypanosoma brucei*. *Molecular and Biochemical Parasitology, 42*, 1–12.
Cubillos, F. A., Billi, E., Zorgo, E., Parts, L., Fargier, P., Omholt, S., et al. (2011). Assessing the complex architecture of polygenic traits in diverged yeast populations. *Molecular Ecology, 20*, 1401–1413.
De Las Penas, A., Pan, S. J., Castano, I., Alder, J., Cregg, R., & Cormack, B. P. (2003). Virulence-related surface glycoproteins in the yeast pathogen *Candida glabrata* are encoded in subtelomeric clusters and subject to RAP1- and SIR-dependent transcriptional silencing. *Genes & Development, 17*, 2245–2258.
Donelson, J. E. (1995). Mechanisms of antigenic variation in *Borrelia hermsii* and African trypanosomes. *Journal of Biological Chemistry, 270*, 7783–7786.
Dore, E., Pace, T., Ponzi, M., Picci, L., & Frontali, C. (1990). Organization of subtelomeric repeats in *Plasmodium berghei*. *Molecular and Cellular Biology, 10*, 2423–2427.
Dore, E., Pace, T., Picci, L., Pizzi, E., Ponzi, M., & Frontali, C. (1994). Dynamics of telomere turnover in *Plasmodium berghei*. *Molecular Biology Reports, 20*, 27–33.
El-Sayed, N. M., Myler, P. J., Blandin, G., Berriman, M., Crabtree, J., Aggarwal, G., et al. (2005). Comparative genomics of trypanosomatid parasitic protozoa. *Science, 309*, 404–409.
Fairhead, C., & Dujon, B. (2006). Structure of *Kluyveromyces lactis* subtelomeres: Duplications and gene content. *FEMS Yeast Research, 6*, 428–441.
Farman, M. L., & Leong, S. A. (1995). Genetic and physical mapping of telomeres in the rice blast fungus, *Magnaporthe grisea*. *Genetics, 140*, 479–492.
Fischer, G., Kyriacou, A., Decaris, B., & Leblond, P. (1997). Genetic instability and its possible evolutionary implications on the chromosomal structure of *Streptomyces*. *Biochimie, 79*, 555–558.
Flint, J., Bates, G. P., Clark, K., Dorman, A., Willingham, D., Roe, B. A., et al. (1997). Sequence comparison of human and yeast telomeres identifies structurally distinct subtelomeric domains. *Human Molecular Genetics, 6*, 1305–1313.
Freitas-Junior, L. H., Bottius, E., Pirrit, L. A., Deitsch, K. W., Scheidig, C., Guinet, F., et al. (2000). Frequent ectopic recombination of virulence factor genes in telomeric chromosome clusters of *P. falciparum*. *Nature, 407*, 1018–1022.
Fukuhara, H., Sor, F., Drissi, R., Dinouel, N., Miyakawa, I., Rousset, S., et al. (1993). Linear mitochondrial DNAs of yeasts: Frequency of occurrence and general features. *Molecular and Cellular Biology, 13*, 2309–2314.
Galagan, J. E., Calvo, S. E., Cuomo, C., Ma, L. J., Wortman, J. R., Batzoglou, S., et al. (2005). Sequencing of *Aspergillus nidulans* and comparative analysis with *A. fumigatus* and *A. oryzae*. *Nature, 438*, 1105–1115.
Ganal, M. W., Lapitan, N. L., & Tanksley, S. D. (1991). Macrostructure of the tomato telomeres. *Plant Cell, 3*, 87–94.
Ganal, M. W., Broun, P., & Tanksley, S. D. (1992). Genetic mapping of tandemly repeated telomeric DNA sequences in tomato (*Lycopersicon esculentum*). *Genomics, 14*, 444–448.
Goffeau, A., Barrell, B. G., Bussey, H., Davis, R. W., Dujon, B., Feldmann, H., et al. (1996). Life with 6000 genes. *Science, 274*, 546, 563–567.
Gottschling, D. E., Aparicio, O. M., Billington, B. L., & Zakian, V. A. (1990). Position effect at *S. cerevisiae* telomeres: Reversible repression of Pol II transcription. *Cell, 63*, 751–762.
Guizetti, J., & Scherf, A. (2013). Silence, activate, poise and switch! Mechanisms of antigenic variation in *Plasmodium falciparum*. *Cellular Microbiology, 15*, 718–726.

Hayashida, K., Hara, Y., Abe, T., Yamasaki, C., Toyoda, A., Kosuge, T., et al. (2012). Comparative genome analysis of three eukaryotic parasites with differing abilities to transform leukocytes reveals key mediators of Theileria-induced leukocyte transformation. *MBio, 3*, e00204–e00212.

Hertz-Fowler, C., Figueiredo, L. M., Quail, M. A., Becker, M., Jackson, A., Bason, N., et al. (2008). Telomeric expression sites are highly conserved in *Trypanosoma brucei*. *PLoS ONE, 3*, e3527.

Hinnebusch, J., Bergstrom, S., & Barbour, A. G. (1990). Cloning and sequence analysis of linear plasmid telomeres of the bacterium *Borrelia burgdorferi*. *Molecular Microbiology, 4*, 811–820.

Horn, D., & Barry, J. D. (2005). The central roles of telomeres and subtelomeres in antigenic variation in African trypanosomes. *Chromosome Research, 13*, 525–533.

Horowitz, H., & Haber, J. E. (1984). Subtelomeric regions of yeast chromosomes contain a 36 base-pair tandemly repeated sequence. *Nucleic Acids Research, 12*, 7105–7121.

Horowitz, H., Thorburn, P., & Haber, J. E. (1984). Rearrangements of highly polymorphic regions near telomeres of *Saccharomyces cerevisiae*. *Molecular and Cellular Biology, 4*, 2509–2517.

Kitten, T., & Barbour, A. G. (1990). Juxtaposition of expressed variable antigen genes with a conserved telomere in the bacterium *Borrelia hermsii*. *Proceedings of the National Academy of Sciences USA, 87*, 6077–6081.

Kuo, H. F., Olsen, K. M., & Richards, E. J. (2006). Natural variation in a subtelomeric region of *Arabidopsis*: Implications for the genomic dynamics of a chromosome end. *Genetics, 173*, 401–417.

Laroche, T., Martin, S. G., Gotta, M., Gorham, H. C., Pryde, F. E., Louis, E. J., et al. (1998). Mutations of yeast Ku genes disrupts the subnuclear organization of telomeres. *Current Biology, 8*, 653–656.

Le, B. S., Korman, S. H., & Van, D. P. L. (1991). Frequent rearrangements of rRNA-encoding chromosomes in *Giardia lamblia*. *Nucleic Acids Research, 19*, 4405–4412.

Leblond, P., Fischer, G., Francou, F. X., Berger, F., Guerineau, M., & Decaris, B. (1996). The unstable region of *Streptomyces ambofaciens* includes 210 kb terminal inverted repeats flanking the extremities of the linear chromosomal DNA. *Molecular Microbiology, 19*, 261–271.

Liti, G., Carter, D. M., Moses, A. M., Warringer, J., Parts, L., James, S. A., et al. (2009). Population genomics of domestic and wild yeasts. *Nature, 458*, 337–341.

Liti, G., & Louis, E. J. (2012). Advances in quantitative trait analysis in yeast. *PLoS Genetics, 8*, e1002912.

Louis, E. J., & Borts, R. H. (1995). A complete set of marked telomeres in *Saccharomyces cerevisiae* for physical mapping and cloning. *Genetics, 139*, 125–136.

Louis, E. J., & Haber, J. E. (1990a). Mitotic recombination among subtelomeric Y' repeats in *Saccharomyces cerevisiae*. *Genetics, 124*, 547–559.

Louis, E. J., & Haber, J. E. (1990b). The subtelomeric Y' repeat family in *Saccharomyces cerevisiae*: An experimental system for repeated sequence evolution. *Genetics, 124*, 533–545.

Louis, E. J., Naumova, E. S., Lee, A., Naumov, G., & Haber, J. E. (1994). The chromosome end in yeast: Its mosaic nature and influence on recombinational dynamics. *Genetics, 136*, 789–802.

Marvin, M. E., Becker, M. M., Noel, P., Hardy, S., Bertuch, A. A., & Louis, E. J. (2009a). The association of yKu with subtelomeric core X sequences prevents recombination involving telomeric sequences. *Genetics, 183*, 453–467, 451SI–413SI.

Marvin, M. E., Griffin, C. D., Eyre, D. E., Barton, D. B., & Louis, E. J. (2009b). *Saccharomyces cerevisiae*, yKu and subtelomeric core X sequences repress homologous recombination near telomeres as part of the same pathway. *Genetics, 183*, 441–451, 441SI–412SI.

Mefford, H. C., & Trask, B. J. (2002). The complex structure and dynamic evolution of human subtelomeres. *Nature Reviews Genetics, 3*, 91–102.

Mefford, H. C., Linardopoulou, E., Coil, D., van den Engh, G., & Trask, B. J. (2001). Comparative sequencing of a multicopy subtelomeric region containing olfactory receptor genes reveals multiple interactions between non- homologous chromosomes. *Human Molecular Genetics, 10*, 2363–2372.

Moraes Barros, R. R., Marini, M. M., Antonio, C. R., Cortez, D. R., Miyake, A. M., Lima, F. M., et al. (2012). Anatomy and evolution of telomeric and subtelomeric regions in the human protozoan parasite *Trypanosoma cruzi*. *BMC Genomics, 13*, 229.

Nosek, J., & Tomaska, L. (2003). Mitochondrial genome diversity: Evolution of the molecular architecture and replication strategy. *Current Genetics, 44*, 73–84.

Nosek, J., Dinouel, N., Kovac, L., & Fukuhara, H. (1995). Linear mitochondrial DNAs from yeasts: Telomeres with large tandem repetitions. *Molecular and General Genetics, 247*, 61–72.

Novo, M., Bigey, F., Beyne, E., Galeote, V., Gavory, F., Mallet, S., et al. (2009). Eukaryote-to-eukaryote gene transfer events revealed by the genome sequence of the wine yeast *Saccharomyces cerevisiae* EC1118. *Proceedings of the National Academy of Sciences USA, 106*, 16333–16338.

Okazaki, S., Ishikawa, H., & Fujiwara, H. (1995). Structural analysis of TRAS1, a novel family of telomeric repeat-associated retrotransposons in the silkworm, *Bombyx mori. Molecular and Cellular Biology, 15*, 4545–4552.

Pace, T., Ponzi, M., Dore, E., Janse, C., Mons, B., & Frontali, C. (1990). Long insertions within telomeres contribute to chromosome size polymorphism in *Plasmodium berghei. Molecular and Cellular Biology, 10*, 6759–6764.

Pace, T., Ponzi, M., Scotti, R., & Frontali, C. (1995). Structure and superstructure of *Plasmodium falciparum* subtelomeric regions. *Molecular and Biochemical Parasitology, 69*, 257–268.

Ponzi, M., Pace, T., Dore, E., Picci, L., Pizzi, E., & Frontali, C. (1992). Extensive turnover of telomeric DNA at a *Plasmodium berghei* chromosomal extremity marked by a rare recombinational event. *Nucleic Acids Research, 20*, 4491–4497.

Prabhu, A., Morrison, H. G., Martinez, C. R, 3rd, & Adam, R. D. (2007). Characterisation of the subtelomeric regions of *Giardia lamblia* genome isolate WBC6. *International Journal for Parasitology, 37*, 503–513.

Pryde, F. E., & Louis, E. J. (1997). *Saccharomyces cerevisiae* telomeres. A review. *Biochemistry-English Translation, 62*, 1232–1241.

Pryde, F. E., & Louis, E. J. (1999). Limitations of silencing at native yeast telomeres. *EMBO Journal, 18*, 2538–2550.

Pryde, F. E., Gorham, H. C., & Louis, E. J. (1997). Chromosome ends: All the same under their caps. *Current opinion in genetics & development, 7*, 822–828.

Restrepo, B. I., Kitten, T., Carter, C. J., Infante, D., & Barbour, A. G. (1992). Subtelomeric expression regions of *Borrelia hermsii* linear plasmids are highly polymorphic. *Molecular Microbiology, 6*, 3299–3311.

Riethman, H. C., Moyzis, R. K., Meyne, J., Burke, D. T., & Olson, M. V. (1989). Cloning human telomeric DNA fragments into *Saccharomyces cerevisiae* using a yeast-artificial-chromosome vector. *Proceedings of the National Academy of Sciences USA, 86*, 6240–6244.

Riethman, H., Ambrosini, A., & Paul, S. (2005). Human subtelomere structure and variation. *Chromosome Research, 13*, 505–515.

Roth, C. W., Kobeski, F., Walter, M. F., & Biessmann, H. (1997). Chromosome end elongation by recombination in the mosquito *Anopheles gambiae. Molecular and Cellular Biology, 17*, 5176–5183.

Saiga, H., & Edstrom, J. E. (1985). Long tandem arrays of complex repeat units in *Chironomus telomeres. EMBO Journal, 4*, 799–804.

Saint, G. I., & Barbour, A. G. (1991). Antigenic variation in *Borrelia. Research in Microbiology, 142*, 711–717.

Scherf, A., Figueiredo, L. M., & Freitas-Junior, L. H. (2001). Plasmodium telomeres: A pathogen's perspective. *Current Opinion in Microbiology, 4*, 409–414.

Sykorova, E., Lim, K. Y., Kunicka, Z., Chase, M. W., Bennett, M. D., Fajkus, J., et al. (2003). Telomere variability in the monocotyledonous plant order Asparagales. *Proceedings of the Biological Sciences, 270*, 1893–1904.

Sykorova, E., Fajkus, J., Meznikova, M., Lim, K. Y., Neplechova, K., Blattner, F. R., et al. (2006). Minisatellite telomeres occur in the family Alliaceae but are lost in *Allium. American Journal of Botany, 93*, 814–823.

Takahashi, H., Okazaki, S., & Fujiwara, H. (1997). A new family of site-specific retrotransposons, SART1, is inserted into telomeric repeats of the silkworm, *Bombyx mori. Nucleic Acids Research, 25*, 1578–1584.

Tomaska, L., McEachern, M. J., & Nosek, J. (2004). Alternatives to telomerase: Keeping linear chromosomes via telomeric circles. *FEBS Letters, 567*, 142–146.

Trask, B. J., Friedman, C., Martingallardo, A., Rowen, L., Akinbami, C., Blankenship, J., et al. (1998). Members of the olfactory receptor gene family are contained in large blocks of DNA duplicated polymorphically near the ends of human chromosomes. *Human Molecular Genetics, 7*, 13–26.

Underwood, A. P., Louis, E. J., Borts, R. H., Stringer, J. R., & Wakefield, A. E. (1996). *Pneumocystis carinii* telomere repeats are composed of TTAGGG and the subtelomeric sequence contains a gene encoding the major surface glycoprotein. *Molecular Microbiology, 19*, 273–281.

Upcroft, P., Chen, N. H., & Upcroft, J. A. (1997). Telomeric organization of a variable and inducible toxin gene family in the ancient eukaryote *Giardia duodenalis*. *Genome Research, 7*, 37–46.

Vershinin, A. V., Schwarzacher, T., & Heslopharrison, J. S. (1995). The large-scale genomic organization of repetitive DNA families at the telomeres of rye chromosomes. *Plant Cell, 7*, 1823–1833.

Wada, M., & Nakamura, Y. (1996). Antigenic variation by telomeric recombination of major-surface-glycoprotein genes of *Pneumocystis carinii*. *Journal of Eukaryotic Microbiology, 43*, S8.

Witmer, K., Schmid, C. D., Brancucci, N. M., Luah, Y. H., Preiser, P. R., Bozdech, Z., et al. (2012). Analysis of subtelomeric virulence gene families in *Plasmodium falciparum* by comparative transcriptional profiling. *Molecular Microbiology, 84*, 243–259.

Young, B. S., Pession, A., Traverse, K. L., French, C., & Pardue, M. L. (1983). Telomere regions in *Drosophila* share complex DNA sequences with pericentric heterochromatin. *Cell, 34*, 85–94.

# Chapter 2
# Subnuclear Architecture of Telomeres and Subtelomeres in Yeast

**Emmanuelle Fabre and Maya Spichal**

**Abstract** Subtelomeres, upstream telomeres, have a very dynamic spatial positioning along the cell cycle. During G1 phase of the mitotic cell growth, subtelomere localisation close to the nuclear periphery results from the so-called Rabl chromosome configuration found in budding yeasts. In this chromosome configuration, centromeres are found clustered at one pole of the cell and chromosome arms lag behind. Subtelomere anchoring to the nuclear envelope relies on partly redundant molecular pathways, involving nuclear envelope components and structural composition of chromosome ends themselves. Subtelomere positioning also depends on chromosome arm length. Characteristic yeast subtelomere clustering thus results from chromosome arm length and location of subtelomeres close to the nuclear edge. During cell cycle progression, subtelomeres dynamics varies and subtelomeres localize towards the nuclear interior. During meiosis, distinct subtelomere positioning result from different spatial regulations. Dynamic spatial positioning of subtelomeres emerges as an important feature for chromosome end regulation and function.

## 2.1 Introduction

The closed mitosis, found in many fungi and in the yeast *Saccharomyces cerevisiae*, implies some basic rules in chromosome end positioning. During closed mitosis, the nuclear envelope never disassembles and in *S. cerevisiae* the mitotic organizing centre, i.e. the spindle pole body (SPB), sits in the double membrane of the nuclear envelope all along the cell cycle. Rigid microtubules

E. Fabre (✉) · M. Spichal
Groupe Régulation Spatiale des Génomes, Institut Pasteur, UMR 3525 CNRS, rue du Dr Roux, 75015 Paris, France
e-mail: emmanuelle.fabre@pasteur.fr

M. Spichal
Univ Pierre et Marie Curie, Cellule Pasteur UPMC, rue du Dr Roux, 75015 Paris, France

E. J. Louis and M. M. Becker (eds.), *Subtelomeres*, DOI: 10.1007/978-3-642-41566-1_2, 

emanate from the SPB, and a single nuclear microtubule binds to each of the kinetochores of the 16 chromosomes of *S. cerevisiae*. All chromosomes are thus attached to the centrosome by their centromere—composed by one to three nucleosomes in budding yeast (Lawrimore et al. 2011)—implying a spatial arrangement named Rabl configuration, after Carl Rabl (1885). This arrangement, which results from the maintenance of chromosome orientation after telophase at the end of mitotic cell division, is in part due to rapid cell cycle division, chromosome attachment by their centromere to the SPB and telomeres anchoring at the nuclear envelope. A rather Rabl-like chromosome configuration was however initially depicted in yeast because centromere clustering is not only a consequence of the anaphase movement of centromeres and the telomere–centromere polarization is moderately relaxed (Jin et al. 2000).

Telomeres correspond to the most distal part of linear chromosomes; subtelomeres, as it is implicit by their denomination, encompass the region immediately upstream to telomeric extremities. The structural definition of subtelomeres, described in detail in Chap. 3, remains a challenge since no clear barrier exists to distinguish a subtelomere from a non-subtelomeric "central" domain. Yet a consensus in the field emerges which refers to subtelomeres as large chromosomal regions in which few non-essential genes, separated by long AT-rich intergenic regions are found. These genes in addition often belong to similar structural and functional gene families involved in adaptive processes. Evolutionary studies in most eukaryotic species demonstrate the large number of chromosomal rearrangements that happen in these domains and converge to the idea that subtelomeres may perform in a function of gene reservoir (next chapters). Understanding spatial positioning of these regions is therefore particularly relevant for its possible role in chromosome end regulation.

## 2.2 Yeast Chromosome Ends: A Structural Definition of Telomeres and Subtelomeres

### *2.2.1 Yeast Telomere Structure*

An important structural distinction exists between telomeres and subtelomeres. Telomeres are repeated nucleoprotein TG-rich regions in which no genes are encoded. In particular, telomeres permit to distinguish a chromosome extremity from a double-strand break (DSB). The protective role of telomeres in chromosome degradation, fusion or recombination events has pushed important studies for telomere–protein identification. Various techniques of biochemical fractionation including development of proteomics of isolated chromatin fragments (PICh) help to delineate the particular composition of this unusual structural chromatin cap (Wright et al. 1992; Dejardin and Kingston 2009). Indeed, telomere chromatin is not predominantly formed by nucleosomes, but rather by non-histone proteins including chromatin assembly factors, replication, repair and telomere components (Enomoto et al. 1997). In addition to ~300-bp double-stranded $TG_{1-3}$ repeats, yeast telomeric DNA shows a 12–14 bases 3′ overhang of a G-rich strand, the G-tail. In vitro, G-tails can form G-quadruplexes that are four-stranded DNA structures between four-stacked Guanines connected through stable non Watson–Crick-based associations (Sundquist and Klug 1989). Because G-quadruplexes are stable, DNA helicases, like Sgs1 of the RecQ family or Pif1, are thought to resolve G quartets by trapping single-stranded G-tails (Paeschke et al. 2011; Huppert 2010). In vivo G-quadruplexes are potentially present at telomeres, but also elsewhere in the genome (Capra et al. 2010). They could play a positive role in telomere stability, but their role is yet incompletely solved and it remains to determine to what extent these structures are present at each telomere (Smith et al. 2011).

At each replication cycle, a specific addition of the 3′G-rich overhang is required to avoid telomere shortening that will inevitably lead to cell ageing and cell death. G-tail addition happens through a specialized telomerase complex, which in *S. cerevisiae* is constitutively expressed and composed by Est1, Est2—the reverse transcriptase catalytic subunit—and the integral RNA component *TLC1*. G-tail formation also involves Rap1, the essential repressor/activator protein 1, found both at ~5 % of polII-driven ribosomal promoters and at repeated double-stranded telomeric DNA, the heterotrimer CST (Cdc13/Stn1/Ten1) that binds the single-stranded G-rich overhang, the Ku heterodimer, the MRX complex (Mre11/Rad50/Xrs2) and many other proteins whose role is yet unclear (Lieb et al. 2001). All these components are critical for a positive or negative access to the telomerase. For instance, Rap1 establishes a negative feedback loop on telomere elongation, by recruiting two factors, Rif1 and Rif2 through its C-terminal domain (Levy and Blackburn 2004; Marcand et al. 1999). Increased binding of Rif1 and Rif2 inhibits Tel1 (ATM) binding at longer telomeres (Hirano et al. 2009). Since Tel1 is also required for telomerase recruitment, extension of long telomeres is less efficient (Goudsouzian et al. 2006). Notably, telomere repeats are not homogeneous among telomeres inside a single cell, and telomerase does not act on every telomere in each cell cycle (Teixeira et al. 2004). The number of repeats and therefore of Rap1 molecules bound to it (see below) might intervene to convert telomeres from closed non-extendible state to open-extendible configurations which in turn influence telomere function (McEachern and Blackburn 1995; Teixeira et al. 2004). For instance, 100–125 repeat length increases telomerase processivity; ~30 bp are not enough for Rap1 to inhibit non-homologous end joining (NHEJ) which ends in telomere fusion (Marcand et al. 1999). The dynamic interplay of Rap1 with each telomere repeat element together with the inherent dynamics of each telomere is likely to influence the regulation of telomere function and localization.

Interestingly, telomeres are transcribed into specific transcripts, referred to as telomeric repeat-containing RNA (TERRA). In metazoans where TERRA have been discovered, these molecules are heterogeneous in size, are exclusively nuclear and colocalize with telomeres (Azzalin et al. 2007; Schoeftner and Blasco

2008). In budding yeast, TERRA are also produced. These molecules, larger than the telomere repeats, contain subtelomeric-derived sequences and are degraded by the 5′ to 3′ RNA exonuclease Rat1. In *rat1-1* mutants, RNA accumulation is linked to telomere shortening leading to the interesting possibility that DNA/RNA or TERRA/TLC1 hybrids inhibit telomerase (Luke et al. 2008). Telomere length regulation occurs hence through multiple pathways.

### *2.2.2 Yeast Subtelomere Structure and the Sir Proteins*

Subtelomeric sequences are also repeated in different complex forms. They contain X elements, subtelomeric repeats (STR) and long tandem Y′ repeats which could origin from transposable elements (Fourel et al. 1999). As described in detail in Chap. 3, one to four copies of tandem Y′ sequences flanked by TG1-3 repeats are present on two-thirds of yeast S288c subtelomeres. S288c is the first fully assembled and sequenced *S. cerevisiae* strain (Goffeau et al. 1996). These elements are however highly variable between strains and species (Liti et al. 2009). For instance, in W303 strain, a recent cross between S288c and other recent lineages, the exact number of Y′ sequences remains partly unknown because of the difficulty to assemble these repeats (Liti et al. 2009). The inherent variability of subtelomeres and of Y′ sequences, often used as a probe for cytological subtelomere position studies, hence predicts some changeability between the different *S. cerevisiae* backgrounds analysed.

Subtelomeres, contrary to telomeres, show a nucleosome composition. In addition to subtelomeric histones however, a number of additional chromatin modifiers, including the silencing insulator (Sir) proteins, are specifically enriched in there (Rusche et al. 2003). Sir proteins were initially discovered as responsible for silencing of *HML* and *HMR* mating-type cassettes each located at one subtelomere of chromosome 3 (Rine and Herskowitz 1987), but Sir-mediated silencing is also found at some other subtelomeres and at the rDNA (Pryde and Louis 1999; Smith et al. 1998). Sir proteins are recruited to silencer sequences at the mating-type loci, while at subtelomeres Sir recruitment occurs through Rap1 binding to double-stranded telomeric repeats and ORC and Abf1 binding to the core X element (Pryde and Louis 1999). Sir3 binds deacetylated amino-terminal residues of histones H3 and H4, and cooperative binding between histone deacetylase Sir2, Sir3 and Sir4 is thought to enable Sir spreading through silenced regions (Hecht et al. 1995). Interestingly, Sir3 overproduction can lead to subtelomeric clustering independently of Sir3 function in silencing, pointing to the architectural role of this protein (see below and Ruault et al.2011). Note that Rap1, as well as yKu, is also found by chromatin immunoprecipitation experiments not only at double-stranded telomeric repeats, but also in several kb of subtelomeric sequences, suggesting that the terminal region of the chromosome may fold into internal subtelomeric sequences (de Bruin et al. 2000; Marvin et al. 2009a).

Only Sir2 is required for silencing of polII-driven genes inserted into the rDNA (Pryde and Louis 1999; Smith et al. 1998). At subtelomeres, all Sir proteins, except Sir1, are recruited to silence similar reporter insertions. These reporters have allowed to distinguish silencing occurring at the so-called truncated ends, in which subtelomeric sequences are deleted and silencing at native ends, in which all subtelomeric sequences are maintained (Gottschling et al. 1990; Pryde and Louis 1999). In the first case, Sir spreading can encompass several kb, while Sir proteins occupancy at native ends is limited to regions of 1–2 kb peaking at X elements and avoiding Y′ sequences (Pryde and Louis 1999). These last data are in agreement with recent kinetic localization mapping by deep sequencing of an overexpressed version of Sir3. A rapid binding covers ~2 kb (i.e. 6 to 10 nucleosomes) around silencers and a slower binding is constrained to subtelomeric PAU genes and highly transcribed euchromatic sites, suggesting that Sir spreading is rather spatially limited and Sir3 recruitment not restricted to subtelomeres (Lynch and Rusche 2009; Radman-Livaja et al. 2011). This experimental evidence is coherent with the extremely variable silencing detected among native ends (Pryde and Louis 1999). Again, it points to a role for Sir proteins distinct to silencing which could rather be architectural, for instance to avoid recombination with internal sequences (Marvin et al. 2009b; Pryde and Louis 1999).

### 2.2.3 Sir Structural Properties: In Vitro Consequences

The chromatin architectural Sir3 protein provides one of the key structural properties of the subtelomeric Sir complex. The structure of the amino-terminal bromo-associated homology (BAH) domain is known (Hou et al. 2006). Even more, the crystal structure of a BAH-mutated version is now solved at 3Å resolution in complex with a nucleosome (Armache et al. 2011). Crystal shows that BAH significantly binds each of the four core histone proteins (Armache et al. 2011). Although the BAH domain itself has weak self-associating properties, extensive BAH binding to nucleosome might contribute to the compaction properties of the full-length Sir3 protein. Indeed, Sir3 alone can compact nucleosomal arrays in vitro and high levels of Sir3 proteins promote oligomeric structure formation between individual nucleosomal arrays and Sir3 oligomers (McBryant et al. 2008). Furthermore, in vitro reconstitution of the Sir complex shows that in these conditions, Sir2, Sir3 and Sir4 maintain a 1:1:1 stoichiometry (Martino et al. 2009). Sedimentation analyses of this reconstituted holocomplex with chromatin estimate that one SIR complex binds two nucleosomes, arguing for a binding between two neighbouring nucleosomes (Martino et al. 2009).

It is interesting to note that in vitro Rap1 binds telomeric repeats at a frequency of 1 per 18 bp, corresponding in theory to 14 to 20 Rap1-binding sites per 250–350-bp-long telomere (Gilson et al. 1993). However, due to the intrinsic heterogeneity of *S. cerevisiae* telomeres, the precise number of Rap1 molecules remains uncertain. It has been recently shown that in solution, Rap1 binds multiple sites as a monomer and can bind multiple sites with a repeat array without obvious cooperativity and regardless of binding site affinity (Williams et al. 2010). Interestingly, a correlation exists between the ability of telomeric repeats to regulate telomere length and in vitro Rap1 affinity to its binding sites (Williams et al. 2010). Relationship between in vitro and in vivo experiments is lacking, but these data suggest that multiple binding sites for Rap1 do not by themselves promote formation of an energetically favourable complex (Williams et al. 2010).

### 2.2.4 *Dynamic Assembly of Subtelomeric and Telomeric Proteins*

In normal conditions, telomere length changes rapidly among cells and in a few generations, suggesting that assembly and disassembly of telomeric and subtelomeric proteins is a dynamic process all along the cell cycle and even during each phase of the cell cycle. For instance, only 2 to 3 telomeres are bound by telomerase per cell in S phase (Teixeira et al. 2004). A number of telomeric or subtelomeric proteins like Rap1, Sir3, Sir4, Rif1 and Rif2 change their occupancy over the cell cycle. Lower at the G2/M phase, occupancy increases at G1 or S phase (Laroche et al. 2000). The C-terminal Rap1 domain interacts with Rif1, Rif2 and Sir proteins (Wotton and Shore 1997). Rap1 multiple partners raise the possibility that these proteins compete for Rap1 binding either along the cell cycle or even between each subtelomere. Consistently these data point to the variability that exists not only in a cell population but also between different chromosome ends in a single cell, lighting the variable position of distinct chromosome ends. Yet, as it is now discussed, rules for subtelomere positioning in the nuclear space are shared by many of them and are probably driven by the intrinsic property of the chromatin fibre.

## 2.3 Subtelomeres: Intranuclear Positioning and Relative Interactions

### 2.3.1 *How to Define Subtelomere Position in the Nuclear Space*

Deciphering telomere and subtelomere positions relative to the nuclear space has been essentially tackled by examining dedicated subtelomeric and/or telomeric DNA sequences or proteins, in fixed or living cells. Relative subtelomeric positions have furthermore benefited from recent outstanding progresses performed to capture frequent interactions between DNA segments, namely the chromosome conformation capture 3C derivatives (Dekker et al. 2002). The technique exists in many variations but can easily be understood as a process based on the cross-linking of DNA followed by restriction enzyme cutting and intramolecular ligation with an adaptor that allows the purification of the circularized DNA. By sequencing the DNA circles, inter- and intrachromosomal interactions existing in the nuclei at the moment of fixation can thus be identified. As this method requires a high number of cells, it is possible to determine the statistical significance of given interactions. These frequencies are currently being used to model chromosome configurations through polymer physics-based models (Kalhor et al. 2012; Lieberman-Aiden et al. 2009). Using Hi-C, one model for 3D organization of the budding yeast genome, in an asynchronous population has been proposed, validating previous knowledge about subtelomere behaviour (Duan et al. 2010; Zimmer and Fabre 2011). A close Hi-C method has similarly been applied to the 3 chromosomes that fission yeast contains. Close spatial proximity between both chromosome ends of chromosomes 2 and 3 is uncovered, chromosome 1 being particular as both of its subtelomeres carry rDNA sequences (Tanizawa et al. 2010).

Yet, it is still difficult to correlate frequencies with real distances. These distances can be reached by single-cell observation of dedicated differently labelled loci. A popular system consists to insert a number of bacterial operator sequences, namely LacO or TetO sequences, respectively, 42 bp and 14 bp long, at chosen positions. Subsequent binding by the respective repressors LacI and TetR fused to different variants of the green fluorescent protein allows locus labelling and detection of a single spot with a good signal to noise ratio over the nucleoplasmic background (Robinett et al. 1996; Straight et al. 1997). Recently, a new generation of fluorescent repressor/repeats operators has been constructed allowing three-colour detection (Lassadi, I and Bystricky K, personal communication). Repressors bind their targets with a certain affinity; for instance, LacI binds its target with a Kd ~ $10^{-11}$ mol/l and TetR with a Kd ~ $5 \times 10^{-9}$ mol/l—for comparison, streptavidin binds biotin with a Kd ~ $10^{-15}$ mol/l (Falcon and Matthews 1999). Therefore, it can be questioned whether this binding can affect global and/or local chromosome biology. Cell cycling remains unchanged in the presence of different repeat insertions bound or not bound by their respective repressors, arguing for unchanged chromosome replication and segregation (Belmont 2001). If the LacI hinge region is mutated, LacI-binding affinity decreases by 500-fold in vitro, i.e. the range of TetR constant affinity (Falcon and Matthews 1999). In vivo, this mutation relieves cooperative Sir4 recruitment observed at silencers when lacO arrays are inserted nearby (Dubarry et al. 2011). In the presence of silencers, it might therefore be prudent to verify expression behaviour of neighbouring genes after lacO repeats insertions. Yet, tetO and lacO insertions have been successfully used to localize many subtelomeres, recapitulating initial observations with subtelomeric probes on fixed cells (Gotta et al. 1996; Hediger et al. 2002b).

To localize subtelomeres towards the nuclear periphery in living cells, a functional GFP fusion of nuclear pore protein Nup49 is often chosen as a marker of the nuclear envelope. Most of the subsequent images captured are then based on 3D imaging, with a number of $z$ stacks whose optimum is derived from the Nyquist criterion (which itself depends on the microscope used). After focal plane selection, close to the equatorial $Z$ stack, spot 2D localization towards the nuclear envelope is defined. Gasser's lab has taken advantage of the focal plane to define three zones of similar surface and determine fluorescent spot distribution in either one of the three zones. By counting ~ hundred cells, the subtelomere is determined as being peripheral if preferentially found in the outmost zone, or randomly located if equally distributed between the three zones (33.3 % of the cells in each zone (Hediger et al. 2002b).

Recently, to overcome limitations due to 2D localization, it has been proposed to detect a second nuclear landmark, the nucleolus. In yeast, the nucleolus forms a stable crescent structure at one pole of the cell, opposite to the spindle pole body during interphase, allowing for segmentation and barycentre definition (Berger et al. 2008). This additional landmark allows defining an axis joining the two centres of mass of the nucleoplasm and the nucleolus (Berger et al. 2008). Orientated alignment of thousands of nuclei through this axis results in dimensional probability maps of a given chromosomal locus (Berger et al. 2008; Therizols et al. 2010). As discussed in the Chap. 3, the number of studies based on either methodology has localized subtelomeres inside the nucleoplasmic space.

### 2.3.2 Subtelomere Position in the Nuclear Space

Subtelomere positioning depends on various aspects, i.e. chromosome arm length, cell cycle stage and telomere replication state. Subtelomere positioning also varies between cell cycle stages during both vegetative and meiotic growth (see Sect. 2.2.4). As structural chromosome end variability anticipates, each chromosome end has a particular behaviour that can in addition depend on the strain background (Table 2.1).

Rap1 detection shows a limited number of spots at proximity of the nuclear envelope when observed in G1 and S phases (Gotta et al. 1996; Hiraga et al. 2008; Klein et al. 1992; Palladino et al. 1993; Schober et al. 2008), suggesting that if Rap1 signal reflects chromosome ends, they tend to cluster. After S phase, Rap1 signal is delocalized and subtelomeres randomly position in the nucleoplasm (Laroche et al. 2000). Individual labelling of subtelomeres also detects subtelomeres close to the nuclear envelope during almost the entire cell cycle, except during the G2/M phase (Bystricky et al. 2005; Hediger et al. 2002b, 2008; Schober et al. 2008; Therizols et al. 2010).

Moreover, two-dimensional probability maps of number of individually labelled subtelomeres show that they are non-randomly situated at the nuclear periphery depending on the size of their corresponding chromosome arm (Therizols et al. 2010). Interestingly, these probability maps based on the nucleolus as nuclear landmark are similar when the SPB is chosen as a reference (unpublished results). Subtelomeres on shorter chromosome arms are close to the SPB with a gradually increasing distance to the SPB corresponding to increasing chromosome arm size. Furthermore, the volume of the nucleolus influences subtelomere position. This region is largely avoided by all subtelomeres. Increase in chromosome arm length does not influence the exclusion of subtelomeres in the nucleolar area. Moreover, reducing the nucleolar mass by rapamycin extends the space subtelomere occupy at nuclear periphery, suggesting that the nucleolus represents a physical barrier (Therizols et al. 2010).

**Table 2.1** Compilation of yeast subtelomeres positions studied in living cells. Subtelomeres are ranked by chromosome arm size

| Chromosome arm | Length (kb) (S288c) | Strain | WT | | | | Mutant | | | | | Method | References |
|---|---|---|---|---|---|---|---|---|---|---|---|---|---|
| | | | Asynchronous (%) | G1 (%) | S (%) | G2 (%) | | Asynchronous (%) | G1 (%) | S (%) | G2 (%) | | |
| **9R** | 85 | S288c | 80 | | | | | | | | | 3D GM | 12 |
| **8L** | 105 | W303 | | 50 | Early S 60, late S 45 | | | | | | | 2D | 1 |
| | | | 52 | | | | *mps3Δ75-150* | **30** | | | | 2D | 1 |
| | | | 81 | | | | *ctf18* | **50** | | | | 2D | 11 |
| | | | | | | | *ctf8* | **45** | | | | 2D | 11 |
| | | | | | | | *dcc1* | **52** | | | | 2D | 11 |
| | | | | | | | *ku70* | **45** | | | | 2D | 11 |
| | | | | 50 | | | *ku70* | **33** | | | | 2D | 5 |
| | | | | 61 | 58 | | *rtt109* | | **42** | **40** | | 2D | 10 |
| | | | | | Early S 60, late S **35** | **35** | *clb5/clb6* | | 50 | Early S 60, late S 50 | **35** | 2D | 4 |
| **3L** | 115 | W303 | | 43 | | | | | | | | 2D | 7 |
| **6R** | 122 | W303 | | 60 | Early S 60, late S **35** | **35** | | | | | | 2D | 4 |
| | | | 45 | 50 | 49 | | *mps3Δ75–150* | **25** | 50 | **35** | | 2D | 1 |
| | | | | 55 | 65 | | *sir4* | | 50 | 42 | | 2D | 2 |
| | | | | 58 | | | *sir4* | | 57 | | | 2D | 7 |
| | | | | | | | *ku70* | | **30** | | | 2D | 7 |
| | | | | 50 | 58 | | *ku70* | | **45** | 60 | | 2D | 8 |
| | | | | | | | *esc1* | | **40** | **42** | | 2D | 8 |
| | | | | | | | *esc1/ku70* | | **37** | **38** | | 2D | 8 |

(continued)

**Table 2.1** (continued)

| Chromosome arm | Length (kb) (S288c) | Strain | WT | | | | Mutant | | | | | Method | References |
|---|---|---|---|---|---|---|---|---|---|---|---|---|---|
| | | | Asynchronous (%) | G1 (%) | S (%) | G2 (%) | | Asynchronous (%) | G1 (%) | S (%) | G2 (%) | | |
| | | | | 58 | Early S 63, late S 48 | | *ku70* | | **30** | Early S **40**, late S **33** | | 2D | 9 |
| | | | | | | | *mlp1/mlp2* | | 63 | Early S 65, late S 52 | | 2D | 9 |
| | | | | 65 | 65 | | *rtt109* | | **45** | **45** | | 2D | 10 |
| | | | 82 | | | | *ctf18* | **55** | | | | 2D | 11 |
| | | | | | | | *ctf8* | **53** | | | | 2D | 11 |
| | | | | | | | *dcc1* | **45** | | | | 2D | 11 |
| | | S288c | 70 | | | | | | | | | 3D GM | 12 |
| **6L** | 148 | | | 49 | | | *ku70* | | **40** | | | 2D | 7 |
| | | | | | | | *sir4* | | **25** | | | 2D | 7 |
| | | S288c | 80 | | | | | | | | | 3D GM | 12 |
| **1L** | 150 | S288c | 70 | | | | | | | | | 3D GM | 12 |
| **5L** | 150 | | | 60 | | | | | | | | 2D | 7 |
| **14R** | 157 | S288c | 70 | | | | | | | | | 3D GM | 12 |
| | | | | 71 | | | | | | | | 2D | 7 |
| **3R** | 200 | | | 51 | | | | | | | | 2D | 7 |
| **10R** | 310 | S288c | 70 | | | | | | | | | 3D GM | 12 |
| **5R** | 430 | S288c | 80 | | | | | | | | | 3D GM | 12 |
| | | W303 | | 40 | | | | | | | | 2D | 5 |
| | | W303 | | 35 | | | | | | | | 2D | 7 |
| **11L** | 440 | S288c | 65 | | | | *nup145* | **55** | | | | 3D | 6 |
| | | | | | | | *nup84* | **50** | | | | 3D | 6 |

(continued)

**Table 2.1** (continued)

| Chromosome arm | Length (kb) (S288c) | Strain | WT | | | | | Mutant | | | | Method | References |
|---|---|---|---|---|---|---|---|---|---|---|---|---|---|
| | | | Asynchronous (%) | G1 (%) | S (%) | G2 (%) | | Asynchronous (%) | G1 (%) | S (%) | G2 (%) | | |
| | | | | | | | *nup133* | **45** | | | | 3D | 6 |
| | | | | | | | *nup120* | **45** | | | | 3D | 6 |
| | | | | | | | *ku70* | **45** | | | | 3D | 6 |
| | | | | | | | *sir4* | **40** | | | | 3D | 6 |
| | | | | | | | *ku70/sir4* | **45** | | | | 3D | 6 |
| | | | | | | | *ku70/nup145* | **40** | | | | 3D | 6 |
| | | | | | | | *esc1* | **50** | | | | 3D | 6 |
| | | | 80 | | | | | | | | | 3D GM | 12 |
| **7L** | 492 | W303 | | 55 | | | | | | | | 2D | 5 |
| **14L** | 628 | W303 | | 50 | | | | | | | | 2D | 5 |
| | | | | 54 | | | | | | | | 2D | 7 |
| | | | | 52 | Early S 63, late S 53 | | | | | | | 2D | 9 |
| | | | 63 | 59 | 55 | | *mps3Δ75–150* | **30** | **45** | | | 2D | 1 |
| | | | 65 | 65 | 65 | | *mps3Δ75–150* | | 50 | | | 2D | 3 |
| | | | | | | | *mps3Δ75–150/heh1* | | **45** | **43** | | 2D | 3 |
| | | | | | | | *esc1/heh1* | | | **45** | | 2D | 3 |
| | | | | | | | *heh1* | | 58 | 60 | | 2D | 3 |
| | | | | | | | *lrs4* | **38** | **45** | **40** | | 2D | 3 |
| | | | | | | | *nur1* | | | 60 | | 2D | 3 |
| | | | | | | | *ku80* | | **33** | | | 2D | 3 |
| | | | | | | | *ku80/lrs4* | | **33** | | | 2D | 3 |
| | | | | | | | *esc1/lrs4* | | | **33** | | 2D | 3 |
| | | | | | | | *rif1* | 70 | | | | 2D | 3 |
| | | | | | | | *lrs4/rif1* | 50 | | | | 2D | 3 |

(continued)

**Table 2.1** (continued)

| Chromosome arm | Length (kb) (S288c) | Strain | WT | | | | | Mutant | | | | Method | References |
|---|---|---|---|---|---|---|---|---|---|---|---|---|---|
| | | | Asynchronous (%) | G1 (%) | S (%) | G2 (%) | | Asynchronous (%) | G1 (%) | S (%) | G2 (%) | | |
| | | | 84 | 60 | 60 | | *ctf18* | **45** | | **40** | 52 | 2D | 11 |
| | | | | | | | *ctf8* | 53 | | | | 2D | 11 |
| | | | | | | | *dcc1* | 58 | | | | 2D | 11 |
| | | | | | | | *ku70* | 58 | | | | 2D | 11 |
| | | | | 50 | Early S 50, late S 40 | 35 | *sir4* | | 50 | Early S 70, late S 40 | **35** | 2D | 4 |
| | | | | | | | *clb5/clb6* | | 50 | Early S 50, late S 50 | **35** | 2D | 4 |
| | | | | | | | *ku70* | | 70 | Early S 70, late S 80 | 70 | 2D | 4 |
| | | | | 50 | 58 | | *esc1* | | 50 | 60 | | 2D | 8 |
| | | | | | | | *ku70* | | **40** | 62 | | 2D | 8 |
| | | | | | | | *ku70/esc1* | | **40** | 38 | | 2D | 8 |
| | | | | 65 | 67 | | *sir4* | | **45** | 57 | | 2D | 10 |
| | | | | | | | *ku70* | | 55 | 57 | | 2D | 10 |
| | | | | | | | *ctf18* | | **42** | 53 | | 2D | 10 |
| | | | | | | | *pgd1* | | **37** | **45** | | 2D | 10 |
| | | | | | | | *rrm3* | | **45** | 37 | | 2D | 10 |
| | | | | | | | *elp4* | | **48** | 63 | | 2D | 10 |
| | | | | | | | *tel1* | | **45** | 73 | | 2D | 10 |
| | | | | | | | *asf1* | | 68 | **43** | | 2D | 10 |

(continued)

**Table 2.1** (continued)

| Chromosome arm | Length (kb) (S288c) | Strain | WT | | | | | Mutant | | | | Method | References |
|---|---|---|---|---|---|---|---|---|---|---|---|---|---|
| | | | Asynchronous (%) | G1 (%) | S (%) | G2 (%) | | Asynchronous (%) | G1 (%) | S (%) | G2 (%) | | |
| | | | | | | | *mre11* | | 58 | **45** | | 2D | 10 |
| | | | | | | | *sic1* | | 75 | **50** | | 2D | 10 |
| | | | | | | | *csm4* | | 55 | 58 | | 2D | 10 |
| | | | | | | | *jjj1* | | 50 | 58 | | 2D | 10 |
| | | | | | | | *sir3* | | 58 | 55 | | 2D | 10 |
| | | | | | | | *rtt109* | | **42** | **39** | | 2D | 10 |
| | | | | | | | *vp75* | | 73 | 67 | | 2D | 10 |
| | | | | | | | *asf1/vps75* | | **47** | **40** | | 2D | 10 |
| | | | | | | | *H3K56R* | | **42** | **40** | | 2D | 10 |
| | | | | | | | *H3K56Q* | | **42** | **37** | | 2D | 10 |
| **13R** | 658 | S288c | 70 | | | | | | | | | 3D GM | 12 |
| **15R** | 705 | S288c | 80 | | | | | | | | | 3D GM | 12 |
| **4R** | 1050 | S288c | 60 | | | | | | | | | 3D GM | 12 |

For historical reasons, many studies are concentrated on subtelomeres of short arms 6L, 6R, 8L and long arm 14L in W303 background. Cell cycle (Asynchronous, G1, S, G2) and mutant effects, calculation mode and strain backgrounds are shown. Mutants considered as having a significant effect on subtelomere position at the periphery by the authors (*P* values $< 10^{-2}$) are in bold. *P* values represent significance relative to random localization, except in study (11), where they are calculated versus the WT position. 2D, the subtelomere focus is considered when in the equatorial z stack. 3D GM, subtelomere 3D position is reconstituted in a statistical map (GM, gene map), and peripheral subtelomeres are in the 50 % outmost volume (6, 12). Note that there is a discrepancy between S288c and W303 backgrounds for subtelomere 5R. (1) Bupp et al. {2007), (2) Schober et al. (2008), (3) Chan et al. (2011), (4) Ebrahimi and Donaldson (2008), (5) Hediger et al. (2006), (6) Therizols et al. (2006), (7) Bystricky et al. (2005), (8) Taddei et al. (2004), (9) Hediger et al. (2002b), (10) Hiraga et al. (2008), (11) Hiraga et al. (2006), (12) Therizols et al. (2010)

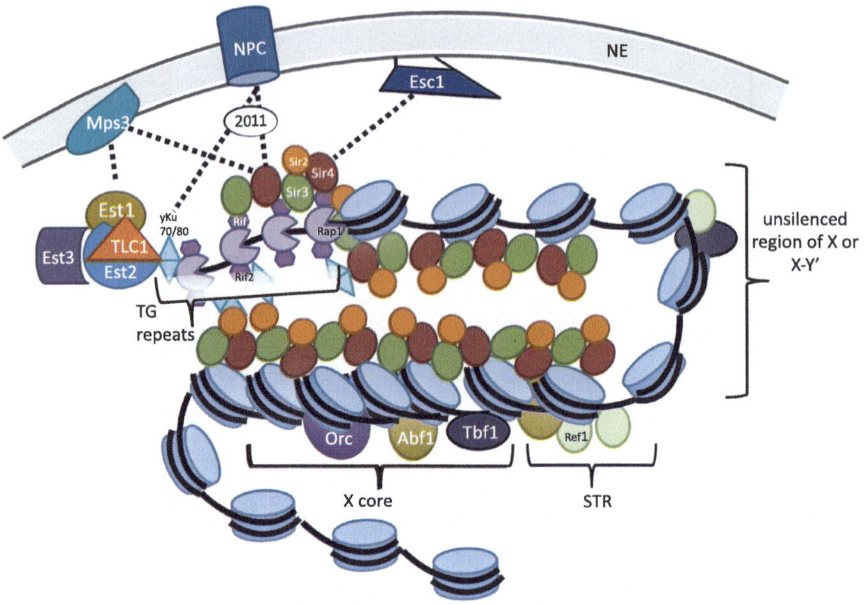


**Fig. 2.1** Model for telomere and subtelomere nuclear architecture and links with molecular components of the nuclear periphery (*dashed lines*)

### *2.3.3 How Subtelomeres are Anchored at the Nuclear Periphery*

A number of molecular pathways, partly redundant, have been discovered which are responsible for the transient anchoring of subtelomeres to the nuclear envelope (Fig. 2.1). They involve both nuclear envelope components and structural composition of chromosome ends.

Initially, two pathways, Sir4 and Ku dependent, were shown to mediate tethering of telomeres to the nuclear envelope (Hediger et al. 2002b; Laroche et al. 1998; Taddei et al. 2004; Tham et al. 2001; Therizols et al. 2006). These two telomeric proteins are evidently not integral membrane proteins, but they do transiently associate with nuclear envelope partners, favouring telomere positioning. Sir4 binds to Esc1, to Mps3, but also to Ku80 (Ribes-Zamora et al. 2007; Taddei et al. 2004). Esc1 (establishes silent chromatin) is a non-abundant acidic protein found in patches on the nuclear membrane distinct from nuclear pores and is thought to associate with the nuclear envelope through post-translational modification of its carboxy terminus (Andrulis et al. 2002; Lewis et al. 2007; Taddei et al. 2004). Esc1 deletion or yku70 mutant on their own has a modest effect on subtelomere 14L (L stands for left, R for right) tethering but when combined, they lead to its defective localization (Taddei et al. 2004). Mps3 is a Sad1/UNC-84 (SUN) domain integral membrane protein essential in SPB duplication. However, simultaneous deletion of *MPS3* and the gene encoding the nuclear pore protein Pom152 and/or Nup157 renders Mps3 redundant even in SPB duplication, implying that changes in the lipid environment of the nuclear membrane can alleviate the lack of the nuclear membrane protein Mps3 (Friederichs et al. 2011; Witkin et al. 2010). The deletion of the nucleoplasmic N-terminal domain of Mps3 allows disconnecting Mps3 from its anchoring function without disturbing its influence on the SPB (Bupp et al. 2007). Ku70 mediates telomeric peripheral anchoring (Laroche et al. 1998). A Ku70 interaction partner could be the nuclear pore complex (Galy et al. 2000; Feuerbach et al. 2002), in agreement with the observation when components of nuclear pore complex are mutated, the Nup84 complex or peripheral Mlp1/2 proteins, subtelomeres are displaced towards the nucleoplasm (Galy et al. 2000; Therizols et al. 2006). Yet the role of Mlp1/2 proteins in subtelomere tethering has been called into question (Hediger et al. 2002a). The fact that Mps3 was recently shown to bind not only to Sir4 but also to the yKu70/yKu80 heterodimer and Est1 (Bupp et al. 2007; Chan et al. 2011; Schober et al. 2009) and the fact that Sir4 was also shown to interact with to the cohibin complex (Lrs4/Csm1) through the integral proteins Heh1/Nur1 (Chan et al. 2011) indicate the intricate multiplicity and dependence of these different pathways. The observation that localization of different subtelomeres is differently compromised according to different mutated contexts additionally points to the variable dependency of each of the 32 chromosome ends towards the nuclear envelope, which is shown Table 2.1.

Chromosome end positioning is cell cycle regulated, mostly peripheral in G1/S phase, and it is displaced towards the nucleoplasm G2/M. In G1, Sir4 seems to be predominant in telomere anchoring in W303, but not in S288c (Table 2.1, Hediger et al. 2002b; Hiraga et al. 2006; Tham et al. 2001). In S phase, Mps3 seems to play a prominent role together with the telomerase and the Ku complex (Bupp et al. 2007; Schober et al. 2009). However, the lack of clear correlation between telomere length and localization behaviour and the fact that different cell cycle specificities are observed between *mre11* and *tel1* mutants (see also Sect. 2.1.1), suggest that telomerase might have a role in telomere anchoring distinct to telomere length regulation (Hiraga et al. 2008). Yet, S-specific events are important for telomere localization. Telomere dislodgement is delayed when S phase is delayed, suggesting that anchoring is probably dependent on DNA replication. Once DNA is fully replicated, telomeres are released (Ebrahimi and Donaldson 2008; Ferreira et al. 2011). This is coherent with the finding that Ctf18-RFC, a subunit of the replication factor required for sister chromatid cohesion, is required for subtelomere positioning (Hiraga et al. 2006). Moreover, the number of proteins involved in chromatin structure, particularly during DNA replication, is shown to be important in telomere positioning (Hiraga et al. 2008). Asf1, the histone chaperone that stimulates acetylation of K56 of newly synthesized histone H3, is one of them (Hiraga et al. 2008).

Sumoylation might be critical in understanding how temporal regulation of these different pathways happens. For instance, Ku80 and Sir4 are found to be sumoylated, and their sumoylation occurs through the PIAS-like SUMO E3 ligase Siz2 (Ferreira et al. 2011). In siz2Δ mutants, telomeres are randomly positioned in both G1 and S phases. The *siz2Δ* mutant phenotype can be antagonized by deletion of the Pif1 helicase, probably through the increase in telomere-bound telomerase. Since only telomeres that are elongated are detached from the nuclear periphery, sumoylation acts as a negative regulator of telomere length, probably through sumoylation of many telomere-bound proteins, as it was shown for Cdc13 (Ferreira et al. 2011; Hang et al. 2011). Interestingly, Cdc13 sumoylation is cell cycle regulated (Hang et al. 2011). On the other hand, the Slx5/Slx8 SUMO-dependent ubiquitin ligase is found at the nuclear pore complex in interaction with the Nup84 complex; the SUMO-protease Ulp1 is also associated with the NPC (Collins et al. 2007; Palancade et al. 2007). Modification of the sumoylated status of proteins that come in association with the NPC, including nuclear pore proteins themselves, might participate in the regulation of telomere positioning as it was proposed to regulate DNA break repair (Nagai et al. 2008; Therizols et al. 2006).

The strength of telomere anchoring to the nuclear periphery is also controlled by the structural composition of the subtelomeric region, i.e. Y′ elements or the STR repeats. Different tethering capacities are indeed related to the presence or absence of Y′ elements (Hediger et al. 2002b, 2006; Hediger and Gasser 2006; Tham et al. 2001). Upon STR repeat deletion, anchoring efficiency of weakly attached telomeres increases. This could be attributed to the absence of the STR-repeat-binding proteins Reb1 and Tbf1 (Hediger et al. 2002b; Lieb et al. 2001). However, the increased anchoring cannot be explained by removal of a Sir4 spreading barrier, as there is no link between the abundance of Sir proteins and anchoring efficiency (Lieb et al. 2001). It rather seems that Reb1 and Tbf1 are implicated in the conformation of telomeres that influences the silencing and anchoring strength to the nuclear periphery (see Sect. 2.1.2).

### *2.3.4 Subtelomere Position During Meiosis*

In contrast to the mitotic cell division, in which chromosomes are separated after DNA replication, meiosis proceeds in two subsequent division processes, meiosis I and II. During the first meiotic cell division, DNA is replicated in the diploid cell and homologous chromosomes undergo recombination, resulting in two genetically unique diploid cells. In meiosis II, the two daughter cells undergo a second division to produce four haploid cells, spores in yeast.

These unique cell cycle events, especially during meiosis I, also require distinctive telomere and subtelomere behaviour. At the beginning of meiotic prophase I, between the leptotene and zygotene transition, telomeres concentrate in a cluster at the centromeres, forming the so-called meiotic "bouquet" (Trelles-Sticken et al. 1999, 2005). The structure is transient but seems to be important for the subsequent recombination, as mutants that are defective for bouquet formation also show malfunctioning or altered steps in crossing over (Chua and Roeder 1997; Kosaka et al. 2008; Wanat et al. 2008). At early zygotene stage, the bouquet formation dissolves through rapid movements of dispersed telomeres (Conrad et al. 2008; Koszul et al. 2008).

The rapid telomere movement along the nuclear envelope in budding yeast meiosis is mediated through a complex consisting of Mps3, Ndj1 and Csm4.

In meiosis, Mps3 is connected to both the actin cytoskeleton by Csm4, a protein that sits in the outer nuclear membrane and to telomeres through Ndj1, a telomeric meiotic protein (Conrad et al. 1997, 2008). In support to an actin skeleton-led telomere movement, actin cable extension occurs at a speed similar to chromosome motion (ca. 0.3 μm/s; Yang and Pon 2002). Furthermore, actin cytoskeleton and Csm4 mutants do not impede telomere attachment to the nuclear periphery but rather impair directed rapid telomere movements (Conrad et al. 1997, 2008). Although Mps3 also connects telomeres to the nuclear membrane in interphase (see above), directed chromosome movements mediated by Mps3 have yet only been observed during meiotic prophase. The role of such a chromosome movement is still unclear, but it could help chromosome pairing resolution of meiotic crossovers (Conrad et al. 2008; Koszul et al. 2008; Sonntag Brown et al. 2011).

### *2.3.5 Subtelomeric Dynamics and Associations*

Contrary to the rapid movements observed for telomeres during meiosis, telomeric movements observed during the G1 and S phases of the mitotic cycle are much slower. Moreover, the dynamic behaviour observed for subtelomere 6R during G1 is more constrained than centromere proximal autonomously replicating origin (ARS) during the same cell cycle phase, with a mean speed of 98 nm/s vs. 118 nm/s for the ARS (Heun et al. 2001). Each studied telomere had a distinct mobility, possibly depending on its chromosome arm size and compaction (Bystricky et al. 2005). Subtelomere loci generally occupy a confinement radius of 0.2–0.4 μm in the about 2-μm-diameter yeast nucleus, due to their attachment to the nuclear periphery (Bressan et al. 2004; Bystricky et al. 2005; Heun et al. 2001; Rosa et al. 2006). For comparison, a 16 kb artificially generated chromatin ring moves within a confinement radius of more than 0.8 μm (Gartenberg et al. 2004). Accordingly, the subtelomere confinement radius increases to 0.5 μm in *sir4* mutants and to more than 0.6 μm in y*Ku70* deficient cells (Bystricky et al. 2005; Hediger et al. 2002b). Note that movement measurements by locus detection methods are restricted to G1 phase of the cell cycle, as the replication of DNA coincides with the presence of two fluorescent spots that can hardly be distinguished from one another.

Subtelomeres move with a speed of 0.1–0.5 μm/s meaning that regions preferably occupied by a particular chromosome end are largely overlapping (Heun et al. 2001; Marshall et al. 1997; Schober et al. 2008). Given that subtelomeres of different chromosome arm lengths show similar probability maps of their localization, it is not surprising to find that subtelomeres on similar chromosome arm sizes interact more frequently, although very transiently (Therizols et al. 2010). Hence, different subtelomere interactions detected in living cells as foci are direct consequences of preferred subtelomere localizations (Schober et al. 2008; Therizols et al. 2010; Zimmer and Fabre 2011).

Subtelomeres rarely exceed a distance of 0.3 μm from the nuclear envelope, in an oscillating movement suggesting reversible interactions with nuclear envelope components (Bystricky et al. 2005; Hediger et al. 2002b; Heun et al. 2001; Therizols et al. 2010). Large movements of over 0.5 μm in a 10.5s interval take place only once every 10 min in G1 phase for chromosome ends, in contrast to other chromosomal loci that come to about 10 large movements in the same time frame (Heun et al. 2001). These large telomere movements in G1 phase point towards a certain directionality, although nature or function of the force that could be responsible for them are missing. A higher temporal resolution is expected to give more information on the nature of these movements (Hajjoul et al. 2009). Yet, abolition of all movements by ATP depletion shows that chromosome movements are not only randomly caused by diffusion (Heun et al. 2001).

## 2.4 Conclusions on Functional Implications of Chromosome End Nuclear Architecture from the Repair Point of View

Double-strand break (DSB) repair efficiency in haploid yeast depends on the chromosomal region affected (Ricchetti et al. 2003). The closer the regions are to the chromosomal ends, the higher the repair efficiency rate. This can be explained by the fact that in yeast haploids G1 cells, central chromosome regions are repaired by NHEJ, while chromosomal ends can also be repaired by other mechanisms like telomere addition and break-induced replication (BIR) (Malkova et al. 1996; Sandell and Zakian 1993). Moreover, subtelomeres position close to the nuclear periphery is important, since mutations that release nuclear envelope tethering of subtelomere 11L, decrease repair efficiency of DSBs induced in there (Therizols et al. 2006). Persistent DNA DSBs are also shown to interact with Cdc13 and the telomerase component Est2. These proteins anchor the slowly repaired or unrepairable DSBs to Mps3 at the nuclear periphery (Oza et al. 2009; Schober et al. 2009). The nuclear periphery thus appears to have evolved as a compartment that helps uptake of these specific DNA DSBs.

Following S phase, DNA is replicated, and therefore, two identical homologues exist for each chromosome allowing DNA repair by homologous recombination. After S phase, telomeres are no longer anchored to the nuclear periphery. May be a consequence of nuclear envelope positioning failure in S phase, a hyper-recombination among telomeres has been observed in mutants for the telomere length maintenance protein Tel1 (Schober et al. 2009). One reason why telomeres are no longer anchored at the nuclear envelope could be that more efficient repair mechanisms make telomere anchoring difficult.

Mechanisms for insulating chromosome ends could have been evolved in order to distinguish chromosomal ends from DSBs. For instance, in budding yeast, different pathways to prevent chromosome fusion inhibit NHEJ. They involve at least Rap1 and Rif2 (Marcand et al. 2008). Sir4 also inhibits NHEJ through an unknown mechanism. Coexistence of several pathways is understandable to ensure that chromosome ends will never fuse, unless they are too short (Pardo and Marcand 2005). In agreement with such a negative role in telomere fusion, is the Ku70-mediated fold-back structure, depicted between telomeres and subtelomeres (Marvin et al. 2009b). Accordingly, telomeres are 1,000-fold less effective in DNA damage response activation than DSBs (Lydall 2009). These mechanisms might have evolved by taking advantage of the proteins enriched in there, as it is the case for Rap1 and Sir4. Therefore, nuclear architecture of telomeres appears to be a major pathway to ensure genome stability, by avoiding chromosome end fusion.

On the other hand, subtelomere positioning in a restricted domain at the nuclear periphery also ensures that breaks happening in there can efficiently be repaired with other neighbouring subtelomeres. This could explain why, from an evolutionary point of view, subtelomeres could behave as gene reservoirs. Their improved genome flexibility could also explain why linear chromosomes and therefore telomeres and subtelomeres might have been evolved. In fact, cells seem to be very adaptable with respect to chromosomal structure; telomeres do not seem to be neither specific to eukaryotes as there are examples of bacteria with linear chromosomes nor required to eukaryotes since yeast *S. pombe pot1* mutants survive with fused telomeres forming circular chromosomes (Baumann and Cech 2001; Jain et al. 2010).

Nuclear architecture of yeast telomeres and subtelomeres in a Rabl-like chromosome organization is a characteristic often associated with fast-dividing cells. Many yeast mutants show that telomere attachment to the nuclear envelope is not required for cell survival. It rather seems that the attachment helps to untangle and sort chromosomes for different chromosome interactions and therefore faster cell division, thus out competing less organized chromosome conformations.

**Acknowledgments** We thank all members of the laboratory for many thoughtful discussions. MS is recipient of Univ Pierre et Marie Curie fellowship and EF is supported by CNRS, Institut Pasteur and ANR-PIRIBIO Grant No. ANR-09-PIRI-0024.

## References

Andrulis, E. D., Zappulla, D. C., Ansari, A., Perrod, S., Laiosa, C. V., Gartenberg, M. R., et al. (2002). Esc1, a nuclear periphery protein required for Sir4-based plasmid anchoring and partitioning. *Molecular and Cellular Biology, 22*, 8292–8301.

Armache, K. J., Garlick, J. D., Canzio, D., Narlikar, G. J., & Kingston, R. E. (2011). Structural basis of silencing: Sir3 BAH domain in complex with a nucleosome at 3.0 Å resolution. *Science, 334*, 977–982.

Azzalin, C. M., Reichenbach, P., Khoriauli, L., Giulotto, E., & Lingner, J. (2007). Telomeric repeat containing RNA and RNA surveillance factors at mammalian chromosome ends. *Science, 318*, 798–801.

Baumann, P., & Cech, T. R. (2001). Pot1, the putative telomere end-binding protein in fission yeast and humans. *Science, 292*, 1171–1175.

Belmont, A. S. (2001). Visualizing chromosome dynamics with GFP. *Trends in Cell Biology, 11*, 250–257.

Berger, A. B., Cabal, G. G., Fabre, E., Duong, T., Buc, H., Nehrbass, U., et al. (2008). High-resolution statistical mapping reveals gene territories in live yeast. *Nature Methods, 5*, 1031–1037.

Bressan, D. A., Vazquez, J., & Haber, J. E. (2004). Mating type-dependent constraints on the mobility of the left arm of yeast chromosome III. *Journal of Cell Biology, 164*, 361–371.

Bupp, J. M., Martin, A. E., Stensrud, E. S., & Jaspersen, S. L. (2007). Telomere anchoring at the nuclear periphery requires the budding yeast Sad1-UNC-84 domain protein Mps3. *Journal of Cell Biology, 179*, 845–854.

Bystricky, K., Laroche, T., van Houwe, G., Blaszczyk, M., & Gasser, S. M. (2005). Chromosome looping in yeast: Telomere pairing and coordinated movement reflect anchoring efficiency and territorial organization. *Journal of Cell Biology, 168*, 375–387.

Capra, J. A., Paeschke, K., Singh, M., & Zakian, V. A. (2010). G-quadruplex DNA sequences are evolutionarily conserved and associated with distinct genomic features in Saccharomyces cerevisiae. *PLoS Computational Biology, 6*, e1000861.

Chan, J. N., Poon, B. P., Salvi, J., Olsen, J. B., Emili, A., & Mekhail, K. (2011). Perinuclear cohibin complexes maintain replicative life span via roles at distinct silent chromatin domains. *Developmental Cell, 20*, 867–879.

Chua, P. R., & Roeder, G. S. (1997). Tam1, a telomere-associated meiotic protein, functions in chromosome synapsis and crossover interference. *Genes & Development, 11*, 1786–1800.

Collins, S. R., Miller, K. M., Maas, N. L., Roguev, A., Fillingham, J., Chu, C. S., et al. (2007). Functional dissection of protein complexes involved in yeast chromosome biology using a genetic interaction map. *Nature, 446*, 806–810.

Conrad, M. N., Dominguez, A. M., & Dresser, M. E. (1997). Ndj1p, a meiotic telomere protein required for normal chromosome synapsis and segregation in yeast. *Science, 276*, 1252–1255.

Conrad, M. N., Lee, C. Y., Chao, G., Shinohara, M., Kosaka, H., Shinohara, A., et al. (2008). Rapid telomere movement in meiotic prophase is promoted by NDJ1, MPS3, and CSM4 and is modulated by recombination. *Cell, 133*, 1175–1187.

de Bruin, D., Kantrow, S. M., Liberatore, R. A., & Zakian, V. A. (2000). Telomere folding is required for the stable maintenance of telomere position effects in yeast. *Molecular and Cellular Biology, 20*, 7991–8000.

Dejardin, J., & Kingston, R. E. (2009). Purification of proteins associated with specific genomic Loci. *Cell, 136*, 175–186.

Dekker, J., Rippe, K., Dekker, M., & Kleckner, N. (2002). Capturing chromosome conformation. *Science, 295*, 1306–1311.

Duan, Z., Andronescu, M., Schutz, K., McIlwain, S., Kim, Y. J., Lee, C., et al. (2010). A three-dimensional model of the yeast genome. *Nature, 465*, 363–367.

Dubarry, M., Loiodice, I., Chen, C. L., Thermes, C., & Taddei, A. (2011). Tight protein-DNA interactions favor gene silencing. *Genes & Development, 25*, 1365–1370.

Ebrahimi, H., & Donaldson, A. D. (2008). Release of yeast telomeres from the nuclear periphery is triggered by replication and maintained by suppression of Ku-mediated anchoring. *Genes & Development, 22*, 3363–3374.

Enomoto, S., McCune-Zierath, P. D., Gerami-Nejad, M., Sanders, M. A., & Berman, J. (1997). RLF2, a subunit of yeast chromatin assembly factor-I, is required for telomeric chromatin function in vivo. *Genes & Development, 11*, 358–370.

Falcon, C. M., & Matthews, K. S. (1999). Glycine insertion in the hinge region of lactose repressor protein alters DNA binding. *Journal of Biological Chemistry, 274*, 30849–30857.

Ferreira, H. C., Luke, B., Schober, H., Kalck, V., Lingner, J., & Gasser, S. M. (2011). The PIAS homologue Siz2 regulates perinuclear telomere position and telomerase activity in budding yeast. *Nature Cell Biology, 13*, 867–874.

Feuerbach, F., Galy, V., Trelles-Sticken, E., Fromont-Racine, M., Jacquier, A., Gilson, E., et al. (2002). Nuclear architecture and spatial positioning help establish transcriptional states of telomeres in yeast. *Nature Cell Biology, 4*, 214–221.

Fourel, G., Revardel, E., Koering, C. E., & Gilson, E. (1999). Cohabitation of insulators and silencing elements in yeast subtelomeric regions. *EMBO Journal, 18*, 2522–2537.

Friederichs, J. M., Ghosh, S., Smoyer, C. J., McCroskey, S., Miller, B. D., Weaver, K. J., et al. (2011). The SUN protein Mps3 is required for spindle pole body insertion into the nuclear membrane and nuclear envelope homeostasis. *PLoS Genetics, 7*, e1002365.

Galy, V., Olivo-Marin, J. C., Scherthan, H., Doye, V., Rascalou, N., & Nehrbass, U. (2000). Nuclear pore complexes in the organization of silent telomeric chromatin. *Nature, 403*, 108–112.

Gartenberg, M. R., Neumann, F. R., Laroche, T., Blaszczyk, M., & Gasser, S. M. (2004). Sir-mediated repression can occur independently of chromosomal and subnuclear contexts. *Cell, 119*, 955–967.

Gilson, E., Roberge, M., Giraldo, R., Rhodes, D., & Gasser, S. M. (1993). Distortion of the DNA double helix by RAP1 at silencers and multiple telomeric binding sites. *Journal of Molecular Biology, 231*, 293–310.

Goffeau, A., Barrell, B. G., Bussey, H., Davis, R. W., Dujon, B., Feldmann, H., et al. (1996). Life with 6000 genes. *Science, 274*(546), 563–567.

Gotta, M., Laroche, T., Formenton, A., Maillet, L., Scherthan, H., & Gasser, S. M. (1996). The clustering of telomeres and colocalization with Rap1, Sir3, and Sir4 proteins in wild-type Saccharomyces cerevisiae. *Journal of Cell Biology, 134*, 1349–1363.

Gottschling, D. E., Aparicio, O. M., Billington, B. L., & Zakian, V. A. (1990). Position effect at S. cerevisiae telomeres: Reversible repression of Pol II transcription. *Cell, 63*, 751–762.

Goudsouzian, L. K., Tuzon, C. T., & Zakian, V. A. (2006). S. cerevisiae Tel1p and Mre11p are required for normal levels of Est1p and Est2p telomere association. *Molecular Cell, 24*, 603–610.

Hajjoul, H., Kocanova, S., Lassadi, I., Bystricky, K., & Bancaud, A. (2009). Lab-on-Chip for fast 3D particle tracking in living cells. *Lab on a Chip, 9*, 3054–3058.

Hang, L. E., Liu, X., Cheung, I., Yang, Y., & Zhao, X. (2011). SUMOylation regulates telomere length homeostasis by targeting Cdc13. *Nature Structural & Molecular Biology, 18*, 920–926.

Hecht, A., Laroche, T., Strahl-Bolsinger, S., Gasser, S. M., & Grunstein, M. (1995). Histone H3 and H4 N-termini interact with SIR3 and SIR4 proteins: A molecular model for the formation of heterochromatin in yeast. *Cell, 80*, 583–592.

Hediger, F., Berthiau, A. S., van Houwe, G., Gilson, E., & Gasser, S. M. (2006). Subtelomeric factors antagonize telomere anchoring and Tel1-independent telomere length regulation. *EMBO Journal, 25*, 857–867.

Hediger, F., Dubrana, K., & Gasser, S. M. (2002a). Myosin-like proteins 1 and 2 are not required for silencing or telomere anchoring, but act in the Tel1 pathway of telomere length control. *Journal of Structural Biology, 140*, 79–91.

Hediger, F., & Gasser, S. M. (2006). Heterochromatin protein 1: Don't judge the book by its cover! *Current Opinion in Genetics & Development, 16*, 143–150.

Hediger, F., Neumann, F. R., Van Houwe, G., Dubrana, K., & Gasser, S. M. (2002b). Live imaging of telomeres. yKu and Sir proteins define redundant telomere-anchoring pathways in yeast. *Current Biology, 12*, 2076–2089.

Heun, P., Laroche, T., Shimada, K., Furrer, P., & Gasser, S. M. (2001). Chromosome dynamics in the yeast interphase nucleus. *Science, 294*, 2181–2186.

Hiraga, S., Botsios, S., & Donaldson, A. D. (2008). Histone H3 lysine 56 acetylation by Rtt109 is crucial for chromosome positioning. *Journal of Cell Biology, 183*, 641–651.

Hiraga, S., Robertson, E. D., & Donaldson, A. D. (2006). The Ctf18 RFC-like complex positions yeast telomeres but does not specify their replication time. *EMBO Journal, 25*, 1505–1514.

Hirano, Y., Fukunaga, K., & Sugimoto, K. (2009). Rif1 and rif2 inhibit localization of tel1 to DNA ends. *Molecular Cell, 33*, 312–322.

Hou, Z., Danzer, J. R., Fox, C. A., & Keck, J. L. (2006). Structure of the Sir3 protein bromo adjacent homology (BAH) domain from *S. cerevisiae* at 1.95 A resolution. *Protein Science, 15*, 1182–1186.

Huppert, J. L. (2010). Structure, location and interactions of G-quadruplexes. *FEBS Journal, 277*, 3452–3458.

Jain, D., Hebden, A. K., Nakamura, T. M., Miller, K. M., & Cooper, J. P. (2010). HAATI survivors replace canonical telomeres with blocks of generic heterochromatin. *Nature, 467*, 223–227.

Jin, Q. W., Fuchs, J., & Loidl, J. (2000). Centromere clustering is a major determinant of yeast interphase nuclear organization. *Journal of Cell Science, 113*(Pt 11), 1903–1912.

Kalhor, R., Tjong, H., Jayathilaka, N., Alber, F., & Chen, L. (2012). Genome architectures revealed by tethered chromosome conformation capture and population-based modeling. *Nature Biotechnology, 30*, 90–98.

Klein, F., Laroche, T., Cardenas, M. E., Hofmann, J. F., Schweizer, D., & Gasser, S. M. (1992). Localization of RAP1 and topoisomerase II in nuclei and meiotic chromosomes of yeast. *Journal of Cell Biology, 117*, 935–948.

Kosaka, H., Shinohara, M., & Shinohara, A. (2008). Csm4-dependent telomere movement on nuclear envelope promotes meiotic recombination. *PLoS Genetics, 4*, e1000196.

Koszul, R., Kim, K. P., Prentiss, M., Kleckner, N., & Kameoka, S. (2008). Meiotic chromosomes move by linkage to dynamic actin cables with transduction of force through the nuclear envelope. *Cell, 133*, 1188–1201.

Laroche, T., Martin, S. G., Gotta, M., Gorham, H. C., Pryde, F. E., Louis, E. J., et al. (1998). Mutation of yeast Ku genes disrupts the subnuclear organization of telomeres. *Current Biology, 8*, 653–656.

Laroche, T., Martin, S. G., Tsai-Pflugfelder, M., & Gasser, S. M. (2000). The dynamics of yeast telomeres and silencing proteins through the cell cycle. *Journal of Structural Biology, 129*, 159–174.

Lawrimore, J., Bloom, K. S., & Salmon, E. D. (2011). Point centromeres contain more than a single centromere-specific Cse4 (CENP-A) nucleosome. *Journal of Cell Biology, 195*, 573–582.

Levy, D. L., & Blackburn, E. H. (2004). Counting of Rif1p and Rif2p on *Saccharomyces cerevisiae* telomeres regulates telomere length. *Molecular and Cellular Biology, 24*, 10857–10867.

Lewis, A., Felberbaum, R., & Hochstrasser, M. (2007). A nuclear envelope protein linking nuclear pore basket assembly, SUMO protease regulation, and mRNA surveillance. *Journal of Cell Biology, 178*, 813–827.

Lieb, J. D., Liu, X., Botstein, D., & Brown, P. O. (2001). Promoter-specific binding of Rap1 revealed by genome-wide maps of protein-DNA association. *Nature Genetics, 28*, 327–334.

Lieberman-Aiden, E., van Berkum, N. L., Williams, L., Imakaev, M., Ragoczy, T., Telling, A., et al. (2009). Comprehensive mapping of long-range interactions reveals folding principles of the human genome. *Science, 326*, 289–293.

Liti, G., Carter, D. M., Moses, A. M., Warringer, J., Parts, L., James, S. A., et al. (2009). Population genomics of domestic and wild yeasts. *Nature, 458*, 337–341.

Luke, B., Panza, A., Redon, S., Iglesias, N., Li, Z., & Lingner, J. (2008). The Rat1p 5′ to 3′ exonuclease degrades telomeric repeat-containing RNA and promotes telomere elongation in Saccharomyces cerevisiae. *Molecular Cell, 32*, 465–477.

Lydall, D. (2009). Taming the tiger by the tail: Modulation of DNA damage responses by telomeres. *EMBO Journal, 28*, 2174–2187.

Lynch, P. J., & Rusche, L. N. (2009). A silencer promotes the assembly of silenced chromatin independently of recruitment. *Molecular and Cellular Biology, 29*, 43–56.

Malkova, A., Ivanov, E. L., & Haber, J. E. (1996). Double-strand break repair in the absence of RAD51 in yeast: A possible role for break-induced DNA replication. *Proceeding of National Academy Science of the United States of America, 93*, 7131–7136.

Marcand, S., Brevet, V., & Gilson, E. (1999). Progressive cis-inhibition of telomerase upon telomere elongation. *EMBO Journal, 18*, 3509–3519.

Marcand, S., Pardo, B., Gratias, A., Cahun, S., & Callebaut, I. (2008). Multiple pathways inhibit NHEJ at telomeres. *Genes & Development, 22*, 1153–1158.

Marshall, W. F., Straight, A., Marko, J. F., Swedlow, J., Dernburg, A., Belmont, A., et al. (1997). Interphase chromosomes undergo constrained diffusional motion in living cells. *Current Biology, 7*, 930–939.

Martino, F., Kueng, S., Robinson, P., Tsai-Pflugfelder, M., van Leeuwen, F., Ziegler, M., et al. (2009). Reconstitution of yeast silent chromatin: Multiple contact sites and O-AADPR binding load SIR complexes onto nucleosomes in vitro. *Molecular Cell, 33*, 323–334.

Marvin, M. E., Becker, M. M., Noel, P., Hardy, S., Bertuch, A. A., & Louis, E. J. (2009a). The association of yeast Ku with subtelomeric core X sequences prevents recombination involving telomeric sequences. *Genetics, 183*, 453–467.

Marvin, M. E., Griffin, C. D., Eyre, D. E., Barton, D. B., & Louis, E. J. (2009b). In *Saccharomyces cerevisiae*, yKu and Subtelomeric Core X sequences repress homologous recombination near telomeres as part of the same pathway. *Genetics, 83*, 441–451.

McBryant, S. J., Krause, C., Woodcock, C. L., & Hansen, J. C. (2008). The silent information regulator 3 protein, SIR3p, binds to chromatin fibers and assembles a hypercondensed chromatin architecture in the presence of salt. *Molecular and Cellular Biology, 28*, 3563–3572.

McEachern, M. J., & Blackburn, E. H. (1995). Runaway telomere elongation caused by telomerase RNA gene mutations. *Nature, 376*, 403–409.

Nagai, S., Dubrana, K., Tsai-Pflugfelder, M., Davidson, M. B., Roberts, T. M., Brown, G. W., et al. (2008). Functional targeting of DNA damage to a nuclear pore-associated SUMO-dependent ubiquitin ligase. *Science, 322*, 597–602.

Oza, P., Jaspersen, S. L., Miele, A., Dekker, J., & Peterson, C. L. (2009). Mechanisms that regulate localization of a DNA double-strand break to the nuclear periphery. *Genes & Development, 23*, 912–927.

Paeschke, K., Capra, J. A., & Zakian, V. A. (2011). DNA replication through G-quadruplex motifs is promoted by the *Saccharomyces cerevisiae* Pif1 DNA helicase. *Cell, 145*, 678–691.

Palancade, B., Liu, X., Garcia-Rubio, M., Aguilera, A., Zhao, X., & Doye, V. (2007). Nucleoporins prevent DNA damage accumulation by modulating Ulp1-dependent sumoylation processes. *Molecular Biology of the Cell, 18*, 2912–2923.

Palladino, F., Laroche, T., Gilson, E., Axelrod, A., Pillus, L., & Gasser, S. M. (1993). SIR3 and SIR4 proteins are required for the positioning and integrity of yeast telomeres. *Cell, 75*, 543–555.

Pardo, B., & Marcand, S. (2005). Rap1 prevents telomere fusions by nonhomologous end joining. *EMBO Journal, 24*, 3117–3127.

Pryde, F. E., & Louis, E. J. (1999). Limitations of silencing at native yeast telomeres. *EMBO Journal, 18*, 2538–2550.

Radman-Livaja, M., Ruben, G., Weiner, A., Friedman, N., Kamakaka, R., & Rando, O. J. (2011). Dynamics of Sir3 spreading in budding yeast: Secondary recruitment sites and euchromatic localization. *EMBO Journal, 30*, 1012–1026.

Ribes-Zamora, A., Mihalek, I., Lichtarge, O., & Bertuch, A. A. (2007). Distinct faces of the Ku heterodimer mediate DNA repair and telomeric functions. *Nature Structural & Molecular Biology, 14*, 301–307.

Ricchetti, M., Dujon, B., & Fairhead, C. (2003). Distance from the chromosome end determines the efficiency of double strand break repair in subtelomeres of haploid yeast. *Journal of Molecular Biology, 328*, 847–862.

Rine, J., & Herskowitz, I. (1987). Four genes responsible for a position effect on expression from HML and HMR in Saccharomyces cerevisiae. *Genetics, 116*, 9–22.

Robinett, C. C., Straight, A., Li, G., Willhelm, C., Sudlow, G., Murray, A., et al. (1996). In vivo localization of DNA sequences and visualization of large-scale chromatin organization using lac operator/repressor recognition. *Journal of Cell Biology, 135*, 1685–1700.

Rosa, A., Maddocks, J. H., Neumann, F. R., Gasser, S. M., & Stasiak, A. (2006). Measuring limits of telomere movement on nuclear envelope. *Biophysical Journal, 90*, L24–L26.

Ruault, M., De Meyer, A., Loiodice, I., and Taddei, A. (2011). Clustering heterochromatin: Sir3 promotes telomere clustering independently of silencing in yeast. *J Cell Biol*, 192, 417–431.

Rusche, L. N., Kirchmaier, A. L., & Rine, J. (2003). The establishment, inheritance, and function of silenced chromatin in Saccharomyces cerevisiae. *Annual Review of Biochemistry 72*, 481–516.

Sandell, L. L., & Zakian, V. A. (1993). Loss of a yeast telomere: Arrest, recovery, and chromosome loss. *Cell, 75*, 729–739.

Schober, H., Ferreira, H., Kalck, V., Gehlen, L. R., & Gasser, S. M. (2009). Yeast telomerase and the SUN domain protein Mps3 anchor telomeres and repress subtelomeric recombination. *Genes & Development, 23*, 928–938.

Schober, H., Kalck, V., Vega-Palas, M. A., Van Houwe, G., Sage, D., Unser, M., et al. (2008). Controlled exchange of chromosomal arms reveals principles driving telomere interactions in yeast. *Genome Research, 18*, 261–271.

Schoeftner, S., & Blasco, M. A. (2008). Developmentally regulated transcription of mammalian telomeres by DNA-dependent RNA polymerase II. *Nature Cell Biology, 10*, 228–236.

Smith, J. S., Brachmann, C. B., Pillus, L., & Boeke, J. D. (1998). Distribution of a limited Sir2 protein pool regulates the strength of yeast rDNA silencing and is modulated by Sir4p. *Genetics, 149*, 1205–1219.

Smith, J. S., Chen, Q., Yatsunyk, L. A., Nicoludis, J. M., Garcia, M. S., Kranaster, R., et al. (2011). Rudimentary G-quadruplex-based telomere capping in Saccharomyces cerevisiae. *Nature Structural & Molecular Biology, 18*, 478–485.

Sonntag Brown, M., Zanders, S., & Alani, E. (2011). Sustained and rapid chromosome movements are critical for chromosome pairing and meiotic progression in budding yeast. *Genetics, 188*, 21–32.

Straight, A. F., Marshall, W. F., Sedat, J. W., & Murray, A. W. (1997). Mitosis in living budding yeast: Anaphase A but no metaphase plate. *Science, 277*, 574–578.

Sundquist, W. I., & Klug, A. (1989). Telomeric DNA dimerizes by formation of guanine tetrads between hairpin loops. *Nature, 342*, 825–829.

Taddei, A., Hediger, F., Neumann, F. R., Bauer, C., & Gasser, S. M. (2004). Separation of silencing from perinuclear anchoring functions in yeast Ku80, Sir4 and Esc1 proteins. *EMBO Journal, 23*, 1301–1312.

Tanizawa, H., Iwasaki, O., Tanaka, A., Capizzi, J. R., Wickramasinghe, P., Lee, M., et al. (2010). Mapping of long-range associations throughout the fission yeast genome reveals global genome organization linked to transcriptional regulation. *Nucleic Acids Research, 38*, 8164–8177.

Teixeira, M. T., Arneric, M., Sperisen, P., & Lingner, J. (2004). Telomere length homeostasis is achieved via a switch between telomerase-extendible and -nonextendible states. *Cell, 117*, 323–335.

Tham, W. H., Wyithe, J. S., Ko Ferrigno, P., Silver, P. A., & Zakian, V. A. (2001). Localization of yeast telomeres to the nuclear periphery is separable from transcriptional repression and telomere stability functions. *Molecular Cell, 8*, 189–199.

Therizols, P., Duong, T., Dujon, B., Zimmer, C., & Fabre, E. (2010). Chromosome arm length and nuclear constraints determine the dynamic relationship of yeast subtelomeres. *Proceeding of National Academy Science of the United States of America, 107*, 2025–2030.

Therizols, P., Fairhead, C., Cabal, G. G., Genovesio, A., Olivo-Marin, J. C., Dujon, B., et al. (2006). Telomere tethering at the nuclear periphery is essential for efficient DNA double strand break repair in subtelomeric region. *Journal of Cell Biology, 172*, 189–199.

Trelles-Sticken, E., Adelfalk, C., Loidl, J., & Scherthan, H. (2005). Meiotic telomere clustering requires actin for its formation and cohesin for its resolution. *Journal of Cell Biology, 170*, 213–223.

Trelles-Sticken, E., Loidl, J., & Scherthan, H. (1999). Bouquet formation in budding yeast: Initiation of recombination is not required for meiotic telomere clustering. *Journal of Cell Science, 112*(Pt 5), 651–658.

Wanat, J. J., Kim, K. P., Koszul, R., Zanders, S., Weiner, B., Kleckner, N., et al. (2008). Csm4, in collaboration with Ndj1, mediates telomere-led chromosome dynamics and recombination during yeast meiosis. *PLoS Genetics, 4*, e1000188.

Williams, T. L., Levy, D. L., Maki-Yonekura, S., Yonekura, K., & Blackburn, E. H. (2010). Characterization of the yeast telomere nucleoprotein core: Rap1 binds independently to each recognition site. *Journal of Biological Chemistry, 285*, 35814–35824.
Witkin, K. L., Friederichs, J. M., Cohen-Fix, O., & Jaspersen, S. L. (2010). Changes in the nuclear envelope environment affect spindle pole body duplication in Saccharomyces cerevisiae. *Genetics, 186*, 867–883.
Wotton, D., & Shore, D. (1997). A novel Rap1p-interacting factor, Rif2p, cooperates with Rif1p to regulate telomere length in Saccharomyces cerevisiae. *Genes & Development, 11*, 748–760.
Wright, J. H., Gottschling, D. E., & Zakian, V. A. (1992). Saccharomyces telomeres assume a non-nucleosomal chromatin structure. *Genes & Development, 6*, 197–210.
Yang, H. C., & Pon, L. A. (2002). Actin cable dynamics in budding yeast. *Proceeding of National Academy Science of the United States of America, 99*, 751–756.
Zimmer, C., & Fabre, E. (2011). Principles of chromosomal organization: Lessons from yeast. *Journal of Cell Biology, 192*, 723–733.

# Chapter 3
# Subtelomeric Regions Promote Evolutionary Innovation of Gene Families in Yeast

**Tim Snoek, Karin Voordeckers and Kevin J. Verstrepen**

**Abstract** Subtelomeres, the regions proximal to telomeres, are extremely dynamic parts of eukaryotic genomes. Gene families that reside in subtelomeres differ profoundly from non-subtelomeric gene families: they show increased recombination and duplication rates and often reflect the lifestyle of the organism under study. In the baker's yeast *Saccharomyces cerevisiae*, subtelomeric gene families can be classified into three broad categories: genes involved in the utilization of alternative substrates, adhesion genes, and lastly, poorly characterized genes. Although the mechanisms shaping these gene families are not yet completely unraveled, studies on two typical subtelomeric gene families exemplify how the dynamic nature of chromosome ends can be exploited to rapidly evolve and diversify. Gene duplication has driven the evolution of the *MAL* gene families and provided closely related yeast species with appropriate, environment-specific alleles to metabolize various disaccharides. A second subtelomeric gene family, the adhesion (*FLO*) genes, shows frequent intergenic recombination between different *FLO* copies, thereby creating new *FLO* alleles with distinct adhesive properties. Moreover, stochastic transcriptional silencing and desilencing of subtelomeric genes could allow cells to 'test' these newly evolved genes without committing all cells in a population to the same fate.

T. Snoek · K. Voordeckers · K. J. Verstrepen (✉)
VIB Laboratory for Systems Biology, Gaston Geenslaan 1, 3001 Leuven, Belgium
e-mail: kevin.verstrepen@biw.vib-kuleuven.be

T. Snoek · K. Voordeckers · K. J. Verstrepen
CMPG Laboratory for Genetics and Genomics, Katholieke Universiteit Leuven, Gaston Geenslaan 1, 3001 Leuven, Belgium

E. J. Louis and M. M. Becker (eds.), *Subtelomeres*, DOI: 10.1007/978-3-642-41566-1_3, 

## 3.1 Introduction

Subtelomeres, the regions located right next to the telomeres on eukaryotic chromosomes, are some of the most intriguing and mysterious parts of the genome. Due to their high content of repetitive sequences, subtelomeric regions are underrepresented in sequencing reads and also notoriously hard to assemble (Eichler 2001; Eichler et al. 2004). Therefore, these regions are often lacking from published whole-genome sequences. As more data become available, it becomes clear that subtelomeres are highly variable and unstable regions that reflect an organism's lifestyle. More specifically, subtelomeric regions seem to serve as breeding grounds and testing laboratories for novel genes, allowing rapid evolutionary innovation and swift adaptation to new niches or changing environments.

The baker's yeast *Saccharomyces cerevisiae* was the first eukaryotic species of which the whole genome was sequenced (Goffeau et al. 1996). The *S. cerevisiae* genome is arguably still the best eukaryotic genome sequence available, as it consists of fully assembled chromosomes, including the complete subtelomeric sequences. With the advent of new sequencing technologies, more and more other genomes are being sequenced. Although these sequences mostly lack fully assembled subtelomeres, the exponentially increasing number of genome sequences and (partially assembled) subtelomeres has made it possible to study different subtelomeric regions. Comparing subtelomeric sequences of different species shows that subtelomeres are highly unstable, displaying elevated levels of mutation and recombination (Brown et al. 2010; Mefford and Trask 2002). This led to the conclusion that genomes generally consist of two domains: the stable majority on the one hand and the plastic subtelomeric regions on the other hand (Pryde et al. 1997).

Although the DNA sequence of subtelomeres diverges rapidly, the overall structure and properties of subtelomeres are remarkably similar (Flint et al. 1997). In general, subtelomeres are characterized by low gene density, the absence of essential genes and the presence of specific gene families, positional gene silencing, high sequence similarity between non-homologous chromosomes, and an increased rate of recombination (Cohn et al. 2006; Linardopoulou et al. 2005). In this chapter, these features of (yeast) subtelomeres will be discussed. In addition, specific examples of how subtelomeric characteristics shape the rapid evolution of gene families are given.

## 3.2 Subtelomeric Structure and Gene Families

### *3.2.1 How to Define a Subtelomeric Region?*

In yeast, the telomere consists of ~300 bp of the imperfect repeat $TG_{1-3}$, but the length of this repeat varies both between chromosome ends and between strains (Gatbonton et al. 2006; Shampay et al. 1984; Walmsley and Petes 1985).

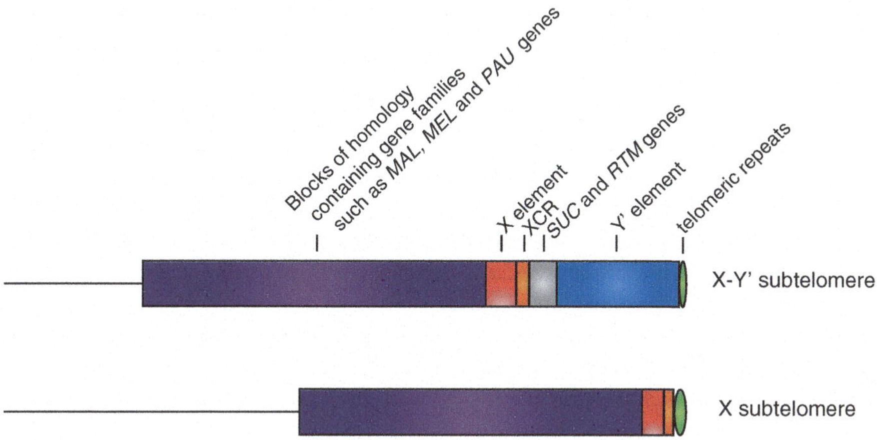


**Fig. 3.1** Subtelomeric structure and location of subtelomeric gene families in *Saccharomyces cerevisiae*. A typical *S. cerevisiae* chromosome end terminates in telomeric repeats and contains a core X element. X element combinatorial repeats (XCR) are often found telomere-proximal to these X elements. Some yeast subtelomeres also have 1–4 copies of a Y′ element and are hence called X–Y′ subtelomeres. Many multigene families, such as the *MAL* and *MEL* genes, reside in large blocks of homology located upstream of the X element. *SUC* and *RTM* genes on the other hand are found between X and Y′ elements

Subtelomeres can be found adjacent to these telomeric repeats and, just like the length of the telomere, the length of subtelomeric regions varies greatly between species (Cohn et al. 2006; Mefford and Trask 2002). The lack of a rigid definition of what a subtelomere exactly is, makes it hard to delineate these regions. Subtelomeres are often defined as the regions next to the telomere that show lower-than-average gene densities and/or that do not contain essential genes (Brown et al. 2010; Pryde and Louis 1997; Teytelman et al. 2008). Using the first definition, yeast subtelomeric regions have been described as varying between 20 and 50 kb from the chromosome end, with 30 kb being the most accurate (Brown et al. 2010). Defining subtelomeres as regions without essential genes causes even greater confusion, as it is not always easy to classify a gene as essential or nonessential.

### *3.2.2 Subtelomeres Contain X and Y′ Elements*

The recent availability of high quality, whole-genome sequences of different (but often closely related) yeast species opened up the possibility to study the variation in subtelomeric regions. One striking observation coming from these comparative genomics studies is that subtelomeres are highly variable, while sharing common structural features. The most prominent elements of yeast chromosome ends are the terminal telomeric repeats, the X elements and the Y′ elements (Fig. 3.1). Every subtelomere harbors an X element, that lies either directly adjacent to the telomere, or is separated from it by one or more Y′ elements (Pryde and Louis 1997). Y′ elements are present at 17 of the 32 chromosome ends in yeast (Chan and Tye 1983; Louis and Haber 1990b, 1992).

X elements show a variable structure, resulting in a total length that ranges from 300 bp to 3 kb. Every X element contains a 473 bp core sequence, which comprises an autonomously replicating sequence consensus sequence (ACS) and usually a binding site for the general regulatory factor Abf1 (Louis et al. 1994; Pryde et al. 1995). At some subtelomeres, telomere-proximal of the core X element, pseudo-repetitive sequences called X element combinatorial repeats (XCRs) are located. There are four types of XCRs (A–D), and different combinations can be found at different chromosome ends, affecting epigenetic silencing (as discussed in Sect. 3.4).

Subtelomeric Y′ elements are either present as a single-copy or tandemly repeated up to four copies, separated by telomeric repeats. Y′ elements exist in two major size classes, long (6.7 kb) and short (5.2 kb), that differ from each other by a series of small insertions/deletions. Several putative open reading frames (ORFs) can be found within Y′ elements, most of which are thought to be non-functional. Interestingly, some ORFs display weak sequence homology to viral helicases (Yamada et al. 1998). Together with the presence of autonomous replication sequences, this suggests that Y′ elements originated from mobile genetic elements that integrated into terminal telomeric repeats (Horowitz and Haber 1985; Louis and Haber 1992).

The function of X and Y′ elements is currently unknown. However, the composition of a subtelomeric region in terms of X and Y′ elements heavily affects positional silencing and nuclear tethering (see Sect. 3.4).

### *3.2.3 Gene Families Found at the Subtelomeres*

Apart from common features such as X and Y′ elements, low gene density and the lack of 'essential' genes, subtelomeres also share another remarkable characteristic. Genes located within subtelomeres are often part of large families that are somehow a telltale sign of the organism's lifestyle (Brown et al. 2010). More specifically, subtelomeres often contain a large number of genes needed to deal with environmental changes. For example, in primates, the genes encoding olfactory receptors are located subtelomerically (Hasin et al. 2008). This large family of receptors allows primates to distinguish a wide variety of aromas and smells, thereby allowing the detection of danger, like fire or rotten food. Similarly, many virulence genes of eukaryotic parasites such as *Plasmodium falciparum* are found near chromosome ends (Roberts et al. 1992). These genes encode many different cell-surface proteins that allow the microbes to elude the host's immune system by constantly switching between different coat proteins (Verstrepen and Fink 2009). Comparative genomics studies underscore the plasticity of

**Table 3.1** Subtelomeric gene families in *S. cerevisiae*

| Category | | Name family | Function | Present in S288c? (number of copies) |
|---|---|---|---|---|
| Carbohydrate consumption | MAL | *MALR* | Regulators that induce *MALR*, *MALS* and *MALT* genes | Yes (4) |
| | MAL | *MALT* | Maltose transporters | Yes (4) |
| | MAL | *MALS* | Maltases (maltose hydrolysis) | Yes (7) |
| | | *MEL* | Alpha-galactosidases (melibiose consumption) | No (0)[a] |
| | | *SUC* | Invertases (sucrose hydrolysis) | No (1)[b] |
| | | *STA* | Glucoamylases (starch consumption) | No (0)[c] |
| Adhesion | | *FLO* | Adhesins | Yes (5)[d] |
| Others (poorly characterized and/or strain-specific) | | *AAD* | Putative aryl-alcohol dehydrogenase | Yes (7)[e] |
| | | *RTM* | Resistance to molasses | No (0) |
| | | *PAU* | Putative role in cell-wall remodeling | Yes (24) |
| | | *COS* | Unknown | Yes (13) |

Representative examples of subtelomeric gene families and their copy number in the standard laboratory strain S288c. Due to the dynamic nature of subtelomeres, some families are completely absent from the S288c genome

[a] *MEL* genes are present in some strains isolated from the wild (Naumov et al. 1995)

[b] S288c only contains *SUC2,* which is non-subtelomeric (Carlson and Botstein 1983)

[c] *STA* genes are found in *S. cerevisiae var. diastaticus* (Yamashita et al. 1985)

[d] *FLO* genes are inactive in S288c due to a mutation in the regulator *FLO8* (Liu et al. 1996)

[e] Distinct *AAD* homologs have been identified in the wine strain AWRI796 (Borneman et al. 2011)

subtelomeres, with a rapid turnover of genes and a remarkable copy number variation in gene families between different species, and often even within one species (Brown et al. 2010).

In the yeast *S. cerevisiae*, gene families located near the end of chromosomes can be roughly divided in three categories: those involved in carbohydrate metabolism, in adhesion, and gene families that are not yet fully characterized (Fig. 3.1; Table 3.1).

#### 3.2.3.1 Gene Families Involved in Carbohydrate Utilization

The *MAL* gene family, encoding genes necessary for the utilization of the sugar maltose, are found in all sequenced *S. cerevisiae* strains to date, but the exact number of *MAL* genes varies greatly between strains. In fact, the *MAL* gene family consists of three subfamilies: *MALT*, encoding transporters: *MALS*, encoding enzymes that hydrolyze maltose (and other sugars); and the *MALR* gene family, encoding transcriptional regulators. These gene families are discussed in more detail in Sect. 3.3.1.1.

Species closely related to *S. cerevisiae* contain dispersed subtelomeric *SUC* loci encoding invertases. These secreted enzymes allow *Saccharomyces*-like yeasts to break down sucrose into glucose and fructose. All *S. cerevisiae SUC* genes are located subtelomerically between X and Y′ elements (except for *SUC2*) (Fig. 3.1). There is a huge variability in number and combinations of *SUC* genes present in different yeast strains (Carlson et al. 1985), with the commonly used laboratory strain S288c only carrying non-subtelomerically located *SUC2*. Similarly, some strains lack *MEL* genes and hence cannot grow on melibiose. Interestingly, wild strains that do posses *MEL* genes often lack *SUC* genes and vice versa (Naumov et al. 1995, 1996).

Another gene family only present in some yeast strains is the *STA* family. The *STA* genes encode secreted glucoamylases that allow *S. cerevisiae var. diastaticus* to utilize starch as a carbon source; something that *S. cerevisiae* is unable to do (Yamashita et al. 1985).

#### 3.2.3.2 Gene Families Conferring Adhesion

Another subtelomeric gene family that has been the subject of many studies is the *FLO* gene family, encoding adhesins that enable cells to adhere to each other and to other (a) biotic surfaces. This gene family is extensively discussed in Sect. 3.5.

#### 3.2.3.3 Poorly Characterized and Strain-Specific Gene Families

A large number of subtelomeric gene families are only partially characterized and their function is mostly inferred based on sequence similarity to genes with known functions. These gene families include *COS* genes (possibly conferring resistance to salt stress) and *PAU* genes (seripauperins, with a putative role in cell-wall remodeling) (Ai et al. 2002; Luo and van Vuuren 2009).

Some strain-specific ORFs were identified with the recent sequencing of feral and industrial strains of *S. cerevisiae*. For example, the wine strain AWRI796 contains a specific subset of subtelomerically located aryl-alcohol dehydrogenase genes (Borneman et al. 2011). These enzymes could contribute significantly to the volatile aromas produced during wine fermentation. Another example is the *RTM1* gene cluster found in strains used in beer production (Ness and Aigle 1995). Multiple copies of this gene confer resistance to an inhibitory substance in molasses, a commonly used substrate in the ethanol production industry. In addition, also the majority of species-specific genes (almost 70 %) are located near telomeres (Kellis et al. 2003).

Taken together, these examples illustrate how subtelomeric gene content reflects the lifestyle of the organism under study. The high degree of plasticity of these genomic regions allows organisms to rapidly adapt to their environment. In the next paragraphs, the different mechanisms that shape subtelomeric genes are discussed in more detail.

## 3.3 Genetic Mechanisms at the Subtelomeres Driving Innovation

Subtelomeres are highly unstable, with increased frequency of gene duplication, recombination, single nucleotide polymorphisms (SNPs), and the occurrence of transposons and unstable repeat sequences. Whereas the first three have already been shown to be statistically significantly enriched at subtelomeres (Brown et al. 2010; Marvin et al. 2009b; Rudd et al. 2007; Schacherer et al. 2009; Teytelman et al. 2008), the situation for transposons and repeat variation is less clear, with only some anecdotal examples available. Below, we discuss these different forms of sequence variability in more detail.

### *3.3.1 Increased Gene Duplication at Subtelomeres and Its Evolutionary Significance*

More than 40 years ago, Susumo Ohno pointed to the key importance of gene duplications for evolutionary innovation (Ohno 1970). He suggested that duplicated genes can provide the raw material needed for the emergence of novel (regulatory) functions in three distinct ways. First, gene duplication creates an extra, dispensable gene copy that can be relieved from purifying selection for its original, pre-duplication function. This copy can acquire mutations that might result in a novel function (*neofunctionalization*), whereas the other paralog keeps performing the original function. Second, duplication can allow for a division-of-labor of the different functions and/or regulatory patterns of a promiscuous pre-duplication gene (*subfunctionalization*). This is particularly interesting for fine-tuning different (conflicting) pre-duplication functions in separate, post-duplication copies. Indeed, the specific molecular composition of an enzyme's active site can prevent optimization of multiple ancestral functions in one single enzyme. Duplication of the ancestral gene can help resolve this adaptive conflict (see for example (Des Marais and Rausher 2008) and the example of the *MALS* gene family discussed below). Third, after duplication, the ancestral function can be preserved in both paralogs, introducing redundancy and/or increasing activity of the gene (*gene dosage effect*).

Duplication events come in two distinct flavors: (1) whole-genome duplications and (2) small-scale duplication events, such as segmental and single-gene duplications. Recent studies have indicated the evolutionary importance of whole-genome duplications (Kellis et al. 2004; Wapinski et al. 2007; Wolfe and Shields 1997). However, compared to segmental and single-gene duplications, such large-scale duplications happen only rarely.

Interestingly, recent studies in yeast demonstrated that small-scale duplications occur much more frequently in subtelomeric regions compared to other genomic regions (Brown et al. 2010; Voordeckers et al. 2012). This elevated level of duplications at subtelomeres is also seen in other species: cytogenic studies have shown that subtelomeres are strikingly polymorphic, with a high degree of copy number variation (CNV) in humans and primates (Linardopoulou et al. 2005; Mefford and Trask 2002; Trask et al. 1998).

*Subtelomeric gene families differ significantly from non-subtelomeric gene families*

Brown and co-workers showed that in yeast, subtelomeric families, i.e., gene families that contain at least one gene located subtelomerically, show significantly more duplication events than non-subtelomeric families ((Brown et al. 2010), see also Fig. 3.2). As a result, subtelomeric families are on average much larger than non-subtelomeric families, containing up to 2–4 times more genes per family. Moreover, subtelomeric genes tend to cluster together in a small number of families: within a specific species, there are far fewer subtelomeric gene families than expected if subtelomeric and non-subtelomeric genes were randomly distributed among families. Additionally, gene families containing a subtelomeric gene are far more likely to contain multiple subtelomeric genes. Taken together, this suggests that subtelomeric genes tend to duplicate frequently, thereby giving rise to additional subtelomeric genes.

Whereas increased duplication rates are observed for most subtelomeric gene families, the duplication events are often lineage-specific (i.e., duplications do not always co-occur in all strains or species). If subtelomeric genes were independently amplified in specific lineages, as opposed to high ancestral copy numbers being lost in many independent lineages, one would expect that subtelomeric members show more sequence similarity to genes within the same species than to genes in other species. Analysis shows that subtelomeric gene families indeed contain more closely related proteins than non-subtelomeric gene families, indicative of recent duplication events.

Moreover, subtelomeric genes can also disappear again. Computational analyses inferred global rates of gene gain and loss and demonstrated that subtelomeric gene families display rapid gene turnover (Brown et al. 2010). Together, these observations portray subtelomeric gene families as large gene families displaying high CNV between (and even within) species. One plausible explanation is the elevated recombination frequencies observed for subtelomeric regions (Louis and Haber 1990a; Louis et al. 1994; Marvin et al. 2009b) (discussed in Sect. 3.3.2).

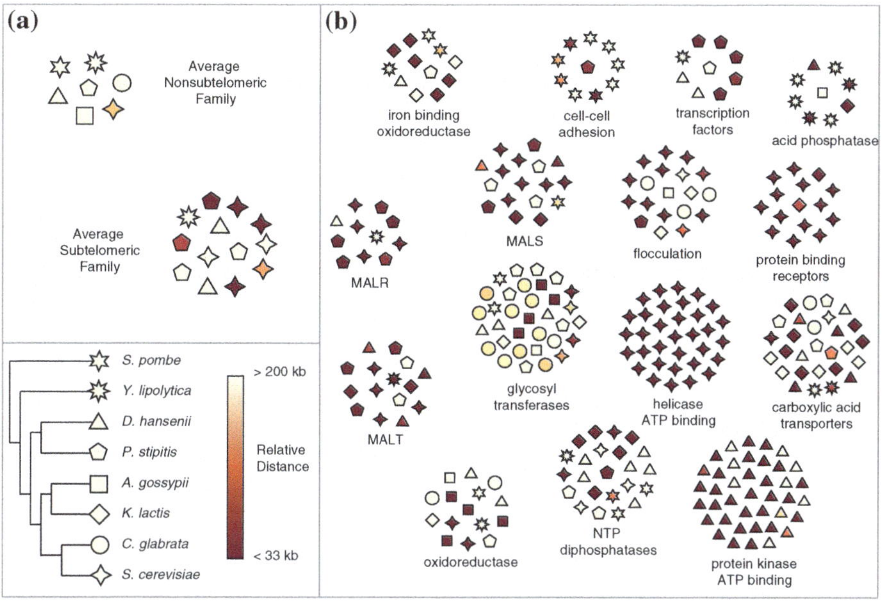


**Fig. 3.2** Subtelomeric gene families differ significantly from non-subtelomeric gene families. Every gene family is represented by a number of polygons (individual genes per gene family). The color of each polygon indicates the closest distance to a chromosome end ('relative distance'), ranging from *pale yellow* (>200 kb) to *dark red* (<33 kb). Each of the eight differently shaped polygons represents one of the individual species (e.g., circle for *Candida glabrata*). **a** The average composition of non-subtelomeric and subtelomeric gene families is represented by two (artificial) clusters. These clusters depict the mean copy number and mean distance from the chromosome end for non-subtelomeric and subtelomeric gene families. Subtelomeric gene families show high copy number variation between species, are generally larger than non-subtelomeric gene families, and have more members close to the chromosome end (especially within 33 kb, indicated by the enrichment for red polygons). **b** Representative subtelomeric gene families, together with their functional annotation. Shared characteristics for these gene families include high copy number variation between species (for example, *Pichia stipitis* has many transcription factors located subtelomerically, whereas *S. cerevisiae* has none) and multiple members close to the nearest chromosome end. Figure from Brown et al. (2010)

Certain functional categories of genes, such as those involved in carbohydrate utilization, genes involved in the response to stress and toxins and genes required for the metabolism of a plethora of substrates are significantly enriched at subtelomeres (Brown et al. 2010; Liti and Louis 2005). On the other hand, typical housekeeping genes, such as those responsible for cell cycle control and ribosomal function, are significantly depleted at subtelomeres. This raised the following question: Is the observed increase in gene turnover and duplication rate for subtelomeric genes an inherent property of their chromosomal location or is it rather a property of the type of genes found at the subtelomeres and the functional category they belong to? Comparing CNV and family size of non-subtelomeric to subtelomeric families belonging to the same functional category clearly points to subtelomeric location as the driving force for increased duplication rates (Brown et al. 2010).

Next to their increased duplication rates (and quite possibly a direct consequence of it), subtelomeric genes also display increased expression divergence and responsiveness (Brown et al. 2010). Together, this further demonstrates the rapid divergence and evolutionary potential of subtelomeric genes. In keep with this hypothesis, duplications of specific subtelomeric regions are commonly found in yeast strains isolated from experimental evolution studies (Gresham et al. 2008).

What mechanism(s) underlie the increased duplication rates in subtelomeric regions? In humans, translocations between different chromosome ends appear to be mainly the result of aberrant repair of double-strand breaks by non-homologous end-joining (Linardopoulou et al. 2005). In this way, subtelomeric blocks can become duplicated. After these translocation events, recombination is possible between blocks on different chromosomal ends, leading to even more variation and new combinations of sequence variants (Rudd et al. 2007) (see also Sect. 3.3.2).

Although the exact mechanism for duplication of subtelomeric genes in *S. cerevisiae* is still unknown, one very likely explanation is the high frequency of ectopic (between copies at different genomic locations) recombination observed in subtelomeric regions (Louis and Haber 1990a; Louis et al. 1994; Marvin et al. 2009a). This is further discussed in Sect. 3.3.2.

#### 3.3.1.1 The *MAL* Gene Families Illustrate the Evolutionary Importance of Subtelomeric Gene Duplications

A key example of subtelomeric duplication and divergence in *S. cerevisiae* are the *MAL* gene families, involved in utilization of the alpha-glucoside maltose (Charron et al. 1989). As will be discussed below, the name *MAL* is actually somewhat misleading, since some of the genes do not respond to maltose at all (Teste et al. 2010).

*Saccharomyces* species contain three related *MAL* gene families, all located subtelomerically and organized as multi-gene loci. The first family, *MALT*, consists of transporters responsible for maltose uptake; the second family, *MALS*, encodes enzymes that hydrolyze maltose into its two constituent glucose molecules; and the third family, *MALR*, contains transcriptional regulators that induce expression of *MALT*, *MALS,* and *MALR* genes when maltose is present (Charron et al. 1989). Each of these gene families can be further subdivided into subfamilies (clades) that group together based on sequence similarity (Brown et al. 2010).

These *MAL* families are typical subtelomeric gene families: they display an extraordinary variability in copy number and chromosomal location between different yeast species (Fig. 3.3). For example, the pathogenic yeast *Candida glabrata* does not contain a single *MAL* locus and thus cannot use maltose as a carbon source. The genome of the *S. cerevisiae* laboratory strain S288c, on the other hand, encodes four *MALT* alleles, seven different *MALS* genes (*MAL12*,

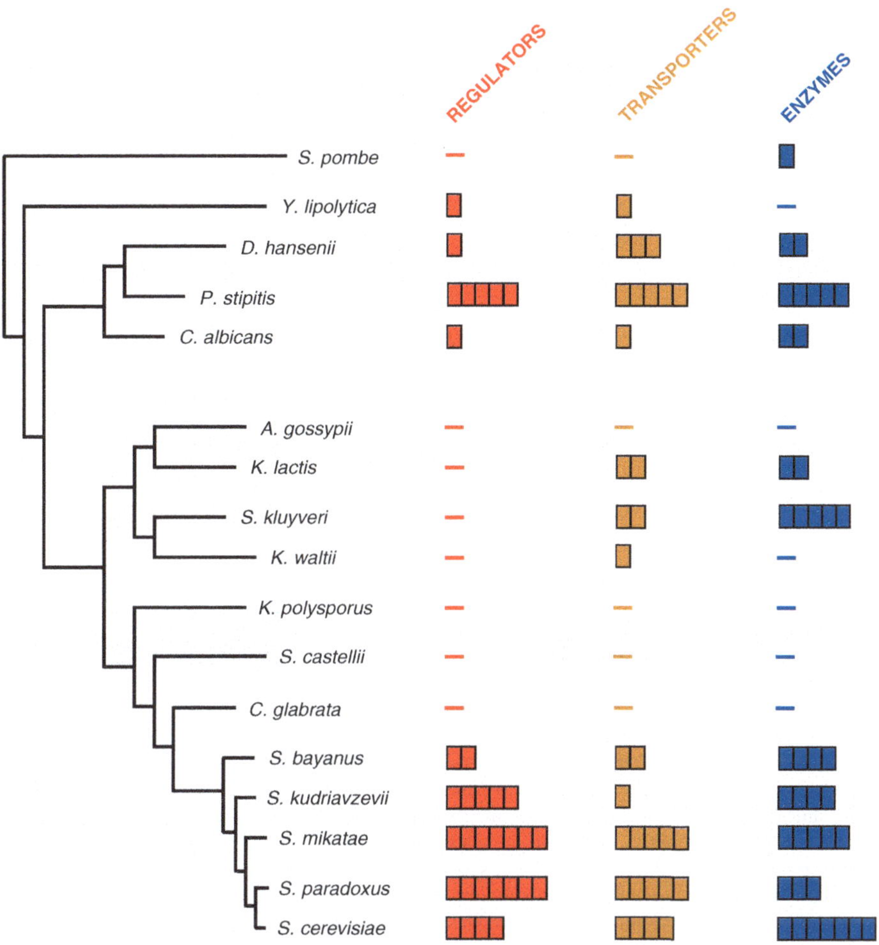


**Fig. 3.3** Copy number variation of the different *MAL* gene families in *Ascomycetes*. The subtelomeric *MAL* gene families show extreme copy number variation in different yeast species. Blocks to the *right* of each species name denote the number of *MAL* regulators (*red*), *MAL* transporter genes (*yellow*) and *MAL* hydrolytic enzymes (*blue*). Figure was adapted from Brown et al. (2010)

*MAL32*, and *IMA1-5*, each with a specific activity profile, see Fig. 3.4) and four *MALR* alleles (Brown et al. 2010; Teste et al. 2010). Interestingly, fluctuations in copy number are also seen within one species. The absence of one specific *MALR* subfamily in the standard laboratory *S. cerevisiae* strain S288c compared to the feral *S. cerevisiae* isolate RM11 explains the different ability of these strains to grow on maltose. Comparing these fully sequenced *S. cerevisiae* strains also reveals many instances of gene duplication and gene loss, as well as differences in chromosomal location of different *MAL* gene family members (Brown et al. 2010).

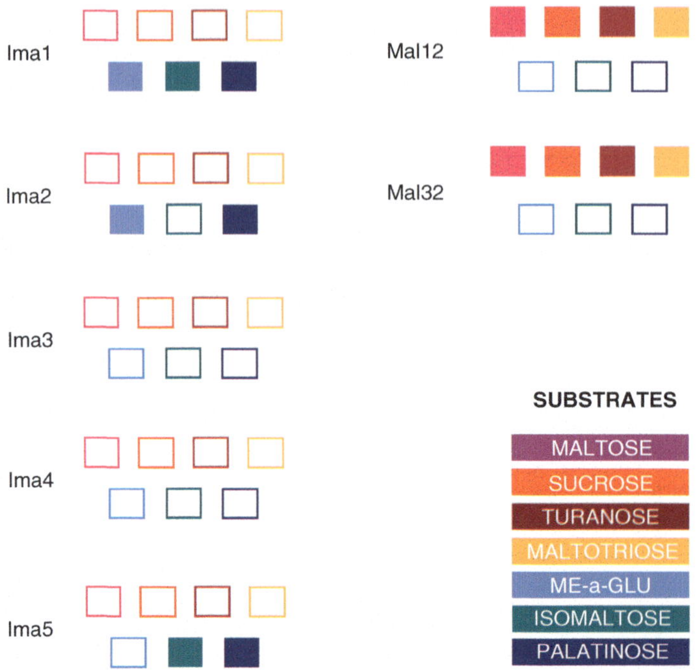


**Fig. 3.4** MalS enzymes in *S. cerevisiae* have different substrate preferences. The standard *S. cerevisiae* laboratory strain has seven different MalS alleles; Ima1–5, Mal12 and Mal32 that can hydrolyze different substrates. Activity toward a specific substrate is indicated by a solid-colored square, white boxes with colored outlines indicate lack of activity for a specific substrate. Mal12 and Mal32 show activity against maltose-like disaccharides often encountered in plant exudates, fruits and cereals, like maltose, maltotriose, sucrose, and turanose (a signaling molecule in plants). Ima1-5 on the other hand shows activity against isomaltose-like sugars including palatinose (found in honey) and isomaltose (an abundant breakdown product of starch). Me-a-glu = methyl-alpha-glucoside. Figure was adapted from Brown et al. (2010)

*Duplications and functional divergence in MAL gene families*

Interestingly, duplication and functional divergence of the different subtelomeric *MAL* gene families seems to underlie the capacity of different yeast species to metabolize a broad spectrum of disaccharides found in plants and fruits. For example, *Saccharomyces paradoxus* strains show amplification of *MAL* alleles that can specifically utilize palatinose and turanose (Brown et al. 2010). These sucrose isomers are commonly found in tree sap and exudates, from which *S. paradoxus* is routinely isolated (Liti et al. 2009). These allele-specific amplifications can thus allow strains to (more efficiently) use sugars found in their ecological niche.

Phylogenetic analyses showed that the common ancestor of the different yeast species investigated contained only a few *MAL* genes (Brown et al. 2010). These ancestral genes were completely lost in some lineages (e.g., in *C. glabrata*) and expanded in others. For example, a single ancestral gene gave rise to the seven present-day *MALS* alleles in the standard laboratory strain S288c through independent duplication events (Voordeckers et al. 2012). These MalS enzymes show activity against a wide range of substrates, with some enzymes having no activity for maltose but instead preferring other alpha-glucosides, such as isomaltose, found in plant exudates (Fig. 3.4). Very recently duplicated alleles, such as Mal12 and Mal32, show an almost identical activity. Here, dosage effects probably provide the necessary selective pressure to preserve these two nearly identical copies in the genome.

Reconstructing the ancestral pre-duplication enzyme shows that this enzyme was multifunctional, hydrolyzing primarily maltose-like substrates but with trace activity for isomaltose-like substrates. Interestingly, the nature of this enzyme's binding pocket prevented optimization of these different activities in one single protein. Homology modeling shows that changes that increase activity toward one substrate class inadvertently compromise activity toward the other substrate class (Voordeckers et al. 2012). Duplication and subsequent divergence allowed optimization of each of these activities in different copies (subfunctionalization). This eventually resulted in a present-day brewer's yeast capable of hydrolyzing a variety of alpha-glucosides much more efficiently than its ancestor. In this way, these duplications may have allowed yeast to colonize new niches containing some of the sugars hydrolyzed by the (novel) MalS alleles. This specific example again illustrates the importance of subtelomeres as breeding grounds for (environment-specific) gene functions.

## *3.3.2 Recombination at Subtelomeric Regions*

### 3.3.2.1 Meiotic Recombination Rates are Low in Subtelomeric Regions

In *S. cerevisiae*, meiotic recombination rates vary over the length of a chromosome, with recombination rates decreasing with increasing distance from the chromosome end (Barton et al. 2008, 2003; Goldman and Lichten 1996). However, little crossing-over is observed in the endmost 10–20 kb of each chromosome. Adjacent euchromatin regions on the other hand exhibit rates that are twofold higher than the genome-wide average (Barton et al. 2008). These results fit with earlier observations that meiotic double-strand break formation is increased 50–100 kb from chromosome ends, the so-called long-range telomere effect (Blitzblau et al. 2007). Since crossovers near chromosome ends have been reported to interfere with proper chromosome segregation (Ross et al. 1996; Su et al. 2000), the decreased meiotic recombination rate observed at (sub) telomeric regions could help to ensure proper chromosome segregation during cell division (Barton et al. 2003).

### 3.3.2.2 Mitotic Recombination Rates are Elevated Near Chromosome Ends

Opposite to what is seen for meiotic recombination rates, rates of mitotic recombination are higher in telomere-proximal regions compared to the average rate observed for the entire yeast genome (Louis and Haber 1990a; Louis et al. 1994;

Marvin et al. 2009b). Studies showed that, whereas the mitotic recombination rate is similar across most of the genome (~$1.5 \times 10^{-7}$ events/mitosis), a significant increase is observed near chromosome ends (Marvin et al. 2009b). For example, recombination rates along the entire right arm of chromosome XV were comparable to the genome average, except for the endmost 5 kb, where recombination rates of up to four times the genome average were reported. It has been suggested that these higher recombination rates are due to telomere clustering during the leptotene stage of the cell cycle, placing the different chromosomal ends in each other's immediate vicinity (Trelles-Sticken et al. 1999).

As discussed earlier, subtelomeric regions in *S. cerevisiae* are all organized in a similar manner (see also Fig. 3.1). A set of elegant experiments demonstrated a high frequency of recombination between different Y′ elements (Louis and Haber 1990a). These ectopic recombination events enable Y′ elements and neighboring sequences to move to chromosome ends that previously did not possess this sequence (thus effectively duplicating this region). This could explain the difference in both copy number and location of Y′ elements between different *S. cerevisiae* strains (Louis and Haber 1990b). No homogenization is seen between the different Y′ size classes, due to the non-random choice in interaction partner for ectopic recombination: long Y′ elements preferentially interact with other long Y′ elements and short Y′ elements opt for other short Y′ elements. The exact reason for this is not yet known, but might be explained by proximity effects during telomere clustering.

Unequal sister chromatid exchange is responsible for the expansion of a single Y′ element into tandem arrays and can thus explain the observed inter-strain Y′ element CNV. This expansion depends on the $(TG_{1-3})_n$ sequences at the junction between X and Y′ elements (Louis et al. 1994). When deprotected, telomeres are recombination hotspots: cells lacking telomerase show increased recombination between these $(TG_{1-3})_n$ tracts and between Y′ elements.

Centromere-proximal sequences are strikingly more divergent than telomere-proximal sequences and X elements have been suggested to act as a recombination barrier (Pryde et al. 1997). Nonetheless, recombination involving X elements is thought to be responsible for the distribution of *SUC* genes in different yeast strains (Carlson et al. 1985). A screen for mutants with increased homologous recombination near telomeres identified yeast Ku proteins as important factors preventing mitotic recombination at these regions (Marvin et al. 2009a, b). Studies suggest that yKu proteins, via association with X elements, help to create a protective fold-back structure at chromosome ends (Marvin et al. 2009a, b). This in turn could help protect these regions from deleterious events, such as aberrant recombination and chromosomal rearrangements.

The increased subtelomeric recombination rate is also evident from comparative genomic studies: subtelomeric gene families, such as the *FLO* and *MALS* gene families, show significant differences in order, number, and orientation between different yeast species and strains (Brown et al. 2010; Van Mulders et al. 2010). Moreover, the *FLO* gene family underwent extensive recombination (Christiaens et al. 2012 and further discussed in Sect. 3.5).

The increased mitotic recombination rate is not the sole explanation for why subtelomeric sequences are more divergent, chromosome ends also display an increased level of nucleotide divergence (Schacherer et al. 2009; Teytelman et al. 2008).

### 3.3.3 Increased SNP Frequency at Subtelomeric Regions

Several comparative genomics studies indicated that telomere-proximal sequences are highly divergent between species and even between strains from the same species (Schacherer et al. 2009). For example, loss and gain of stop codons by nucleotide substitutions occur more frequently near chromosome ends (Kellis et al. 2003) and subtelomerically located transcription factor binding sites are less conserved between species (Francesconi et al. 2011).

Analysis of intergenic regions in *Saccharomyces* sensu stricto species showed that (single-copy) subtelomeric regions display an increased SNP frequency (Teytelman et al. 2008). This frequency decreases with increasing distance from the chromosome end. Interestingly, this hyperdivergence appears to be a shared characteristic of silenced regions and is more pronounced in constitutively silenced regions compared to transiently silenced regions, with subtelomeric regions being an example of the latter (see also Sect. 3.4).

Elevated SNP frequencies reflect the impact of both the strength of selection and the rate of nucleotide exchange. Selection appears to play a negligible role, since also synonymous coding positions in silenced regions display a higher SNP frequency. Increased SNP frequencies could be due to a higher substitution rate or could point to DNA repair issues. Although non-transcribed (and hence non-expressed) genes are targeted less frequently by DNA repair machinery (Svejstrup 2002), no correlation was observed between expression and SNP frequency, indicating that other mechanisms are at play here. One possibility requiring further investigation is that silenced DNA might have a reduced replication fidelity.

What are the potential benefits of hyperdivergence of these genomic regions? First, it is important to note that it remains unsure whether subtelomeres really show elevated rates of mutation events. Subtelomeres have a low gene density (Pryde et al. 1997), so substitutions could merely be tolerated here since chances of hitting an (essential) gene are low. It is possible that non-subtelomeric regions show similar mutation rates, but the mutations may be removed from the population due to stronger negative (purifying) selection associated with the higher density of active and essential genes. Although studies show variation in mutation rates across the genome (Hawk et al. 2005; Ito-Harashima et al. 2002; Lang and Murray 2011), a clear link between subtelomeres and (increased) mutation rate has not been established yet.

Since subtelomeric regions are also hotspots for Ty5 transposon integration (Kim et al. 1998) (see also Sect. 3.3.4), hyperdivergence could help inactivate them after subtelomeric integration. Rapid sequence diversification can also inhibit potential detrimental recombination between chromosome ends. Lastly, given that the type of genes residing at subtelomeres are important for a proper response to changes in the environment (Brown et al. 2010; Liti and Louis 2005), creating diversity at these regions could provide organisms with a significant advantage.

### 3.3.4 Transposable Elements Found at the Subtelomeres

Transposable elements can contribute significantly to genome evolution, for example by promoting chromosomal rearrangements (Zou et al. 1996b). The *S. cerevisiae* genome contains five different classes of retrotransposons (mobile genetic elements that replicate through reverse transcription of an mRNA intermediate), Ty1–5 (Kim et al. 1998; Voytas and Boeke 1992). Only the latter class is often found near chromosome ends: characterization of endogenous Ty5 elements mapped these elements to subtelomeres, and to silenced mating-type loci (Zou et al. 1995). In fact, silenced chromatin appears to act as a 'homing device' for Ty5 integration: newly inserted Ty5 transposons preferentially insert in subtelomeric regions or near silent mating loci (Zou et al. 1996a, b). Subsequent studies demonstrated that Ty5 is recruited to these sites via its interaction with the Sir4 protein (Zhu et al. 2003).

Noteworthy, subtelomeric Y′ elements have a possible mobile element origin (Louis and Haber 1992). However, these elements have never been documented to actually transpose. In the fruitfly *Drosophila melanogaster*, some type of transposons can actually serve as telomeres (Biessmann et al. 1990; Levis et al. 1993).

### 3.3.5 Repeat Variation

Tandem repeats represent another source of genetic variation. These unstable genetic elements consist out of short DNA sequences ('units') repeated head-to-tail (for a recent review, see (Gemayel et al. 2010)). Originally considered to be mainly junk DNA, recent studies demonstrated the importance of tandem repeats for rapid evolution (Verstrepen et al. 2005; Vinces et al. 2009). Changes in the number of repeats occur frequently and are mainly due to strand-slippage replication and recombination.

Chromosome ends are prime examples of repeats, consisting of tracts of ($C_{1-3}$ A/$TG_{1-3}$) repeats (Zakian 1996). Just like other repeats, these simple telomeric repeats are dynamic, with the total length of each repeat tract varying between different chromosomes and even between strains (Gatbonton et al. 2006; Walmsley and Petes 1985). Despite this variability, a minimal length is required for genome stability: telomeres protect chromosome ends from degradation and fusion. Cells defective in telomere maintenance display a shortening of chromosomes and subsequent senescence (Lundblad and Szostak 1989).

A second type of repeats are found distal to these telomeric repeats and comprise two major groups (also discussed in Sect. 3.2.2), X and Y′ sequences. When Y′ minisatellites are excluded, genome-wide studies on tandem repeat variation did not identify subtelomeric regions as being enriched in tandem repeats (Legendre M. personal communication).

Taken together, this enrichment of various types of mutations at the subtelomeres illustrates their extensive plasticity, underscoring the evolutionary potential of these genomic regions. In addition to these genetic differences, also epigenetic effects contribute to this.

## 3.4 Epigenetic Silencing of Subtelomeric Gene Families

Apart from the high degree of DNA sequence variation, another hallmark of subtelomeres is that genes residing in subtelomeres are often transcriptionally silenced in a position-dependent manner (Gottschling et al. 1990; Pryde and Louis 1999; Wyrick et al. 1999). In yeast, positional-silenced regions can be divided into three categories: (1) ribosomal DNA tandem repeats, (2) silent mating-type loci, and (3) subtelomeric regions and telomeric repeats (Sherman and Pillus 1997). The mechanisms involved in silencing these spots are similar, although there are some pronounced differences (extensively covered by Rusche et al. (2003)). Subtelomeric silencing is unstable and seems to have a more stochastic nature: a silenced gene can switch to an active state after a number of generations. Subtelomeric silencing, also called 'Telomere Position Effect' (TPE), is discussed in more detail below.

### *3.4.1 Telomere Position Effect in Yeast*

#### 3.4.1.1 Telomere Position Effect at Truncated Ends Leads to Unstable and Stochastic Expression

Positional silencing at the subtelomeres was first elucidated in *S. cerevisiae* by Gottschling and colleagues (Gottschling et al. 1990). A reporter gene was inserted at a chromosome end between the most distal gene and the telomeric repeats, thereby deleting all native sequence in between (containing X and possibly Y′ elements). Three main observations were made. First, a reporter gene inserted near telomeres is often transcriptionally repressed. Second, in a clonal population of cells, silencing of the reporter gene is heterogeneous; some cells express the gene and others do not. Third, the silencing of the marker gene is mitotically inherited, but the gene can switch to an active state after a number of generations.

Later studies pointed out that also native subtelomeric genes display this stochastic behavior. For example, *FLO11*, a member of the subtelomeric flocculin gene family (which is discussed in more detail in Sect. 3.5), shows variegated expression in a clonal population of diploid cells (Halme et al. 2004) (see Fig. 3.5). Natural chromosome ends show some pronounced differences in TPE compared to truncated chromosome termini.

#### 3.4.1.2 TPE at Natural Chromosome Ends Shows a Pronounced Silencing Pattern

In an elegant study, Pryde and Louis elucidated the extent of the TPE at native yeast subtelomeres. Interestingly, they found that silencing at native subtelomeres is remarkably different from TPE at truncated chromosome ends. The widely used marker gene *URA3* (see Box 1) was inserted in various positions at different chromosome ends, keeping all native subtelomeric elements intact (Pryde and Louis 1999). Silencing was measured as the fraction of cells that were both able to grow on plates containing 5-FOA and on medium lacking uracil after subculturing (thereby only including cells that had reversibly silenced *URA3*). The repressed region in a subtelomeric region is very limited, with the highest level of repression in the proximity of the ARS consensus sequence in an X element. Furthermore, the repression level dropped with increasing distance from the chromosome end. On the other hand, Y′ elements were found to be resistant to silencing (Pryde and Louis 1999) and, even more strikingly, at some chromosome ends, no silencing was observed at all. These findings are in line with observations that some subtelomeric genes are indeed transcribed (Pryde and Louis 1997, 1999).

How is subtelomeric silencing established? Two mechanisms of subtelomeric silencing are present in yeast, both involving histone deacetylation. The first one is heavily dependent on Sir proteins and acts on regions close to the chromosome termini, whereas the second one relies on Hda1 and exerts its effect on more distal regions.

**Box 1 *URA3* System**

The URA3 system is a main driver of genetic studies in S. cerevisiae since the same marker can be used to screen for gene activity and inactivity (positive and negative selection). Because URA3 encodes an enzyme involved in the biosynthesis of pyrimidine ribonucleotides, the URA3 gene is essential for yeast when uracil is not present in the growth medium, allowing selection for cells harboring active URA3. Since the same enzyme will convert 5-FOA to a toxic compound, only cells with an inactive URA3 gene will survive on a medium supplemented with 5-FOA (Boeke et al. 1987).

## *3.4.2 Sir-Dependent Silencing Acts Close to Chromosome Termini and Establishes Silent but Dynamic Heterochromatin*

The first mechanism depends on the histone deacetylase Sir2 and affects genes located up to 6–8 kb from the telomere (Gottschling et al. 1990; Wyrick et al. 1999). Extensive biochemical studies have identified the key players that are indispensable for Sir-mediated gene silencing in *S. cerevisiae*: Sir2, Sir3, and Sir4 (Aparicio et al. 1991); yKu70 and yKu80 (Boulton and Jackson 1998; Laroche et al. 1998); and the C-terminal domain of Rap1 (Kyrion et al. 1992, 1993). TPE in higher organisms is often driven by homologs of these factors, underscoring the evolutionary importance of this mechanism (for a review, see (Ottaviani et al. 2008)). The current model describing Sir-mediated subtelomeric silencing consists of two main steps: nucleation at the telomere followed by polymerization toward the subtelomere.

In the nucleation step, Rap1 binds the telomeric repeats directly (Shore and Nasmyth 1987) and recruits Sir3 and Sir4 through its C-terminal domain (Jeppesen 1997; Moretti et al. 1994; Wotton and Shore 1997). The Ku proteins bind to the end of the telomeres and also recruit Sir4 (Bertuch and Lundblad 2003; Tsukamoto et al. 1997). Bound Sir4 recruits the $NAD^+$-dependent histone deacetylase Sir2. This protein deacetylates the N-terminal tails of histones H3 and H4. The deacetylated histones are bound by the Sir3/Sir4 complex, which will recruit new Sir2/Sir4, resulting in a continuous deacteylation and binding cyclus that leads to spreading of the silencing complex in the direction of the subtelomeres (polymerization).

TPE is counteracted by Sas2, a protein involved in acetylation of H4-K16 (Suka et al. 2002). In this way, a boundary between the silenced subtelomeric region and the adjacent euchromatin is established (Kimura et al. 2002).

Although this silencing mechanism establishes a highly condensed structure, yeast heterochromatin is quite dynamic. For instance, a variety of transcriptional activators can still bind the DNA (Andrau et al. 2006; Chen and Widom 2005; Sekinger and Gross 2001), and also base-pair substitutions can be found in these regions (Teytelman et al. 2008). A recent study elucidated that this dynamic nature can partly be ascribed to acetylation of H4K12, which suppresses abnormal condensation of heterochromatin by Sir proteins (Zhou et al. 2011).

#### 3.4.2.1 Some Subtelomeric Regions Escape Sir-Dependent Silencing

Y′ elements escape from this silencing cascade, whereas X elements undergo maximum silencing (Pryde and Louis 1999; Zhu and Gustafsson 2009). X elements can actually serve as proto-silencers, which means they cannot induce silencing by themselves, but rather relay the silencing activity that originates from the telomeres. Binding of Sir2, Sir3, and Rap1 is maximal at X elements and this decreases in the direction of the centromere. In addition, XCRs, the composition of which varies per chromosome end, can form chromatin domain boundaries which protect Y′ elements from silencing. Physicochemical studies additionally showed that telomeres fold on themselves, thereby bringing telomeres and subtelomeres in close contact and looping out Y′ elements (Strahl-Bolsinger et al. 1997). Since yeast subtelomeres vary in their composition of X and Y′ elements, the degree of silencing at each subtelomere is different.

#### 3.4.2.2 Subtelomeric Silencing is Influenced by Nuclear Architecture

Localization of the chromosome ends in the nucleus contributes to positional silencing. The 32 telomeres of a haploid yeast cell are clustered into 3–8 foci, involving Ku proteins to anchor telomeres to the nuclear envelope (Hediger et al. 2002; Taddei et al. 2004, 2009). It is thought that this clustering in space favors silencing through an increase in the local concentration of Sir proteins. In addition, clustering of silenced regions avoids Sir proteins from binding promiscuously to other sites in the genome (Taddei et al. 2009). It has been argued that gene silencing might be a side-effect resulting from the sequences and proteins involved in nuclear localization of the telomeres (Pryde and Louis 1999).

### *3.4.3 Distal Subtelomeric Regions are Silenced by Hda1*

The second subtelomeric silencing mechanism acts independently from Sir2 and affects regions that are located at ~10–25 kb from the telomere with a length between ~4 and ~34 kb. These silenced regions are long continuous stretches of chromatin that are acetylated upon deletion of the histone deacetylase Hda1 (Robyr et al. 2002), and therefore go by the name of Hda1-affected subtelomeric (HAST) domains. In total, about 149 ORFs reside in HAST domains, including the *MAL* genes and the majority of the *FLO* genes (see Sects. 3.2 and 3.5). A pronounced difference with Sir-mediated subtelomeric silencing is that many HAST-located genes become active when yeast encounters stress or alternative carbon sources (Hughes et al. 2000; Robyr et al. 2002).

### *3.4.4 Subtelomeric Genes Can Be Regulated Dynamically*

Although subtelomeric genes are generally silenced (Wyrick et al. 1999), some genes can be activated by environmental cues. A range of stress conditions can cause inactivation of Sir3, resulting in decreased subtelomeric silencing (Ai et al. 2002). The resistance of cells to chlorpromazine (which stretches the plasma membrane) could be ascribed to the induction of subtelomeric *PAU* genes. Counting 24 members, the *PAU* genes form the largest gene family in *S. cerevisiae*, yet not much is known about their function (Luo and van Vuuren 2009). These genes are highly induced during alcoholic fermentation (Marks et al. 2008; Rossignol et al. 2003) and show homology to proteins involved in maintaining cell-wall integrity during stress (Abramova et al. 2001; Alimardani et al. 2004).

Recent studies reported an enrichment of transcription factors localized to the subtelomeres, some of which only bind these regions under particular conditions (Mak et al. 2009; Smith et al. 2011). For instance, *PAU* genes were induced upon oxidative stress, and this induction correlated with the recruitment of the transcription factor Aft2 to upstream sequences of those genes (Mak et al. 2009).

Could there be any advantage in silencing subtelomeric genes? It has been postulated that the high variability of subtelomeric gene sequences can partially be explained by the fact that silenced genes are invisible to selection (Rando and Verstrepen 2007). Under stress, by inactivating the Sir-silencing complex, cells gain access to this reservoir of genetic variation, which can help to cope with a changing environment. In addition, stochastic expression of a subtelomerically located gene can allow a population to 'bet-hedge,' which is nicely illustrated by the *FLO* gene family, discussed in the next section.

## 3.5 Example of Variability in a Subtelomeric Gene Family: *FLO* Genes

In *S. cerevisiae*, the *FLO* ('flocculation') genes encode adhesins that enable cells to stick to other cells or to (a)biotic surfaces. They are one of the best-characterized subtelomeric gene families [for reviews, see (Verstrepen and Fink 2009; Verstrepen and Klis 2006; Verstrepen et al. 2004)]. Adhesion is an important trait for yeast since it allows colonization of different niches and can protect cells from various stresses (Palkova and Vachova 2006; Smukalla et al. 2008).

*S. cerevisiae* flocculins, like other fungal adhesins, have a basic modular structure with an N-terminal, lectin-like globular domain conferring adhesion to carbohydrates, a highly glycosylated repeat-rich middle domain that acts as a spacer and a C-terminal domain that can be covalently linked to the cell wall via a GPI anchor [reviewed by Verstrepen et al. (2004)]. The standard laboratory strain S288c contains five *FLO* genes. *FLO1*, *FLO5*, *FLO9*, and *FLO10* are located within 40 kb from their respective telomere within a HAST domain (Robyr et al. 2002; Verstrepen et al. 2004), whereas *FLO11* resides at 46 kb from a chromosome end and is therefore sometimes considered not to be subtelomeric (Lo and Dranginis 1996). Each Flo protein confers distinct adhesive properties (Govender et al. 2010). For instance, *FLO10* and *FLO11* are important for both invasive growth and flocculation, whereas *FLO1* is only involved in flocculation (Guo et al. 2000).

Variability in these genes can be generated in various ways (Fig. 3.5). First, *FLO* genes contain intragenic tandem repeats. Expansion and contraction of these

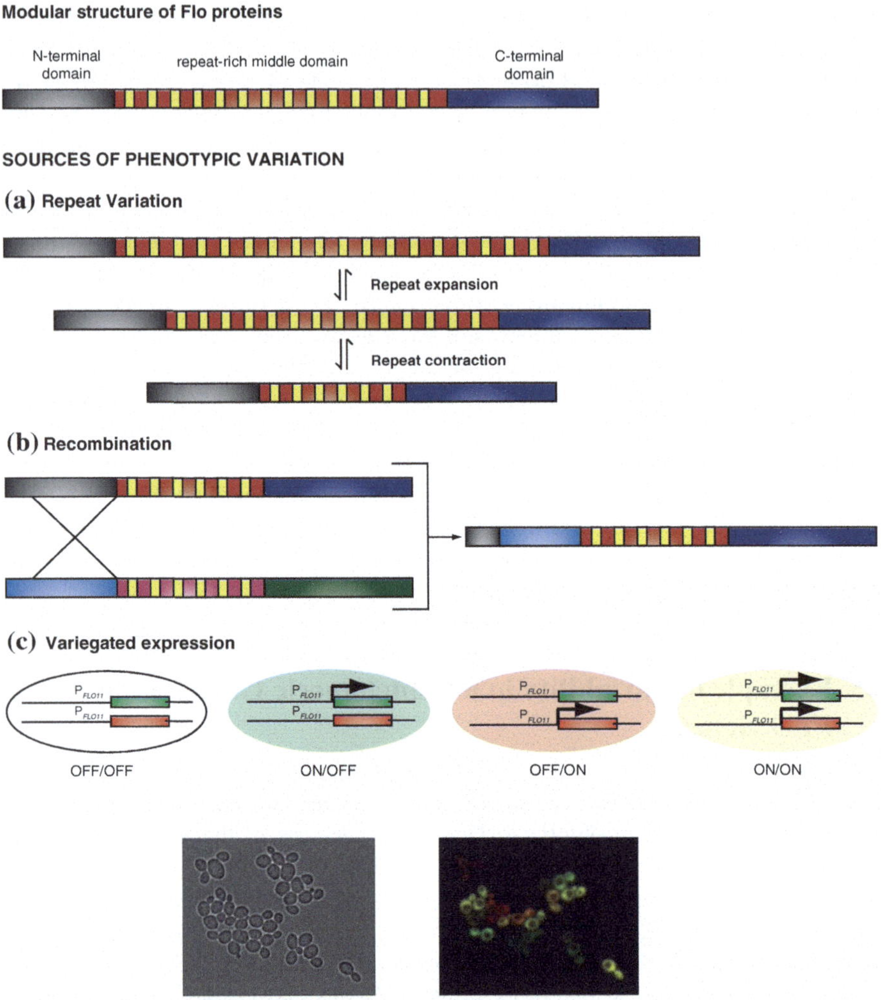


**Fig. 3.5** Sources of phenotypic variation in *FLO* gene family. Flo proteins show a modular structure, consisting of an N-terminal domain involved in adhesion, a repeat-rich middle domain that serves as a spacer and a C-terminal domain that anchors the protein to the cell wall. Both variations in structure (**a** and **b**) and expression (**c**) of *FLO* genes allow for quick changes in phenotype. **a** Unstable intragenic tandem repeats in central domain. The central domain of *FLO* genes contains unstable tandemly repeated DNA. The number of tandem repeats can change rapidly (repeat expansion and repeat contraction), creating genetic and phenotypic variations. **b** Recombination between *FLO* genes generates novel alleles with different characteristics. Recombination events are observed frequently between different *FLO* genes. Recombination can shuffle the domains of existing *FLO* genes and generate a new allele with different adhesive properties. **c** Expression driven by the *FLO11* promoter is variegated and acts as an epigenetic switch. Fluorescent marker genes under control of the *FLO11* promoter ($P_{FLO11}$ in figure) show variegated expression. Octavio and co-workers replaced *FLO11* ORFs in diploid cells with two different fluorescent markers and visualized expression. Four different expression states could be observed (OFF/OFF, ON/OFF, OFF/ON, and ON/ON) in a clonal population of cells, demonstrating variegated expression. Microscopy pictures were kindly provided by Narendra Maheshri and were previously published in Octavio et al. (2009)

repeats is frequently observed and allows cells to rapidly adapt to changes in their environment (Verstrepen et al. 2005). Second, recombination of *FLO* genes is an important driver of variability (Christiaens et al. 2012). Third, epigenetic and genetic regulation of the *FLO* genes allows cells to 'bet-hedge' and quickly adapt to a changing environment.

### *3.5.1 Intragenic Tandem Repeats in FLO Genes Generate Variability in Adhesion*

Just like many other cell-wall (associated) proteins, *FLO* genes contain intragenic tandem repeats. Originally thought of as mainly junk DNA, these tandem repeats have proven to be important sources of genetic and phenotypic variability. Due to recombination, repeat numbers can change rapidly, allowing swift adaptation. The number of tandem repeats in *FLO* genes varies from strain to strain, and this can have functional consequences. For example, increasing the number of repeats in *FLO1* results in increased flocculation (Verstrepen et al. 2005).

### *3.5.2 Recombination of FLO Alleles Drives Innovation*

Anecdotal examples already suggested recombination between different FLO alleles. For instance, *Saccharomyces pastorianus* possesses the gene Lg-*FLO1*, which probably resulted from recombination between the S288c alleles *FLO5* and *YAL065c* (a pseudogene with homology to *FLO1*) (Kobayashi et al. 1999; Ogata et al. 2008).

Analysis of different *Saccharomyces* sensu stricto yeasts revealed that recombination between *FLO* alleles is indeed very common and appears to be a shared characteristic of cell-surface proteins and subtelomerically located genes (Christiaens et al. 2012). Zooming in on the regions where recombination occurred showed that a sequence similarity of around ten nucleotides is sufficient to drive this recombination. For the different *FLO* alleles examined, recombinations were found across the central repeats, in the N-terminal region and in the C-terminal part. Recombination between the N-terminal regions of two *FLO* alleles, for example, can alter the sugar-binding properties and preference of this domain. Indeed, wet laboratory experiments mimicking some of these recombination events demonstrated that recombining different *FLO* alleles results in new, functional adhesins. These engineered proteins conferred adhesive properties to cells that differed from those conferred by the parental alleles (Christiaens et al. 2012). Together with the increased variability generated by the intragenic tandem repeats, these recombination events result in every *S. cerevisiae* strain carrying a set of *FLO* alleles that differ from the five alleles described in the standard laboratory strain S288c.

The increased recombination rate is not limited to *Saccharomyces* adhesins, but is also seen in the *C. glabrata EPA* gene family and the *Candida albicans ALS* gene family, both located subtelomerically (Christiaens et al. manuscript in preparation). Recombination shuffles the domains of existing copies and can thus generate adhesins with novel functional properties. In this way, a relatively small number of genes provides cells with ample raw material for adaptation to novel conditions.

## *3.5.3 Epigenetic Regulation of FLO Genes Generates Variability*

Most *FLO* genes are silenced, yet it has been postulated that silent *FLO* genes form a reservoir of cell-surface variation that can be accessed under particular conditions (Halme et al. 2004). In the common laboratory strain S288c, all *FLO* genes are silenced as a result of a mutation in *FLO8*, encoding a transcriptional regulator of the *FLO* genes (Liu et al. 1996). Hence, most studies on *FLO*-controlled phenotypes are carried out in a different genetic background, such as ∑1278b, where the *FLO8* regulator is not inactivated.

### 3.5.3.1 *FLO11* Expression Drives a Developmental Switch

In a clonal population of ∑1278b cells starved for nitrogen, *FLO11* shows a variegated expression pattern, with only some cells expressing *FLO11* (Halme et al. 2004). Just like genes subjected to TPE (see also Sect. 3.4), *FLO11* expression is heritable and reversible: a single ON cell will give rise to a population of both ON and OFF cells. Most importantly, there is a tight correlation between the presence of Flo11p and the developmental fate of the cell. Cells expressing *FLO11* will form pseudohyphal filaments, where cells remain attached to each other after division (Lambrechts et al. 1996; Lo and Dranginis 1998). On the other hand, cells that do not express *FLO11* will stay in the regular, planktonic form (Halme et al. 2004). For yeast, pseudohyphal growth is a strategy to invade the agar and to forage for nutrients (Gimeno et al. 1992). In this way, variegated *FLO11* expression allows a clonal population to test two strategies without committing all cells to the same developmental form.

Extensive research has been performed to characterize the signaling pathways and regulatory proteins controlling *FLO11* expression. The *FLO11* promoter is large (3.5 kb) and its regulation is very complex, involving many regulatory factors (Octavio et al. 2009). In addition, genetic studies pointed out that *FLO11* silencing is regulated both in a position-dependent and in a promoter-specific way. In line with its genomic location, *FLO11* silencing is not mediated by TPE, but rather relies on specific recruitment of the histone deacetylase Hda1. Moreover, both copies of the *FLO11* promoter in a diploid cell switch in a slow, random, and independent fashion (see Fig. 3.5), which might be an additional mechanism to generate variability in adhesion (Octavio et al. 2009).

#### 3.5.3.2 Benefits of Stochastic Gene Expression

What advantage could stochastic silencing of subtelomeric genes have? Many pathogens, like the malaria pathogen *P. falciparum*, have evolved mechanisms, similar to *FLO* regulation in *S. cerevisiae*, to generate variety at their cell surface to evade the host immune response (Roberts et al. 1992). Several modeling studies demonstrated that stochastic switching is preferred over transcriptional regulation when the environment fluctuates randomly over timescales that are more or less equal to the rate of the phenotypic switch (Jablonka et al. 1995; Kussell and Leibler 2005; Wolf et al. 2005). In this way, cells can explore different lifestyles, without committing the entire population to a specific developmental fate (Halme et al. 2004).

## 3.6 Concluding Remarks

'Exceptional regions underlie exceptional biology': this phrase by Evan Eichler perfectly captures the importance of subtelomeric regions (Eichler 2001). Chromosome ends display numerous features that distinguish them from the rest of the genome: subtelomeric gene families evolve more rapidly, driven by frequent duplication events, elevated recombination and mutation rates, and stochastic transcriptional silencing and desilencing (Brown et al. 2010).

Subtelomeric regions have been proposed to function as 'gene nurseries' or 'gene laboratories', where loci can diversify into novel genes and be further tinkered with (Linardopoulou et al. 2005). Most subtelomeric genes are (transiently) silenced (Gottschling et al. 1990; Pryde and Louis 1999; Wyrick et al. 1999). This positional silencing could prevent immediate expression of possible Frankenstein genes and temporarily relieve genes of negative (purifying) selection, allowing the accumulation of various mutations. Together with the low gene density, increased genetic variation at the subtelomeres thus has fewer detrimental consequences compared to other genomic regions. The stochastic nature of subtelomeric gene silencing can allow cells to explore this hidden genetic variation, for example under stressful conditions (Rando and Verstrepen 2007). This in turn can enable cells to deal with changes in their environment.

In addition, subtelomeric genes show elevated expression divergence and responsiveness (Brown et al. 2010). Although the evolution of gene expression is relatively understudied compared to the divergence of coding regions, swift divergence of the transcriptional regulation of duplicated subtelomeric genes might be an additional mechanism generating variation (Tirosh et al. 2009).

Although some subtelomeric gene families have been intensively studied, they represent only the tip of the iceberg. Most of the genes residing at chromosome ends are poorly characterized. Studying these genes and their dynamics at the subtelomeres will help us to better understand these enigmatic regions of the genome.

**Acknowledgments** We thank Joaquin Christiaens for useful suggestions to improve this manuscript. We are grateful to Narendra Maheshri for providing the microscopy pictures used in Fig. 3.5. All Verstrepen Laboratory members are thanked for useful discussions. We apologize for the omission of several relevant studies we could not cite due to space limitations. Research in the laboratory of KJV is supported by NIH Grant P50GM068763, HumanFrontier Science ProgramHFSPRGY79/2007, ERC Young Investigator Grant 241426, VIB, KU Leuven, the FWO-Odysseus program, and the AB InBev Baillet-Latour. KV is a postdoctoral fellow of the Fonds voor Wetenschappelijk Onderzoek (FWO) Vlaanderen.

## References

Abramova, N., Sertil, O., Mehta, S., & Lowry, C. V. (2001). Reciprocal regulation of anaerobic and aerobic cell wall mannoprotein gene expression in *Saccharomyces cerevisiae*. *Journal of Bacteriology, 183*, 2881–2887.

Ai, W., Bertram, P. G., Tsang, C. K., Chan, T. F., & Zheng, X. F. (2002). Regulation of subtelomeric silencing during stress response. *Molecular Cell, 10*, 1295–1305.

Alimardani, P., Regnacq, M., Moreau-Vauzelle, C., Ferreira, T., Rossignol, T., Blondin, B., et al. (2004). SUT1-promoted sterol uptake involves the ABC transporter Aus1 and the mannoprotein Dan1 whose synergistic action is sufficient for this process. *The Biochemical Journal, 381*, 195–202.

Andrau, J. C., van de Pasch, L., Lijnzaad, P., Bijma, T., Koerkamp, M. G., van de Peppel, J., et al. (2006). Genome-wide location of the coactivator mediator: Binding without activation and transient Cdk8 interaction on DNA. *Molecular Cell, 22*, 179–192.

Aparicio, O. M., Billington, B. L., & Gottschling, D. E. (1991). Modifiers of position effect are shared between telomeric and silent mating-type loci in *S. cerevisiae*. *Cell, 66*, 1279–1287.

Barton, A. B., Su, Y., Lamb, J., Barber, D., & Kaback, D. B. (2003). A function for subtelomeric DNA in *Saccharomyces cerevisiae*. *Genetics, 165*, 929–934.

Barton, A. B., Pekosz, M. R., Kurvathi, R. S., & Kaback, D. B. (2008). Meiotic recombination at the ends of chromosomes in *Saccharomyces cerevisiae*. *Genetics, 179*, 1221–1235.

Bertuch, A. A., & Lundblad, V. (2003). The Ku heterodimer performs separable activities at double-strand breaks and chromosome termini. *Molecular and Cellular Biology, 23*, 8202–8215.

Biessmann, H., Mason, J. M., Ferry, K., d'Hulst, M., Valgeirsdottir, K., Traverse, K. L., et al. (1990). Addition of telomere-associated HeT DNA sequences "heals" broken chromosome ends in Drosophila. *Cell, 61*, 663–673.

Blitzblau, H. G., Bell, G. W., Rodriguez, J., Bell, S. P., & Hochwagen, A. (2007). Mapping of meiotic single-stranded DNA reveals double-stranded-break hotspots near centromeres and telomeres. *Current Biology, 17*, 2003–2012.

Boeke, J. D., Trueheart, J., Natsoulis, G., & Fink, G. R. (1987). 5-Fluoroorotic acid as a selective agent in yeast molecular genetics. *Methods in Enzymology, 154*, 164–175.

Borneman, A. R., Desany, B. A., Riches, D., Affourtit, J. P., Forgan, A. H., Pretorius, I. S., et al. (2011). Whole-genome comparison reveals novel genetic elements that characterize the genome of industrial strains of *Saccharomyces cerevisiae*. *PLoS Genetics, 7*, e1001287.

Boulton, S. J., & Jackson, S. P. (1998). Components of the Ku-dependent non-homologous end-joining pathway are involved in telomeric length maintenance and telomeric silencing. *EMBO Journal, 17*, 1819–1828.

Brown, C. A., Murray, A. W., & Verstrepen, K. J. (2010). Rapid expansion and functional divergence of subtelomeric gene families in yeasts. *Current Biology, 20*, 895–903.

Carlson, M., & Botstein, D. (1983). Organization of the SUC gene family in *Saccharomyces*. *Molecular and Cellular Biology, 3*, 351–359.

Carlson, M., Celenza, J. L., & Eng, F. J. (1985). Evolution of the dispersed SUC gene family of *Saccharomyces* by rearrangements of chromosome telomeres. *Molecular and Cellular Biology, 5*, 2894–2902.

Chan, C. S., & Tye, B. K. (1983). A family of *Saccharomyces cerevisiae* repetitive autonomously replicating sequences that have very similar genomic environments. *Journal of Molecular Biology, 168*, 505–523.

Charron, M. J., Read, E., Haut, S. R., & Michels, C. A. (1989). Molecular evolution of the telomere-associated MAL loci of *Saccharomyces*. *Genetics, 122*, 307–316.

Chen, L., & Widom, J. (2005). Mechanism of transcriptional silencing in yeast. *Cell, 120*, 37–48.

Christiaens, J. F., Van Mulders, S. E., Duitama, J., Brown, C. A., Ghequire, M. G., De Meester, L., & Verstrepen, K. J. (2012). Functional divergence of gene duplicates through ectopic recombination.

Cohn, M., Liti, G., & Barton, D. (2006). Telomeres in fungi. In P. Sunnerhagen, J. Piskur (Eds.), *Comparative genomics using fungi as a model* (pp. 101–130). Heidelberg: Springer.

Des Marais, D. L., & Rausher, M. D. (2008). Escape from adaptive conflict after duplication in an anthocyanin pathway gene. *Nature, 454*, 762–765.

Eichler, E. E. (2001). Segmental duplications: What's missing, misassigned, and misassembled–and should we care? *Genome Research, 11*, 653–656.

Eichler, E. E., Clark, R. A., & She, X. (2004). An assessment of the sequence gaps: Unfinished business in a finished human genome. *Nature Reviews Genetics, 5*, 345–354.

Flint, J., Bates, G. P., Clark, K., Dorman, A., Willingham, D., Roe, B. A., et al. (1997). Sequence comparison of human and yeast telomeres identifies structurally distinct subtelomeric domains. *Human Molecular Genetics, 6*, 1305–1313.

Francesconi, M., Jelier, R., & Lehner, B. (2011). Integrated genome-scale prediction of detrimental mutations in transcription networks. *PLoS Genetics, 7*, e1002077.

Gatbonton, T., Imbesi, M., Nelson, M., Akey, J. M., Ruderfer, D. M., Kruglyak, L., et al. (2006). Telomere length as a quantitative trait: Genome-wide survey and genetic mapping of telomere length-control genes in yeast. *PLoS Genetics, 2*, e35.

Gemayel, R., Vinces, M. D., Legendre, M., & Verstrepen, K. J. (2010). Variable tandem repeats accelerate evolution of coding and regulatory sequences. *Annual Review of Genetics, 44*, 445–477.

Gimeno, C. J., Ljungdahl, P. O., Styles, C. A., & Fink, G. R. (1992). Unipolar cell divisions in the yeast *S. cerevisiae* lead to filamentous growth: Regulation by starvation and RAS. *Cell, 68*, 1077–1090.

Goffeau, A., Barrell, B. G., Bussey, H., Davis, R. W., Dujon, B., Feldmann, H., et al. (1996). Life with 6000 genes. *Science, 274*(546), 547–563.

Goldman, A. S., & Lichten, M. (1996). The efficiency of meiotic recombination between dispersed sequences in *Saccharomyces cerevisiae* depends upon their chromosomal location. *Genetics, 144*, 43–55.

Gottschling, D. E., Aparicio, O. M., Billington, B. L., & Zakian, V. A. (1990). Position effect at *S. cerevisiae* telomeres: Reversible repression of Pol II transcription. *Cell, 63*, 751–762.

Govender, P., Bester, M., & Bauer, F. F. (2010). FLO gene-dependent phenotypes in industrial wine yeast strains. *Applied Microbiology and Biotechnology, 86*, 931–945.

Gresham, D., Desai, M. M., Tucker, C. M., Jenq, H. T., Pai, D. A., Ward, A., et al. (2008). The repertoire and dynamics of evolutionary adaptations to controlled nutrient-limited environments in yeast. *PLoS Genetics, 4*, e1000303.

Guo, B., Styles, C. A., Feng, Q., & Fink, G. R. (2000). A *Saccharomyces* gene family involved in invasive growth, cell–cell adhesion, and mating. *Proceedings of the National Academy of Sciences of the United States of America, 97*, 12158–12163.

Halme, A., Bumgarner, S., Styles, C., & Fink, G. R. (2004). Genetic and epigenetic regulation of the FLO gene family generates cell-surface variation in yeast. *Cell, 116*, 405–415.

Hasin, Y., Olender, T., Khen, M., Gonzaga-Jauregui, C., Kim, P. M., Urban, A. E., et al. (2008). High-resolution copy-number variation map reflects human olfactory receptor diversity and evolution. *PLoS Genetics, 4*, e1000249.

Hawk, J. D., Stefanovic, L., Boyer, J. C., Petes, T. D., & Farber, R. A. (2005). Variation in efficiency of DNA mismatch repair at different sites in the yeast genome. *Proceedings of the National Academy of Sciences of the United States of America, 102*, 8639–8643.

Hediger, F., Neumann, F. R., Van Houwe, G., Dubrana, K., & Gasser, S. M. (2002). Live imaging of telomeres: yKu and Sir proteins define redundant telomere-anchoring pathways in yeast. *Current Biology, 12*, 2076–2089.

Horowitz, H., & Haber, J. E. (1985). Identification of autonomously replicating circular subtelomeric Y′ elements in *Saccharomyces cerevisiae. Molecular and Cellular Biology, 5*, 2369–2380.

Hughes, T. R., Marton, M. J., Jones, A. R., Roberts, C. J., Stoughton, R., Armour, C. D., et al. (2000). Functional discovery via a compendium of expression profiles. *Cell, 102*, 109–126.

Ito-Harashima, S., Hartzog, P. E., Sinha, H., & McCusker, J. H. (2002). The tRNA-Tyr gene family of *Saccharomyces cerevisiae*: Agents of phenotypic variation and position effects on mutation frequency. *Genetics, 161*, 1395–1410.

Jablonka, E., Oborny, B., Molnar, I., Kisdi, E., Hofbauer, J., & Czaran, T. (1995). The adaptive advantage of phenotypic memory in changing environments. *Philosophical Transactions of the Royal Society of London. Series B, Biological sciences, 350*, 133–141.

Jeppesen, P. (1997). Histone acetylation: A possible mechanism for the inheritance of cell memory at mitosis. *BioEssays, 19*, 67–74.

Kellis, M., Birren, B. W., & Lander, E. S. (2004). Proof and evolutionary analysis of ancient genome duplication in the yeast *Saccharomyces cerevisiae. Nature, 428*, 617–624.

Kellis, M., Patterson, N., Endrizzi, M., Birren, B., & Lander, E. S. (2003). Sequencing and comparison of yeast species to identify genes and regulatory elements. *Nature, 423*, 241–254.

Kim, J. M., Vanguri, S., Boeke, J. D., Gabriel, A., & Voytas, D. F. (1998). Transposable elements and genome organization: A comprehensive survey of retrotransposons revealed by the complete *Saccharomyces cerevisiae* genome sequence. *Genome Research, 8*, 464–478.

Kimura, A., Umehara, T., & Horikoshi, M. (2002). Chromosomal gradient of histone acetylation established by Sas2p and Sir2p functions as a shield against gene silencing. *Nature Genetics, 32*, 370–377.

Kobayashi, O., Yoshimoto, H., & Sone, H. (1999). Analysis of the genes activated by the FLO8 gene in *Saccharomyces cerevisiae. Current Genetics, 36*, 256–261.

Kussell, E., & Leibler, S. (2005). Phenotypic diversity, population growth, and information in fluctuating environments. *Science, 309*, 2075–2078.

Kyrion, G., Boakye, K. A., & Lustig, A. J. (1992). C-terminal truncation of RAP1 results in the deregulation of telomere size, stability, and function in *Saccharomyces cerevisiae. Molecular and Cellular Biology, 12*, 5159–5173.

Kyrion, G., Liu, K., Liu, C., & Lustig, A. J. (1993). RAP1 and telomere structure regulate telomere position effects in *Saccharomyces cerevisiae. Genes and Development, 7*, 1146–1159.

Lambrechts, M. G., Bauer, F. F., Marmur, J., & Pretorius, I. S. (1996). Muc1, a mucin-like protein that is regulated by Mss10, is critical for pseudohyphal differentiation in yeast. *Proceedings of the National Academy of Sciences of the United States of America, 93*, 8419–8424.

Lang, G. I., & Murray, A. W. (2011). Mutation rates across budding yeast chromosome VI are correlated with replication timing. *Genome Biology and Evolution, 3*, 799–811.

Laroche, T., Martin, S. G., Gotta, M., Gorham, H. C., Pryde, F. E., Louis, E. J., et al. (1998). Mutation of yeast Ku genes disrupts the subnuclear organization of telomeres. *Current Biology, 8*, 653–656.

Levis, R. W., Ganesan, R., Houtchens, K., Tolar, L. A., & Sheen, F. M. (1993). Transposons in place of telomeric repeats at a Drosophila telomere. *Cell, 75*, 1083–1093.

Linardopoulou, E. V., Williams, E. M., Fan, Y., Friedman, C., Young, J. M., & Trask, B. J. (2005). Human subtelomeres are hot spots of interchromosomal recombination and segmental duplication. *Nature, 437*, 94–100.

Liti, G., & Louis, E. J. (2005). Yeast evolution and comparative genomics. *Annual Review of Microbiology, 59*, 135–153.

Liti, G., Carter, D. M., Moses, A. M., Warringer, J., Parts, L., James, S. A., et al. (2009). Population genomics of domestic and wild yeasts. *Nature, 458*, 337–341.

Liu, H., Styles, C. A., & Fink, G. R. (1996). *Saccharomyces cerevisiae* S288C has a mutation in FLO8, a gene required for filamentous growth. *Genetics, 144*, 967–978.

Lo, W. S., & Dranginis, A. M. (1996). FLO11, a yeast gene related to the STA genes, encodes a novel cell surface flocculin. *Journal of Bacteriology, 178*, 7144–7151.

Lo, W. S., & Dranginis, A. M. (1998). The cell surface flocculin Flo11 is required for pseudohyphae formation and invasion by *Saccharomyces cerevisiae*. *Molecular Biology of the Cell, 9*, 161–171.

Louis, E. J., & Haber, J. E. (1990a). Mitotic recombination among subtelomeric Y′ repeats in *Saccharomyces cerevisiae*. *Genetics, 124*, 547–559.

Louis, E. J., & Haber, J. E. (1990b). The subtelomeric Y′ repeat family in *Saccharomyces cerevisiae*: An experimental system for repeated sequence evolution. *Genetics, 124*, 533–545.

Louis, E. J., & Haber, J. E. (1992). The structure and evolution of subtelomeric Y′ repeats in *Saccharomyces cerevisiae*. *Genetics, 131*, 559–574.

Louis, E. J., Naumova, E. S., Lee, A., Naumov, G., & Haber, J. E. (1994). The chromosome end in yeast: Its mosaic nature and influence on recombinational dynamics. *Genetics, 136*, 789–802.

Lundblad, V., & Szostak, J. W. (1989). A mutant with a defect in telomere elongation leads to senescence in yeast. *Cell, 57*, 633–643.

Luo, Z., & van Vuuren, H. J. (2009). Functional analyses of PAU genes in *Saccharomyces cerevisiae*. *Microbiology, 155*, 4036–4049.

Mak, H. C., Pillus, L., & Ideker, T. (2009). Dynamic reprogramming of transcription factors to and from the subtelomere. *Genome Research, 19*, 1014–1025.

Marks, V. D., Ho Sui, S. J., Erasmus, D., van der Merwe, G. K., Brumm, J., Wasserman, W. W., et al. (2008). Dynamics of the yeast transcriptome during wine fermentation reveals a novel fermentation stress response. *FEMS Yeast Research, 8*, 35–52.

Marvin, M. E., Becker, M. M., Noel, P., Hardy, S., Bertuch, A. A., & Louis, E. J. (2009a). The association of yKu with subtelomeric core X sequences prevents recombination involving telomeric sequences. *Genetics, 183*, 453–467 (451SI–413SI).

Marvin, M. E., Griffin, C. D., Eyre, D. E., Barton, D. B., & Louis, E. J. (2009b). In *Saccharomyces cerevisiae*, yKu and subtelomeric core X sequences repress homologous recombination near telomeres as part of the same pathway. *Genetics, 183*, 441–451 (441SI–412SI).

Mefford, H. C., & Trask, B. J. (2002). The complex structure and dynamic evolution of human subtelomeres. *Nature Reviews Genetics, 3*, 91–102.

Moretti, P., Freeman, K., Coodly, L., & Shore, D. (1994). Evidence that a complex of SIR proteins interacts with the silencer and telomere-binding protein RAP1. *Genes and Development, 8*, 2257–2269.

Naumov, G. I., Naumova, E. S., & Korhola, M. P. (1995). Chromosomal polymorphism of MEL genes in some populations of *Saccharomyces cerevisiae*. *FEMS Microbiology Letters, 127*, 41–45.

Naumov, G. I., Naumova, E. S., Turakainen, H., & Korhola, M. (1996). Identification of the alpha-galactosidase MEL genes in some populations of *Saccharomyces cerevisiae*: A new gene MEL11. *Genetical Research, 67*, 101–108.

Ness, F., & Aigle, M. (1995). RTM1: A member of a new family of telomeric repeated genes in yeast. *Genetics, 140*, 945–956.

Octavio, L. M., Gedeon, K., & Maheshri, N. (2009). Epigenetic and conventional regulation is distributed among activators of FLO11 allowing tuning of population-level heterogeneity in its expression. *PLoS Genetics, 5*, e1000673.

Ogata, T., Izumikawa, M., Kohno, K., & Shibata, K. (2008). Chromosomal location of Lg-FLO1 in bottom-fermenting yeast and the FLO5 locus of industrial yeast. *Journal of Applied Microbiology, 105*, 1186–1198.

Ohno, S. (1970). *Evolution by gene duplication*. Berlin: Springer.

Ottaviani, A., Gilson, E., & Magdinier, F. (2008). Telomeric position effect: From the yeast paradigm to human pathologies? *Biochimie, 90*, 93–107.

Palkova, Z., & Vachova, L. (2006). Life within a community: Benefit to yeast long-term survival. *FEMS Microbiology Reviews, 30*, 806–824.

Pryde, F. E., Huckle, T. C., & Louis, E. J. (1995). Sequence analysis of the right end of chromosome XV in *Saccharomyces cerevisiae*: An insight into the structural and functional significance of sub-telomeric repeat sequences. *Yeast, 11*, 371–382.

Pryde, F. E., Gorham, H. C., & Louis, E. J. (1997). Chromosome ends: All the same under their caps. *Current Opinion in Genetics and Development, 7*, 822–828.

Pryde, F. E., & Louis, E. J. (1997). *Saccharomyces cerevisiae* telomeres. A review. *Biochemistry (Mosc), 62*, 1232–1241.

Pryde, F. E., & Louis, E. J. (1999). Limitations of silencing at native yeast telomeres. *EMBO Journal, 18*, 2538–2550.

Rando, O. J., & Verstrepen, K. J. (2007). Timescales of genetic and epigenetic inheritance. *Cell, 128*, 655–668.

Roberts, D. J., Craig, A. G., Berendt, A. R., Pinches, R., Nash, G., Marsh, K., et al. (1992). Rapid switching to multiple antigenic and adhesive phenotypes in malaria. *Nature, 357*, 689–692.

Robyr, D., Suka, Y., Xenarios, I., Kurdistani, S. K., Wang, A., Suka, N., et al. (2002). Microarray deacetylation maps determine genome-wide functions for yeast histone deacetylases. *Cell, 109*, 437–446.

Ross, L. O., Maxfield, R., & Dawson, D. (1996). Exchanges are not equally able to enhance meiotic chromosome segregation in yeast. *Proceedings of the National Academy of Sciences of the United States of America, 93*, 4979–4983.

Rossignol, T., Dulau, L., Julien, A., & Blondin, B. (2003). Genome-wide monitoring of wine yeast gene expression during alcoholic fermentation. *Yeast, 20*, 1369–1385.

Rudd, M. K., Friedman, C., Parghi, S. S., Linardopoulou, E. V., Hsu, L., & Trask, B. J. (2007). Elevated rates of sister chromatid exchange at chromosome ends. *PLoS Genetics, 3*, e32.

Rusche, L. N., Kirchmaier, A. L., & Rine, J. (2003). The establishment, inheritance, and function of silenced chromatin in *Saccharomyces cerevisiae*. *Annual Review of Biochemistry, 72*, 481–516.

Schacherer, J., Shapiro, J. A., Ruderfer, D. M., & Kruglyak, L. (2009). Comprehensive polymorphism survey elucidates population structure of *Saccharomyces cerevisiae*. *Nature, 458*, 342–345.

Sekinger, E. A., & Gross, D. S. (2001). Silenced chromatin is permissive to activator binding and PIC recruitment. *Cell, 105*, 403–414.

Shampay, J., Szostak, J. W., & Blackburn, E. H. (1984). DNA sequences of telomeres maintained in yeast. *Nature, 310*, 154–157.

Sherman, J. M., & Pillus, L. (1997). An uncertain silence. *Trends Genet, 13*, 308–313.

Shore, D., & Nasmyth, K. (1987). Purification and cloning of a DNA binding protein from yeast that binds to both silencer and activator elements. *Cell, 51*, 721–732.

Smith, J. J., Miller, L. R., Kreisberg, R., Vazquez, L., Wan, Y., & Aitchison, J. D. (2011). Environment-responsive transcription factors bind subtelomeric elements and regulate gene silencing. *Molecular Systems Biology, 7*, 455.

Smukalla, S., Caldara, M., Pochet, N., Beauvais, A., Guadagnini, S., Yan, C., et al. (2008). FLO1 is a variable green beard gene that drives biofilm-like cooperation in budding yeast. *Cell, 135*, 726–737.

Strahl-Bolsinger, S., Hecht, A., Luo, K., & Grunstein, M. (1997). SIR2 and SIR4 interactions differ in core and extended telomeric heterochromatin in yeast. *Genes and Development, 11*, 83–93.

Su, Y., Barton, A. B., & Kaback, D. B. (2000). Decreased meiotic reciprocal recombination in subtelomeric regions in *Saccharomyces cerevisiae*. *Chromosoma, 109*, 467–475.

Suka, N., Luo, K., & Grunstein, M. (2002). Sir2p and Sas2p opposingly regulate acetylation of yeast histone H4 lysine16 and spreading of heterochromatin. *Nature Genetics, 32*, 378–383.

Svejstrup, J. Q. (2002). Mechanisms of transcription-coupled DNA repair. *Nature Reviews Molecular Cell Biology, 3*, 21–29.

Taddei, A., Hediger, F., Neumann, F. R., Bauer, C., & Gasser, S. M. (2004). Separation of silencing from perinuclear anchoring functions in yeast Ku80, Sir4 and Esc1 proteins. *EMBO Journal, 23*, 1301–1312.

Taddei, A., Van Houwe, G., Nagai, S., Erb, I., van Nimwegen, E., & Gasser, S. M. (2009). The functional importance of telomere clustering: Global changes in gene expression result from SIR factor dispersion. *Genome Research, 19*, 611–625.

Teste, M. A., Francois, J. M., & Parrou, J. L. (2010). Characterization of a new multigene family encoding isomaltases in the yeast *Saccharomyces cerevisiae*, the IMA family. *Journal of Biological Chemistry, 285*, 26815–26824.

Teytelman, L., Eisen, M. B., & Rine, J. (2008). Silent but not static: Accelerated base-pair substitution in silenced chromatin of budding yeasts. *PLoS Genetics, 4*, e1000247.

Tirosh, I., Barkai, N., & Verstrepen, K. J. (2009). Promoter architecture and the evolvability of gene expression. *J Biol, 8*, 95.

Trask, B. J., Friedman, C., Martin-Gallardo, A., Rowen, L., Akinbami, C., Blankenship, J., et al. (1998). Members of the olfactory receptor gene family are contained in large blocks of DNA duplicated polymorphically near the ends of human chromosomes. *Human Molecular Genetics, 7*, 13–26.

Trelles-Sticken, E., Loidl, J., & Scherthan, H. (1999). Bouquet formation in budding yeast: Initiation of recombination is not required for meiotic telomere clustering. *Journal of Cell Science, 112*(Pt 5), 651–658.

Tsukamoto, Y., Kato, J., & Ikeda, H. (1997). Silencing factors participate in DNA repair and recombination in *Saccharomyces cerevisiae*. *Nature, 388*, 900–903.

Van Mulders, S. E., Ghequire, M., Daenen, L., Verbelen, P. J., Verstrepen, K. J., & Delvaux, F. R. (2010). Flocculation gene variability in industrial brewer's yeast strains. *Applied Microbiology and Biotechnology, 88*, 1321–1331.

Verstrepen, K. J., & Fink, G. R. (2009). Genetic and epigenetic mechanisms underlying cell-surface variability in protozoa and fungi. *Annual Review of Genetics, 43*, 1–24.

Verstrepen, K. J., Jansen, A., Lewitter, F., & Fink, G. R. (2005). Intragenic tandem repeats generate functional variability. *Nature Genetics, 37*, 986–990.

Verstrepen, K. J., & Klis, F. M. (2006). Flocculation, adhesion and biofilm formation in yeasts. *Molecular Microbiology, 60*, 5–15.

Verstrepen, K. J., Reynolds, T. B., & Fink, G. R. (2004). Origins of variation in the fungal cell surface. *Nature Reviews Microbiology, 2*, 533–540.

Vinces, M. D., Legendre, M., Caldara, M., Hagihara, M., & Verstrepen, K. J. (2009). Unstable tandem repeats in promoters confer transcriptional evolvability. *Science, 324*, 1213–1216.

Voordeckers, K., Brown, C. A., Vanneste, K., van der Zande, E., Voet, A., Maere, S., & Verstrepen, K. J. (2012). Reconstruction of Ancestral Metabolic Enzymes Reveals Molecular Mechanisms Underlying Evolutionary Innovation through Gene Duplication. *PLoS biology, 10*(12), e1001446.

Voytas, D. F., & Boeke, J. D. (1992). Yeast retrotransposon revealed. *Nature, 358*, 717.

Walmsley, R. M., & Petes, T. D. (1985). Genetic control of chromosome length in yeast. *Proceedings of the National Academy of Sciences United States of America, 82*, 506–510.

Wapinski, I., Pfeffer, A., Friedman, N., & Regev, A. (2007). Natural history and evolutionary principles of gene duplication in fungi. *Nature, 449*, 54–61.

Wolf, D. M., Vazirani, V. V., & Arkin, A. P. (2005). Diversity in times of adversity: Probabilistic strategies in microbial survival games. *Journal of Theoretical Biology, 234*, 227–253.

Wolfe, K. H., & Shields, D. C. (1997). Molecular evidence for an ancient duplication of the entire yeast genome. *Nature, 387*, 708–713.

Wotton, D., & Shore, D. (1997). A novel Rap1p-interacting factor, Rif2p, cooperates with Rif1p to regulate telomere length in *Saccharomyces cerevisiae*. *Genes and Development, 11*, 748–760.

Wyrick, J. J., Holstege, F. C., Jennings, E. G., Causton, H. C., Shore, D., Grunstein, M., et al. (1999). Chromosomal landscape of nucleosome-dependent gene expression and silencing in yeast. *Nature, 402*, 418–421.

Yamada, M., Hayatsu, N., Matsuura, A., & Ishikawa, F. (1998). Y′-Help1, a DNA helicase encoded by the yeast subtelomeric Y′ element, is induced in survivors defective for telomerase. *Journal of Biological Chemistry, 273*, 33360–33366.

Yamashita, I., Maemura, T., Hatano, T., & Fukui, S. (1985). Polymorphic extracellular glucoamylase genes and their evolutionary origin in the yeast *Saccharomyces* diastaticus. *Journal of Bacteriology, 161*, 574–582.

Zakian, V. A. (1996). Structure, function, and replication of *Saccharomyces cerevisiae* telomeres. *Annual Review of Genetics, 30*, 141–172.

Zhou, B. O., Wang, S. S., Zhang, Y., Fu, X. H., Dang, W., Lenzmeier, B. A., et al. (2011). Histone H4 lysine 12 acetylation regulates telomeric heterochromatin plasticity in *Saccharomyces cerevisiae*. *PLoS Genetics, 7*, e1001272.

Zhu, X., & Gustafsson, C. M. (2009). Distinct differences in chromatin structure at subtelomeric X and Y′ elements in budding yeast. *PLoS ONE, 4*, e6363.

Zhu, Y., Dai, J., Fuerst, P. G., & Voytas, D. F. (2003). Controlling integration specificity of a yeast retrotransposon. *Proceedings of the National Academy of Sciences of the United States of America, 100*, 5891–5895.

Zou, S., Wright, D. A., & Voytas, D. F. (1995). The *Saccharomyces* Ty5 retrotransposon family is associated with origins of DNA replication at the telomeres and the silent mating locus HMR. *Proceedings of the National Academy of Sciences of the United States of America, 92*, 920–924.

Zou, S., Ke, N., Kim, J. M., & Voytas, D. F. (1996a). The Saccharomyces retrotransposon Ty5 integrates preferentially into regions of silent chromatin at the telomeres and mating loci. *Genes and Development, 10*, 634–645.

Zou, S., Kim, J. M., & Voytas, D. F. (1996b). The *Saccharomyces* retrotransposon Ty5 influences the organization of chromosome ends. *Nucleic Acids Research, 24*, 4825–4831.

# Chapter 4
# Subtelomere Organization, Evolution, and Dynamics in the Rice Blast Fungus *Magnaporthe oryzae*

**Mark Farman, Olga Novikova, John Starnes and David Thornbury**

## 4.1 Introduction

*Magnaporthe oryzae* is a filamentous, ascomycete fungus best known as the causal agent of a devastating disease of rice known as blast. In addition to being a pathogen of rice, it also causes diseases on other important crops, including wheat, millets, and forage grasses, and on turf grasses such as perennial ryegrass and St. Augustinegrass. Despite the species' broad host range, fungal isolates from one host genus usually are unable to infect other host genera. Such specificity can also be manifested at the subspecies level, such that an isolate from one rice cultivar is often unable to infect other cultivars. Genetic analyses of numerous examples of host and cultivar specificity have revealed that these traits are inherited in a simple Mendelian fashion (Murakami et al. 2000, 2003; Tosa et al. 2006), and the subsequent cloning of several host- and cultivar-specificity genes has shown that avirulence (the inability to infect) is the dominant trait (Farman and Leong 1998; Orbach et al. 2000; Sweigard et al. 1995). Most avirulence genes code for small, secreted, effector proteins (Orbach et al. 2000; Peyyala and Farman 2006; Sweigard et al. 1995) that are translocated into the host cell cytoplasm (Khang

M. Farman (✉) · O. Novikova · J. Starnes · D. Thornbury
Department of Plant Pathology, University of Kentucky, Lexington, KY 40546, USA
e-mail: mark.farman@uky.edu

O. Novikova
Department of Biology, Life Sciences Building, University at Albany,
1400 Washington Avenue, Albany, NY 12222, USA

J. Starnes
Somerset Community College, 808 Monticello Street, Somerset, KY 42501, USA

D. Thornbury
Department of Plant Pathology, University of Kentucky, 1724 Appomatox Rd,
Lexington, KY 40504, USA

E. J. Louis and M. M. Becker (eds.), *Subtelomeres*, DOI: 10.1007/978-3-642-41566-1_4, 

et al. 2010) where it is believed they interfere with plant metabolism to the benefit of the pathogen. However, some plant genotypes are capable of recognizing these foreign proteins and mount a powerful defense response that prevents infection.

The specificity conferred by AVR:R-gene interactions serves as the basis for breeding disease resistance into rice. However, a major constraint to the durable control of rice blast through the use of natural host resistance is that the fungus is rapidly able to defeat resistant plants (Bonman 1992). This indicates that avirulence genes are highly mutable and/or highly polymorphic within the rice pathogen population. In contrast, most other loci that have been examined show very little polymorphism among rice-infecting isolates (Couch et al. 2005; Couch and Kohn 2002).

A possible explanation for the increased variation in Magnaporthe avirulence genes is that a large proportion of them (~50 %) map very close to telomeres (Farman 2007). Indeed, one of the first avirulence genes to be cloned, the highly mutable *Avr-Pita* gene, was located just 48 bp from the telomere repeats (Kang et al. 2001; Orbach et al. 2000). An analysis of several spontaneous virulent mutants revealed that they arose through chromosome truncations, point mutations, and transposon insertions (Kang et al. 2001; Orbach et al. 2000). Parallel studies of Magnaporthe telomeres and subtelomeres suggested that these chromosome regions are highly variable and experience rearrangement at much higher frequencies than internal chromosome regions (Farman and Kim 2005; Farman and Leong 1995; Gao et al. 2002). Here, we describe our current state of knowledge on the molecular basis for subtelomere/subterminal chromosome variability in *M. oryzae*.

## 4.2 *M. oryzae* Strains from Perennial Ryegrass have Hypervariable Chromosome Ends

DNA fingerprinting using various repeat sequences as probes has been widely used to classify Magnaporthe strains from different host plants. These probes have the power to distinguish among host-specific pathogen populations (Farman 2002) and have even identified well over a hundred discrete genetic lineages within the global population of rice-infecting strains (Levy et al. 1991). Rice-pathogenic isolates that belong to the same lineage tend to have similar telomere fingerprints to one another (Farman 2007). This is true even for isolates collected several years apart and indicates that the chromosome ends of the rice pathogens are fairly stable. In striking contrast, strains from perennial ryegrass (*prg*) usually have completely different telomere profiles even when there is little or no detectable polymorphism at internal repeat loci (Farman and Kim 2005). This suggests that the telomeres of the *prg* pathogens are more unstable than their counterparts in the rice-infecting strains.

As a formal test of this idea, we directly compared the stability of telomere profiles during growth *in planta* for a rice pathogen, 70-15, and a strain from *prg*, LpKY97-1A (Starnes et al. 2012). First, the strains were genetically purified by culturing isolates that had been single-spored. The resulting cultures were allowed

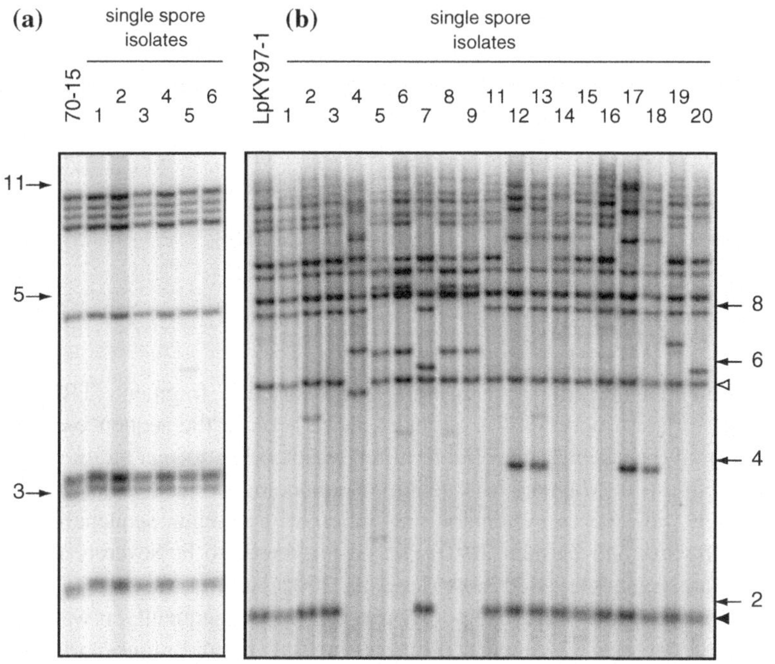


**Fig. 4.1** Telomere stability during growth *in planta*. *M. oryzae* isolates 70-15 and LpKY97-1A were genetically purified by single-spore isolation and used to inoculate their respective hosts (rice and *prg*). After two rounds of infection, single spores were collected and DNA was extracted from the resulting cultures. Shown are phosphor images obtained from Southern blots of *Pst*I-digested DNAs that were hybridized with a telomere probe. **a** shows the telomere profiles of the 70-15 starting culture and a subset of single-spore isolates. **b** telomere profiles of LpKY97-1A and 19 single spores. Molecular sizes (in kb) are shown on the sides of each image. The *open arrowhead* marks a highly stable telomere that was rarely rearranged. The *closed arrowhead* marks the highly unstable rDNA telomere

to sporulate, and the new spores were used to inoculate the respective hosts (rice and *prg*). After 7 days, spores were harvested from leaf lesions and used immediately to inoculate a second batch of plants. Another 7 days later, spores were collected from leaf lesions, subjected to single-spore isolation, and the telomere profiles of 19 representative isolates were examined.

For 70-15, only one of the single-spore cultures (#5) exhibited a telomere profile change (see Fig. 4.1a). This involved the appearance of a single novel, but weak, hybridization signal, which most likely represented a telomere variant that emerged in the culture as it was being grown for DNA extraction. In contrast, all the single-spore cultures from LpKY97-1A exhibited telomere alterations, with the number of visible changes per isolate ranging from one to eight (Fig. 4.1b). Spores sampled immediately prior to the first round of inoculation exhibited very few rearrangements, suggesting that most of the alterations had occurred during *in planta* growth. Furthermore, parallel culturing of an aliquot of the inoculum on agar plates yielded spores with few telomere alterations. Considering that plate-grown cultures typically undergo many more rounds of nuclear division than do colonies in leaf lesions, this suggests that *in planta* growth significantly enhances telomere change.

## 4.3 Rice and *Prg* Pathogens have Major Differences in the Organization of their Chromosome Ends

### *4.3.1 Subtelomere Structure in 70–15*

We surmised that the difference in telomeric restriction fragment (TRF) stability was due to variation between 70–15 and LpKY97-1A in the sequence content and organization of their subtelomere/subterminal regions. The former strain was the one selected for the first Magnaporthe genome sequencing project (Dean et al. 2005). Unfortunately, however, a systematic bias inherent in shotgun sequencing methodologies (Schwartz and Farman 2010) caused telomeres to be severely underrepresented in fungal genome assemblies (Li et al. 2005), and in the case of *M. oryzae*, the subtelomere regions were also poorly assembled. Consequently, it was necessary to acquire the necessary sequence information from fosmid clones containing the 14 chromosome ends (Rehmeyer et al. 2006). Analysis of these sequences revealed that 11 of the chromosome ends contain a highly conserved sequence consisting of a telomere-linked helicase (TLH) gene flanked by numerous short, tandem, helicase-associated repeat (HAR) motis [Fig. 4.2a; (Rehmeyer et al. 2006)]. Based on its presence at multiple chromosome ends, this sequence qualifies as a canonical subtelomere, as defined by Pryde et al. 1997. In most cases, the subtelomere sequence extended all the way to the telomere repeats, but it was truncated to various extents at different chromosome ends and was often interrupted by transposon insertions and other sequences (Fig. 4.2a). Three chromosome ends lacked the subtelomere region—telomeres 2, 3, and 12. Telomere 3 is comprised of telomere repeats attached to 28S ribosomal RNA gene sequences, which constitute the distal end of the major rDNA repeat (Fig. 4.2b). Aside from a short stretch of sequence that is duplicated near telomeres 2 and 12, the corresponding subterminal regions lack features that differentiate them from generic chromosomal sequences (Fig. 4.2c).

The TLH genes are interesting because they are widespread in fungi and intact gene copies are almost always located within a few kilobases of a telomere (Novikova et al., in preparation). As such, they represent the only known example of a gene family whose members' chromosomal positions are conserved across an entire kingdom. The reason for this telomere proximity is unknown, but one possibility is that it is important for their function (Rehmeyer et al. 2009). However, herein lies a conundrum because if their function is so important that their chromosomal position is conserved, then why are they present only in certain fungal strains (see below)? One reason may be that they function only under certain conditions; alternatively, perhaps they are mobilized to chromosome ends through some kind of transposon activity.

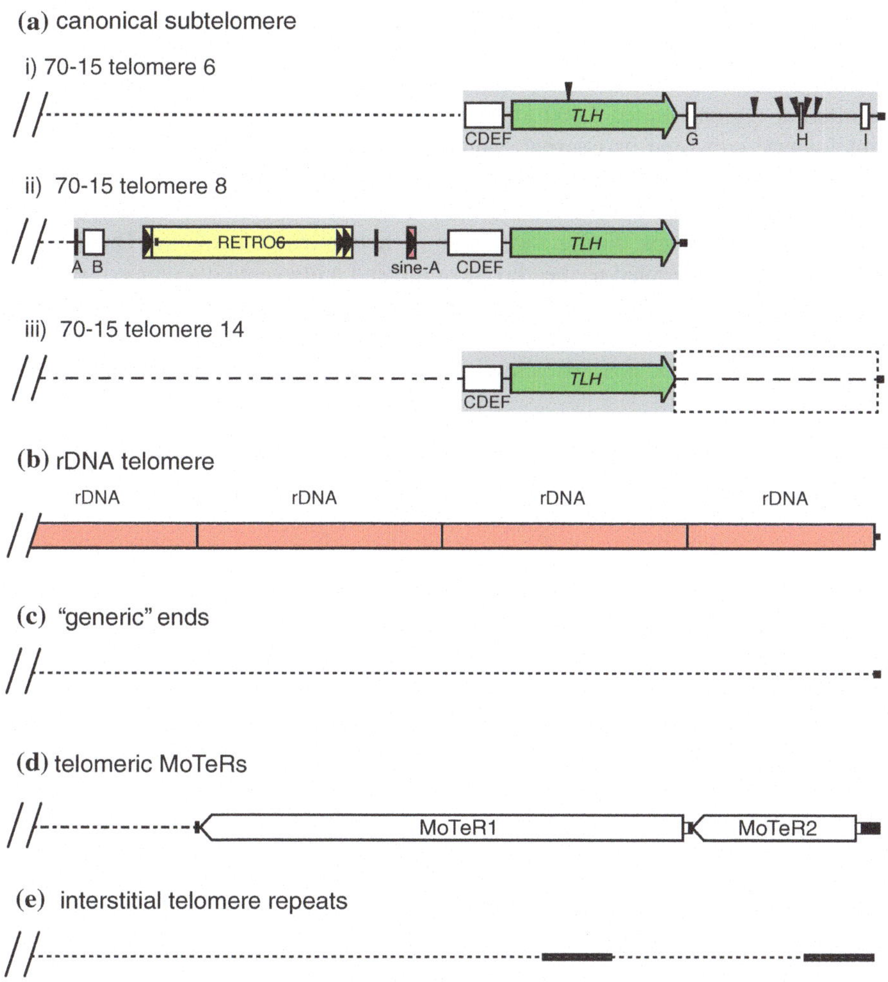


**Fig. 4.2** Summary of subtelomere/subterminal organization. The figure shows the five basic structures that have been identified to date. The telomeres are shown on the *right-hand side* and are depicted as *black boxes*. **a** shows the canonical subtelomere region that is present in most rice pathogens. Subtelomere sequences are enclosed in a *gray box*. The *dotted lines* to the *left* represent chromosome-unique sequences. Three basic subtelomere forms have been identified: (i) standard; (ii) extended; and (iii) subtelomere core separated from the telomere by a foreign (internal) sequence. The green arrows represent telomere-linked helicase gene, and motifs A through H denote helicase-associated repeats (HARs). RETRO6 and sine-A are retroelements that occur within the conserved extended subtelomere domain. *Arrowheads* in (i) show points where the subtelomere sequence is truncated at other chromosome ends. **b** shows the distal end of the tandem rDNA array, which terminates in a telomere. In **c**, the telomere repeats are attached to "normal" genomic sequences, which include genes and repeats that are neither telomere specific nor are they duplicated at different chromosome ends. **d** depicts MoTeRs embedded within the telomere repeats. The *arrows* indicate the direction of MoTeR transcription, and the small *white boxes* at the 5′ ends of each element represent the variant telomere repeats. **e** shows a telomere repeat tract embedded in the subterminal region. These repeats are often present at multiple chromosome ends and therefore qualify as true subtelomeric sequences. The *scale* varies between panels so that different features can be adequately highlighted

Another interesting feature shown by 70-15 was the presence of a "foreign" (i.e., non-subtelomeric) sequence between telomere 14 and its subtelomere region (Rehmeyer et al. 2006). This sequence apparently was derived via duplication of an internal region located ~500 kb from the telomere. Comparison between the subtelomeric copy and its internal counterpart revealed that they diverge at the 5′ boundary of a truncated insertion of MGL, a non-LTR retroelement. Presumably, MGL inserted in subtelomere 14 and then recombined with the MGL copy at the internal locus, perhaps during the repair of a terminal truncation. The internal sequence was then copied to the chromosome end and eventually capped with telomere repeats, giving rise to the structure seen today. This particular chromosome end illustrates the potential for chromosome ends to capture sequences from internal genomic locations, possibly allowing for accelerated evolution of the captured sequences.

### *4.3.2 Subterminal Structure in LpKY97-1A*

Gao and coworkers' (2002) surveyed TLH gene distribution among different host-specific forms of Magnaporthe and showed them to be restricted to rice-pathogenic isolates and absent from other host-specific populations. This told us that the other host-specific forms must have a different subtelomeric composition. To gain insight into the alternative organization of chromosome ends in LpKY97-1A, we sequenced TRFs that were cloned either by screening telomere-enriched random mini-libraries (Arkhipova and Morrison 2001) or by targeted cloning of specific telomeric fragments (Farman 2011). This revealed that most of the telomeres in LpKY97-1A contain insertions of two related non-LTR retrotransposons (NLRs), which are wholly contained within the TTAGGG repeats (Fig. 4.2d). Southern hybridization studies using probes from these elements indicated that intact copies occur almost exclusively in telomeres, while truncated copies tend also to be found at internal chromosome locations (see Sect. 4.7).

The first transposon is ~5 kb in length and codes for a putative reverse transcriptase (RT) with N-terminal zinc finger motifs and a C-terminal restriction enzyme-like (REL) endonuclease domain (Fig. 4.3a). The second element is only 1.7 kb long and has ~800 bp from the 5′ terminus and ~80 bp from the 3′ terminus in common with the larger element. It is distinguished from the larger element by ~800 bp of unique sequence in the middle, which potentially codes for a 232 amino acid protein (Fig. 4.3b). We have named these elements MoTeR1 and MoTeR2, respectively (for *M. oryzae* telomeric retrotransposons). Some telomeres contained single insertions of MoTeR1 or MoTeR2 in full-length or truncated form, while others contained tandem arrays in which adjacent MoTeR copies are separated by short tracts of telomere repeats. Interestingly, intact copies of both elements have variant telomere repeats at their 5′ ends with the consensus sequence, 5′-(CCCGAA)$_2$(CCCAAA)$_8$CCCGAA-3′ (Fig. 4.3c). We believe these variant repeats to be instrumental in the insertion of full-length elements (Starnes et al. 2012).

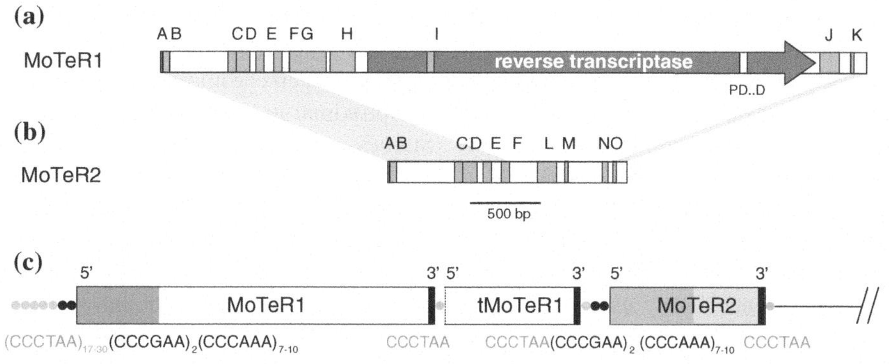


**Fig. 4.3** MoTeR1 and MoTeR2 organization. Putative coding regions are depicted with *dark gray arrows*. Blocks of tandem repeats are shown as medium *gray boxes*. *Light gray shading* connecting the termini of MoTeR1 and MoTeR2 shows regions of significant sequence identity between the two elements. The positions of the restriction enzyme-like endonuclease (REL-endo) domains are shown

BLASTx searches indicated that the MoTeRs are most closely related to hypothetical proteins in the filamentous fungi *Nectria haematococca*, *Cryptococcus neoformans*, *Cryptococcus gattii*, and *Fusarium oxyxporum*. Further investigation proved these hypothetical proteins also to be encoded by telomeric NLRs, although in some instances, the elements no longer resided at chromosome ends. The *C. neoformans* NLR corresponds to the previously described Cnl1 element, which was reported as inserting into copies of itself (Goodwin and Poulter 2001). Closer inspection revealed that Cnl1, in fact, inserts into telomeres in the same manner as the MoTeRs.

A phylogenetic analysis of the RT protein placed MoTeR1 in a clade along with the SLACS, CZAR, and CRE1 retrotransposons found in various protozoan parasites (Aksoy et al. 1990; Gabriel et al. 1990; Peacock et al. 2007; Teng et al. 1995; Villanueva et al. 1991). These elements belong to the oldest known group of non-LTR retrotransposons and are unusual because, rather than inserting at random genomic locations, they target specific chromosomal sequences, namely the spliced leader RNA genes. It is believed that these elements recognize and cleave their targets using a REL endonuclease present in the C-terminus of their respective RT proteins (Mcclure et al. 2002). Based on the phylogenetic placement of the MoTeR1 RT and the presence of the REL endodomain, it seems likely that the MoTeRs are site-specific transposons that target telomere repeats. Interestingly, however, the MoTeRs are only distantly related to other transposons that insert in telomeric sequences, which include TRAS and SART from the silkworm *Bombyx mori* (Okazaki et al. 1995; Takahashi et al. 1997) and Penelope-like elements present in animals, plants, and fungi (Gladyshev and Arkhipova 2007).

Clearly, the presence of transposons inserted in the telomere repeats presents a major structural difference between the telomeres of LpKY97-1A and those of 70-15. Likewise, the absence of TLH genes—as revealed by Southern hybridization—pointed to a major departure in subtelomere organization. This was confirmed by sequencing a number of cosmids containing LpKY97-1A sequences that were homologous to specific 70-15 chromosome ends. Alignment of the homologous ends suggested that TLH gene absence in the *prg* pathogen chromosomes was due to terminal truncations that occurred at positions proximal to where the subtelomeres started in the rice pathogen homologues (Starnes et al. 2012). These comparative analyses also revealed that, unlike their 70-15 counterparts, the chromosome ends of the *prg* pathogens have very few insertions of other types of transposons. Indeed, aside from the MoTeRs, there were no additional sequence features that could explain the enhanced instability of the *prg* pathogen telomeres. In addition, there was no evidence of any kind of conserved subtelomere sequence, although some of the telomere-adjacent sequences were duplicated at non-telomeric locations.

## 4.4 MoTeR Insertions Promote Telomere Instability

It was tempting to speculate that the MoTeRs somehow contribute to the frequent rearrangements of terminal fragments in Magnaporthe strains from *prg*. However, we also considered the possibility that the genetic background of the *prg* pathogens predisposes them to telomere instability. For example, they might possess an inefficient telomerase enzyme, leading to frequent telomere "uncapping" and subsequent rearrangements. To distinguish between these competing hypotheses, we crossed FH—a prg pathogen with unstable, MoTeR-containing telomeres—with 2539 a laboratory strain whose telomeres are comparatively stable. Progeny were collected, and their telomere profiles were examined for stability over several generations of vegetative growth. All progeny exhibited instability at multiple telomeres (representative examples are shown in Fig. 4.4), essentially ruling out the possibility that telomere instability was determined by a simple Mendelian factor and pointing instead to multiple dominant factors. In this regard, it was significant that all of the progeny inherited several MoTeR copies, which is consistent with the idea that the MoTeRs cause instability. Further evidence in support of this hypothesis came from the observation that instability was restricted almost entirely to the telomeres containing MoTeR insertions (Fig. 4.4 and Starnes et al. 2012). Most of the telomeric fragments that were inherited from 2539 were faithfully transmitted through several generations of vegetative growth. The only exceptions were a telomere that contains MoTeR insertions (something we discovered after the fact) and the rDNA telomere, which, as discussed below, tends to be inherently unstable in itself.

## 4.5 The Molecular Basis for Telomere Rearrangements in MoTeR-Containing Strains

To gain insight into the molecular events underlying the frequent MoTeR-induced telomere rearrangements, we used chromosome-unique, telomere-adjacent probes to follow alterations at specific chromosome ends. The altered telomeric

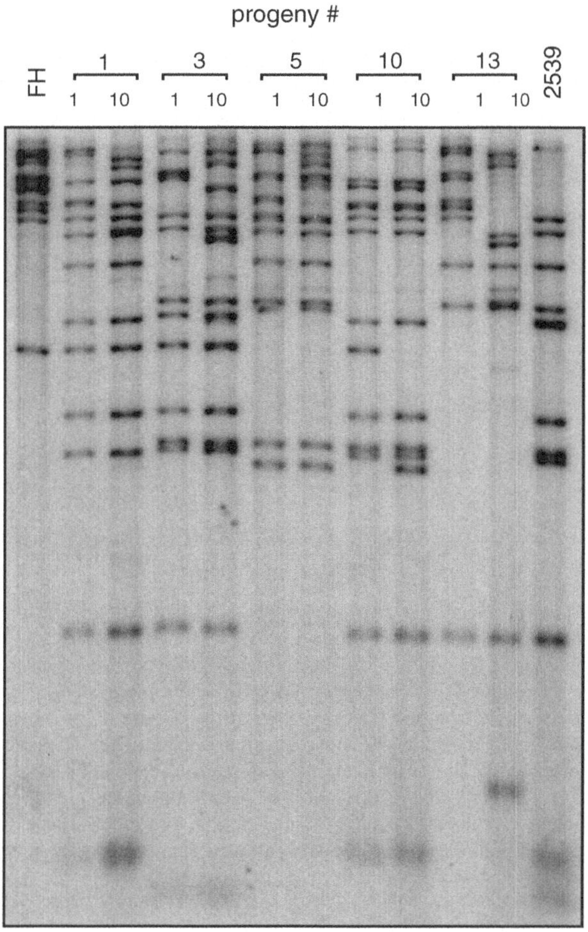


**Fig. 4.4** Telomere instability in progeny from a cross between perennial ryegrass pathogen FH and rice pathogen 2539. Ascospores were collected and used to start plate cultures that were serially transferred to fresh plates a total of 5 times. DNA was extracted from "generations" 1 and 5, digested with *Pst*I, fractioned by agarose gel electrophoresis, blotted to membranes, and hybridized with a telomere probe. The figure shows the resulting phosphor image. DNA from parent FH is on the *left-hand* lane, and 2539 DNA is on the *right*. Note that all progeny show telomere instability, as evidenced by alterations in telomere profiles between generations 1 and 5. Note also that telomeres inherited from 2539 were unaltered, yet those from FH were frequently rearranged (i.e., they were present in generation 1 but absent in generation 5)

fragments were then cloned using a targeted cloning method (Farman 2011), followed by restriction mapping and end-sequencing to identify the natures of the rearrangements.

### *4.5.1 MoTeR Array Expansion and Contraction Caused by Breaks Within Internal Telomere Repeat Tracts*

The use of chromosome-unique probes and subsequent telomere cloning showed that many chromosome ends experienced simple expansions and contractions of MoTeR arrays. LpKY97-1A TEL4 is a good example because it showed remarkable instability. Among the 19 single spores isolated from leaf lesions, four different variants of the TEL4 TRF were detected (Fig. 4.5b). Furthermore, most of

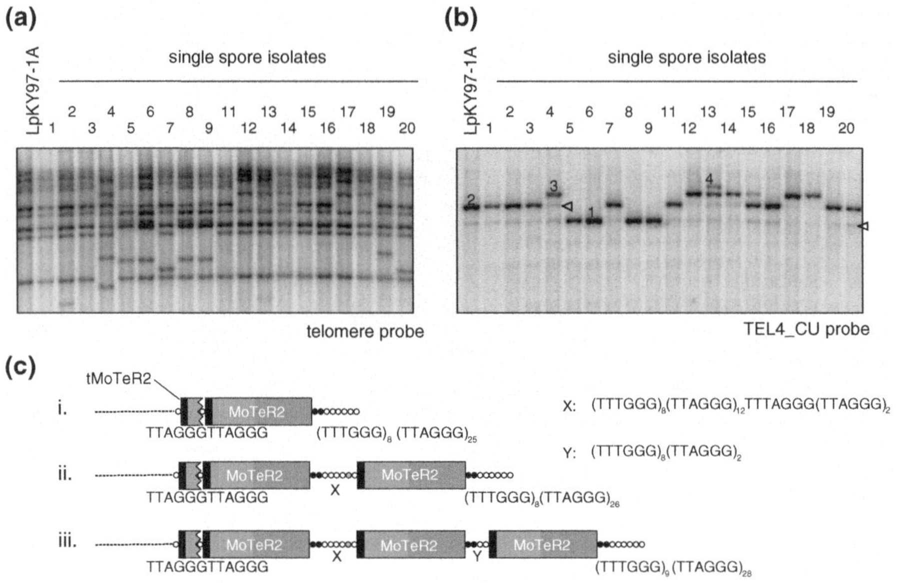


**Fig. 4.5** Recurrent rearrangements at an individual chromosome end. DNA samples from single-spore isolates of strain LpKY97-1A were digested with *Pst*I, fractioned by agarose gel electrophoresis, blotted to membranes, and hybridized with **a** telomere probe and **b** a chromosome-unique (CU) probe derived from a terminal *Pst*I fragment carrying a telomeric MoTeR2 array. DNA from the LpKY97-1A starting culture is loaded in the *left-hand* lane. Four different variants of the telomere were detected in panel (**b**) (labeled *1*, *2*, *3*, and *4*). Note that all cultures containing longer telomere variants also possess some copies of the truncated versions (*open arrowheads*). **b** Structures of variant telomeres identified in panel (**a**). Three different telomere variants (*i*, *ii*, and *iii*) were cloned, and their structures were determined by sequencing and restriction mapping. x and y show the sequences of the internal telomere tracts separating the MoTeR2 copies depicted in **c**

the spore cultures contained more than one variant (one culture even had signals from all four), indicating that rearrangement is extremely frequent and an ongoing process. The sizes of the telomeric variants differed by multiples of ~1.7 kb, which corresponds to the unit length of MoTeR2. Therefore, it appeared that this particular telomere underwent rearrangements involving simple expansions and contractions of a MoTeR2 array.

Cloning and characterization of the highly unstable LpKYTEL4 revealed that the original telomere (the one in the starting culture) contained two full-length MoTeR2 insertions and a drastically truncated copy which was positioned immediately adjacent to the chromosome-unique sequence (Fig. 4.5c). Two new variants—one shorter and one longer—also were characterized. The shorter one contained a single MoTeR2 distal to the truncated copy, while the longer one contained three intact distal MoTeR2s (Fig. 4.5c). The organization of these clones confirmed our suspicion that the observed rearrangements were due to MoTeR array expansions and contractions. Moreover, the truncation points corresponded precisely to the positions of internal telomere tracts.

In theory, it should have been possible for the array to be contracted further to versions containing the single truncated element or zero elements, yet such variants were not observed, nor were there faint hybridization signals of the appropriate sizes. A possible explanation is that the telomere tracts flanking the truncated MoTeR on both sides consist of only one TTAGGG repeat, while the other interstitial tracts are longer—$(TTTGGG)_8(TTAGGG)_{12}TTTAGGG(TTAGGG)_2$ and $(TTTGGG)_8(TTAGGG)_2$ (Fig. 4.5c). We propose that the longer tracts experience frequent breakage, resulting in deprotected ends that terminate in short tracts of telomeric sequence. Repair by telomere addition would produce truncations, while ectopic recombination with MoTeRs at other chromosome ends could produce array truncations or expansions, depending on the organization of the donor array and the recombination point. It is worth noting that internal telomere repeats promote chromosome breaks in fungi, humans, and plants (Itzhaki et al. 1992; Nelson et al. 2011; Wu et al. 2009; Yu et al. 2006), although in these cases, the normal route to repair appears to be capping with telomere repeats.

The Magnaporthe strains with the longer MoTeR arrays always exhibited faint signals from shorter variants (Fig. 4.5b), such that, in any given culture, an estimated one in five nuclei contained a truncated version. This indicates that breakage of internal arrays is an extremely frequent event. In striking contrast, the shortest variant of the MoTeR array appears to be very stable. Single-spore cultures that inherited this variant showed no hybridization signals from longer arrays, and no segregation was observed upon further single-spore isolation. This demonstrates that the MoTeR sequences themselves are not prone to breakage and that the terminal repeats rarely become deprotected.

### *4.5.2 MoTeR Terminal Truncations*

Cloning of additional TRFs and their rearranged derivatives identified a number of terminal truncations in which the 5′ end of a MoTeR was missing and the remaining sequence was capped with a canonical telomere minus the variant repeats (Fig. 4.6a.iii). Most of our data suggest that the telomere proper is not easily compromised so it seems unlikely that these truncated elements arose through erosion of terminal sequences. In one case, we have evidence that a chromosome end acquired a truncated MoTeR that was already present (at a different end) in the starting culture. It is also possible that some of these instances result from *de novo* transposition of 5′ truncated elements.

### *4.5.3 MAGGY Insertions*

Sequencing of two newly arisen TRFs identified copies of the MAGGY LTR retrotransposon (Farman et al. 1996) inserted in the middle of MoTeR1. MAGGY contains the *Pst*I recognition sequence in each of its long terminal repeats, so insertion

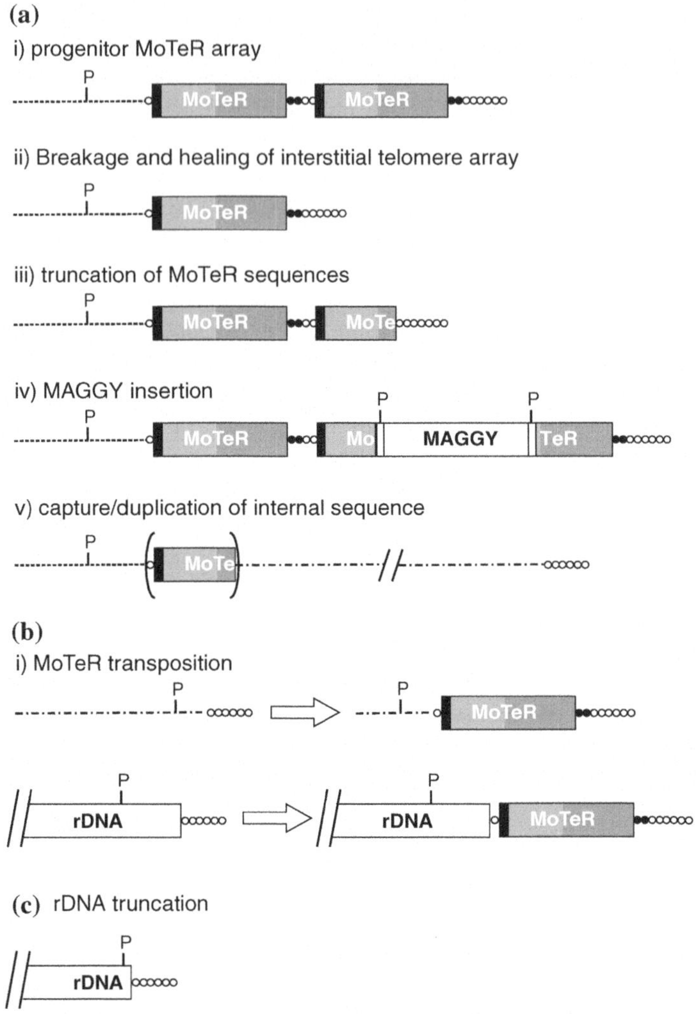


**Fig. 4.6** Schematics of telomere rearrangement types detected in MoTeR-containing strains. Types of rearrangements were determined by cloning novel telomere fragments and characterizing their inserts by restriction mapping and sequencing. **a** MoTeR array alterations. (i) shows an representative telomeric MoTeR array; (ii) through (v) show various rearranged versions. Note the truncated MoTeR in (v) is shown in parentheses because, although we suspect that the internal sequence was captured following MoTeR array breakage/truncation, we have no direct evidence to support this. **b** MoTeR transposition events; **c** rDNA truncation. Open circles represent canonical telomere repeats, and closed circles represent variant repeats. The lengths of telomere repeats are not drawn to scale. *P PstI* restriction site

of this element effectively introduces sites within the MoTeR array, resulting in a shortening of the corresponding TRF (Fig. 4.6a.iv). To determine how frequently MAGGY promoted TRF changes, we developed a diagnostic Southern blot test for *de novo* insertions. This revealed that most of the newly formed TRFs less than 5 kb in length were caused by MAGGY insertions, occurring in either orientation.

The MAGGY integrase protein contains a chromodomain which directs integration to heterochromatin (Gao et al. 2008) and is essential for efficient transposition (Nakayashiki et al. 2005). Their frequent acquisition of MAGGY insertions suggests that the MoTeR arrays are rich in heterochromatin, possibly the result of a telomere position effect.

### 4.5.4 Capture/Duplication of an Internal Sequence

One of the newly formed TRFs that was cloned had the telomere attached to a sequence that was non-telomeric in the progenitor strain. Furthermore, the original, non-telomeric locus was still present. This appears to be another example of the capture and duplication of an internal sequence by a chromosome end, as was previously observed at 70-15 telomere 14 (see Fig. 4.2a.iii). The duplication junction has not been characterized so the catalyst for this rearrangement is unknown. However, as depicted in Fig. 4.6a.v, we speculate that an inter-MoTeR breakage was possibly involved. Alternatively, a MAGGY insertion may have precipitated a subsequent ectopic recombination event.

### 4.5.5 MoTeR Transposition

To date, we have detected two instances in which "naked" telomeres lacking MoTeR insertions spontaneously gained MoTeR1 sequences (see Fig. 4.6b). In a third case, a telomere containing only MoTeR1 gained a truncated copy of MoTeR2 (not shown). For two of these putative transposition events, there are fairly long tracts of telomeric sequence (6 and 9 TTAGGGs) proximal to the MoTeR 3′ end. Thus, it is possible that the MoTeR sequences were acquired during recombinational repair of telomeres that somehow became deprotected. However, considering our evidence that terminal telomere repeats are resistant to loss of integrity, we favor the transposition hypothesis. The third putative transposition event cannot easily be explained by recombination because the newly acquired MoTeR was separated from the chromosome-unique sequence by just two TTAGGG repeats, which presumably is too short a substrate for homologous recombination. Current experiments are focused on the development of a retrotransposition assay to confirm MoTeR1 and MoTeR2 mobility.

### 4.5.6 Ribosomal Array Truncations

In the LpKY97-1A progenitor strain, the rDNA telomere lacks MoTeR insertions, and yet, it is highly unstable. Indeed, throughout our studies, we have identified many single-spore isolates in which the 1.9-kb rDNA fragment is absent (for example, see Fig. 4.1b). As in many fungi, the *M. oryzae* ribosomal RNA genes occur in a large (~2 Mb) tandem array, which in *M. oryzae* strain 70-15 occupies a single locus on chromosome 2. At least four of the Magnaporthe strains that we have studied have the rDNA array extending all the way to the chromosome end where it is capped by telomeric repeats. Cloning of a number of rearranged rDNA telomere fragments suggests that the majority arise through simple truncation and recapping (Fig. 4.6c), although we cannot rule out the possibility that expansions also occur.

Instability of the rDNA telomere was also detected when analyzing the raw sequence data generated in the 70-15 genome sequencing project. Most of the sequence reads derived from TEL3 contained TTAGGG repeats attached at position 7,807 in the rDNA array. However, there were six singleton traces representing variously truncated versions of the chromosome end, with the telomere attached at rDNA positions 46, 999, 1,817, 1,908, 4,391, and 8,105. Thus, it appears that neighboring ribosomal sequences somehow compromise the telomere's protective function.

Telomere-healed breaks have been documented in the rDNA of other organisms, including *Giardia lamblia* (Arkhipova and Morrison 2001) and *Neurospora crassa* (Butler 1992). In the latter case, the breaks were apparently due to the particular strain being a partial diploid, which carried two copies of the rDNA locus. More recently, however, we identified truncated rDNA telomere variants in the *N. crassa* genome sequence data (Wu et al. 2009), indicating that truncations also occur in haploid cells. Expansion and contraction of the rDNA array is one of the most common causes of chromosome length polymorphisms in fungi (Zolan 1995). In *S. cerevisiae*, the major mechanisms underlying size changes in the rDNA are gene conversion (Gangloff et al. 1996) and unequal crossing-over (Petes 1980). Our results indicate that truncation of rDNA arrays may also be a major contributor to rDNA size variation, in organisms where the array occupies a terminal location.

In summary, the MoTeRs promote terminal rearrangements in three basic ways: through their initial insertion into telomeres; by generating extended and, hence, unstable interstitial telomeres upon integration; and by altering the chromatin environment of the regions in which they reside. Some of the rearrangements that occur have major impacts on genome organization, including duplication of internal sequences at termini and, presumably, the attendant movement of subterminal sequence to the interior. As such, it is clear that the MoTeRs have major impacts on *M. oryzae* genome organization.

## 4.6 Subtelomere/Subterminal Organization in Other Host-Specific Forms of Magnaporthe

The organization of the chromosome ends in 70-15 appears to be representative of subtelomere structure in the rice pathogen population as a whole because previous Southern hybridization studies showed the TLH genes to be widely present among

Magnaporthe strains from rice, although their copy numbers varied widely among isolates (Gao et al. 2002). In addition, sequencing of representative cosmids from other rice pathogens identified canonical, TLH-containing subtelomeres (see Sect. 4.9.1). TLH genes are almost universally absent from non-rice pathogens (Gao et al. 2002).

Equivalent studies using MoTeR probes showed the MoTeR1 RT sequence to be present at a high copy number in all of the *prg*, wheat and millet pathogens examined, intermediate copy number in isolates from goosegrass, single copy in the St. Augustinegrass pathogens and only sporadically present among isolates from rice and foxtails. RT sequences were universally absent from the crabgrass pathogens (*Magnaporthe grisea*). When present, the RT sequence was usually located on a TRF and was, therefore, probably telomeric. MoTeR2 tended to occur at a lower copy number than MoTeR1 in all of the populations analyzed and was frequently absent.

Considering that TLH genes are present only in rice pathogens and the MoTeRs were largely restricted to the *prg*, wheat and Eleusine (millets and goosegrass) pathogen populations, this begged the question as to what the subtelomeres/subterminal regions of the other host-specific forms look like. Do they have simple, generic ends, or are there other types of conserved subtelomere sequences? We have started addressing this question using a combination of telomere cloning and genome sequence analysis.

### 4.6.1 Foxtail Pathogens

Sequencing of several cloned TRFs from a strain collected in Lexington, KY (Arcadia 2), revealed no evidence of TLH-associated subtelomere sequences or MoTeR sequences in the telomeres. Of the nine chromosome ends that were sequenced, one contained rDNA sequence adjacent to the telomere repeats, two had "generic" (i.e., non-telomeric) transposon sequences next to the telomere, and five contained novel sequence with no matches to the 70-15 genome. Interestingly, three of the Arcadia telomeric clones contained telomere tracts in the subterminal regions. These tracts, which contained 4, 5, and 25 TTAGGG repeats, are highly unusual because telomeric sequences with more than two repeats have not been found at internal genome locations in any of the other Magnaporthe genomes that have been sequenced.

### 4.6.2 Crabgrass Pathogens

Magnaporthe isolates from crabgrasses belong to a separate species, *M. grisea* (Couch and Kohn 2002). Characterization of 13 telomeric clones from strain 217DC showed that the telomeres were comprised of uninterrupted TTAGGG tracts and, therefore, lacked MoTeR insertions. The subterminal regions in clones 1 and 11 had similar sequences to one another, as did the corresponding regions in clones 4 and 13. However, in both cases, the positions of the telomere repeats varied. Although this organization is reminiscent of the subtelomeres in the rice pathogen 70-15, there was no sequence match. This suggests that 217DC might have one or more novel subtelomeric sequences that differ from the one found in the rice pathogens.

In general, the non-TLH- and non-MoTeR-containing strains had generic chromosomal sequences immediately adjacent to the telomere repeats. This was true also for the foxtail pathogens. However, these strains were unique in having extended interstitial telomere in the subterminal regions. Formation of extended internal tracts is a common response to telomerase deletion in *M. oryzae* (B. Wang, B. Peppers, and M. Farman, in preparation). Thus, we propose that the interstitial telomeres in the foxtail pathogens are generated via occasional failure of the telomere maintenance machinery.

## 4.7 Internal MoTeR Relics are Widely Present in Magnaporthe

Occasionally, the MoTeR RT probe hybridized to restriction fragments that were non-telomeric. In an effort to characterize these loci, we used inverse PCR to amplify a number of 3′ insertion junctions. Unexpectedly, in one strain, we amplified more junctions than there were hybridization signals. This implied the presence of elements with severe 5′ truncations—a suspicion that was confirmed by using a 3′ MoTeR1 probe to reprobe the DNA samples from all of the host-specific forms. The majority of Magnaporthe strains analyzed were found to contain short MoTeR relics, indicating that these elements were present in the common ancestor to *M. grisea* and *M. oryzae* and have experienced recurrent losses from the various host-specialized populations.

In contrast to the full-length elements detected with the RT probe, most of the truncated elements were not telomeric. Searches of genome sequences for isolates from rice, wheat, *prg*, and crabgrass identified over 50 severely truncated MoTeR 3′ ends at internal genome locations. All these internal relics had intact 3′ ends, which were separated from the adjacent chromosomal sequences by short telomere repeat tracts. Considering the evidence that MoTeRs insert into telomeres, this suggested that the relics are vestiges of formerly telomeric insertions that have become internalized. Evidence in support of this hypothesis comes from an internal MoTeR relic in a foxtail pathogen. The vestigial telomere repeat at the 3′ insertion junction of this particular relic corresponds to a true telomere identified in a rice pathogen. This discovery of former telomeres at internal genome locations points to a flow of genetic information from the chromosome tips to the rest of the genome, thus reciprocating the movement of internal sequences to telomeric locations described above.

**Table 4.1** Frequency of telomere rearrangement in Magnaporthe strains isolated from different host plants

| Host plant | Number of strains | Number spore cultures | Total number of novel telomeres[a] | Average # new telomeres/culture |
|---|---|---|---|---|
| Rice | 10 | 89 | 14 | 0.16 |
| Perennial ryegrass | 7 | 141 | 75 | 0.53 |
| Annual ryegrass | 1 | 10 | 2 | 0.2 |
| Foxtail | 8 | 79 | 115 | 1.46 |
| Crabgrass | 3 | 30 | 9 | 0.3 |
| St. Augustinegrass | 1 | 5 | 7 | 1.4 |
| Ginger | 1 | 10 | 0 | 0 |
| Para grass | 1 | 10 | 1 | 0.1 |
| Chinese Sprangletop | 1 | 10 | 13 | 1.3 |

[a] Total number of novel telomere fragments in all single-spore isolates analyzed. Faint hybridization signals corresponding to "incipient" novel telomeres were counted. Similar-sized fragments present in more than one isolate were counted only once

## 4.8 Telomere Instability in Different Host-Specialized Isolates of Magnaporthe

Our current data suggest that the telomere instability exhibited by LpKY97-1A largely is due to breakage of internal telomere repeat tracts that are generated as a consequence of MoTeR insertion. This leads to the prediction that strains with high MoTeR copy numbers should possess unstable telomeres, while those lacking MoTeRs should have relatively stable ones (with the possible exception of the rDNA telomere). To test this, we have explored relative telomere stability in a collection of strains, including additional ones from rice and *prg* as well as representatives of other host-specialized forms. Because the earlier *in planta* growth experiments had proved to be technically challenging, for these further investigations, we simply cultured the single-spored strains on oatmeal agar plates before sampling from the subsequent spore generation. The results are summarized in Table 4.1.

### *4.8.1 Rice Pathogens*

In general, the strains from rice had quite stable telomeres. Of the 10 strains analyzed, two showed no obvious variation in telomere profile among the single-spore cultures, five had only one band change, and three had two or more changes (overall average = 0.16 changes/culture). In strains with a single telomere alteration, the rDNA telomere was most likely the culprit. Note that we could not use an rDNA probe to test this directly because the single telomeric signal tends to be obscured by smearing of the signal from the more than 100 tandemly arranged rDNA copies. Interestingly, the strain experiencing the most band changes (5) was 2539 a recombinant laboratory strain with MoTeR insertions in three telomeres.

### 4.8.2 *Prg* Pathogens

All the strains from *prg* and the relative annual ryegrass (*arg*) showed telomere alterations during vegetative growth, although the frequency of rearrangement varied considerably, ranging from just two changes among 10 spore cultures in strain PL1-1 (*arg*) to 32 changes in 28 cultures of LpKY97-1A (overall average = 0.53 changes/culture). We suspect that this wide variation in stability is because some strains lack the extended internal telomere tracts known to be present in LpKY97-1A. Interestingly, separate cultures of LpKY97-1A maintained by different laboratory personnel exhibited significant differences in telomere stability. Most likely, this difference arose because the repair of unstable, internal telomere tracts had resulted in the formation of stable arrays, either through shortening or terminalization of interstitial telomere tracts.

### 4.8.3 Foxtail Pathogens

All strains analyzed experienced multiple rearrangements, with three yielding an average of more than 2 novel telomeres per single-spore isolate. This was surprising because the majority of foxtail pathogens lack MoTeRs and even those that have them possess just one copy. A possible clue to the rampant telomere alterations in these strains is held by the Arcadia2 subterminal sequences, which contained at least one very long internal telomere tract. According to our data, this sequence should make the corresponding chromosome end prone to frequent breakage, followed by healing or rearrangement.

### 4.8.4 Pathogens of St. Augustinegrass and Chinese Sprangletop

Analysis of single spores from strains infecting these two hosts revealed several novel telomeric fragments. However, in most cases, the new hybridization signals were very faint, which suggests that the rate of telomere rearrangements is low or that counterselection prevented nuclei with rearranged variants from gaining abundant representation within the colony.

### 4.8.5 Pathogens of Crabgrass, Para Grass, and Ginger

Three strains from crabgrass were analyzed, one of which produced single-spore cultures showing several faint hybridization signals from novel telomeric fragments. The remaining strain, along with a single representative each from para grass and ginger, yielded just two, one, and zero novel TRFs among 10, 10, and 8 single spores, respectively.

### *4.8.6 Wheat Pathogens*

Unfortunately, strict regulations governing culturing and manipulation of the exotic wheat blast pathogen on US soil prevented us from working with live cultures. Therefore, we were unable to monitor telomere stability in the wheat-infecting strains. However, an analysis of telomere segregation in a cross between Magnaporthe isolates from wheat and Setaria performed by a Japanese group found frequent non-Mendelian segregation of the wheat telomere "alleles," with three out of 12 visible TRFs exhibiting rearrangements in just five tetrads (Chuma et al. 2011).

In summary, with the exception of the foxtail pathogens, it appears that Magnaporthe strains with abundant MoTeRs tend to have more unstable TRFs than the strains with one or fewer copies. Not counting the strains from foxtails, the MoTeR-containing *prg* and *arg* pathogens experienced TRF alterations approximately four times more frequently than other strains. Along with strains from Chinese sprangletop and St. Augustinegrass, the *M. oryzae* foxtail pathogens were the champions of telomere instability. Intriguingly, the foxtail strains with the most unstable telomere profiles had many more than 14 telomeric hybridization signals, and several of the signals were unusually intense. Based on telomere organization in the Arcadia2 strain, we suspect that this reflects an abundance of long, interstitial telomeres in the subterminal regions of the foxtail pathogens. As noted above, we believe that these to be created through occasional failure of the telomere maintenance machinery. Given the potential for internal breaks to generate terminal rearrangements, this raises the intriguing possibility that there is an adaptive advantage to such failure and, furthermore, that it is genetically programmed. Indeed, this may be an explicit example of "adaptive telomere failure"—a phenomenon that had been postulated to exist (McEachern 2007), but, until now, there have been no known examples.

## 4.9 Comparative Analysis of Subtelomere/Subterminal Regions

Based on our limited sequencing, it appears that Magnaporthe has at least six distinct telomere/subtelomere/subterminal structures: (1) Rice-infecting strains are unique in having a conserved subtelomere sequence containing a telomere-linked helicase gene; (2) strains from perennial ryegrass, wheat, weeping lovegrass, and millets have numerous MoTeR insertions within their telomeres; (3) the foxtail pathogen Arcadia2 has extensive telomere tracts in its subterminal regions; (4) the crabgrass pathogen 217DC possesses novel subtelomere sequences; (5) several strains have rDNA sequences in the subterminal regions; and (6) all strains examined had one or more telomeres that adjoined generic chromosomal sequence lacking notable features. To gain insight into how these various structures may have evolved, we have started to compare homologous chromosome segments among the various host-specific forms.

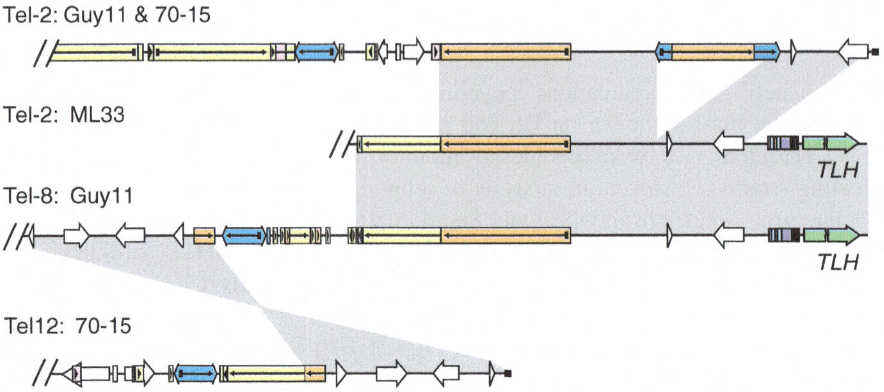


**Fig. 4.7** Evolution of telomeres 2 and 12. A probe derived from sequences immediately adjacent to telomere 2 of 70-15 was used to screen a cosmid library of Guy11 genomic DNA. This resulted in the identification of subtelomere 8, which contains a copy of the sequence adjacent to telomere 2 (Farman and Leong 1995). Subtelomere 2 of strain ML33 was identified in a cosmid library by screening with a TLH gene probe. The telomeric ends of each clone are on the *right*. Regions of sequence identity are connected with *gray shading*. Single and low copy genes are depicted as *white arrows*. Repetitive elements are represented by *colored boxes*. *Boxes* with inverted *arrowheads* are inverted repeat transposons, while those bound by *small boxes* containing *arrowheads* are LTR retrotransposons. Small boxes with *arrowheads* represent solo LTRs. Boxes lacking *arrowheads* or tails represent truncated elements. *Yellow* RETRO6, *blue* Pot2/Pot4, *orange* MGL, *brown* Pyret, *red* MAGGY

### *4.9.1 "Generic" Chromosome Ends in 70-15*

The origin of the generic telomeres in 70-15 was illuminated by cloning homologous chromosome ends from two other rice pathogens, Guy11 and ML33. Guy11 was a recurrent parent in the backcrossing scheme that produced 70-15 (Chao and Ellingboe 1991) and contributed at least 10 telomeres to the latter strain, including telomere 2. In Guy11, the sequence adjacent to telomere 2 is duplicated near telomere 8. To understand the nature of this duplication, we cloned the corresponding region from a cosmid library of Guy11 DNA. Interestingly, in the duplicated region, the sequence found adjacent to telomere 2 in 70-15 was bordered by a canonical subtelomere (Fig. 4.7)—an organization that was mirrored in a subtelomere-containing cosmid clone from a third strain, ML33 (Fig. 4.7). From this, we conclude that telomere 2 in Guy11 and 70-15 was derived via truncation of a chromosome end that once possessed a canonical subtelomere. Interestingly, Guy11 subtelomere 8 also contained the sequence that occurs immediately adjacent to 70-15 telomere 12, albeit in an inverted orientation (Fig. 4.7). Thus, it appears that the subtelomere region capped by telomere 12 in 70-15 could also be a truncated derivative of a chromosome end that once contained a TLH gene.

On reflection, it is not surprising to find that subtelomeres 2 and 12 arose through terminal truncations because the general organization of the 70-15 subtelomeres indicates an evolutionary history replete with such events. One would expect recurrent truncations eventually to cause TLH gene elimination. However, losses of terminal sequence appear to be counterbalanced by the occasional duplication of entire chromosome ends, as reflected by 70-15 subtelomeres 5 and 10, which have identical organizations, right down to the specific nucleotide position where the telomere repeats are attached (Rehmeyer et al. 2006). The presence of doubly and triply, etc. intense TLH gene hybridization signals in Southern blots of DNAs from numerous rice-infecting strains (Gao et al. 2002) points to similar duplication events.

### *4.9.2 Evolution of the Conserved Subtelomere Domain Found in Rice Pathogens*

Gao and coworkers have previously shown that TLH gene sequences are restricted to Magnaporthe isolates from rice (Gao et al. 2002). These findings were based on Southern hybridization analysis using a probe containing only a part of the TLH open reading frame (ORF). Consequently, it was not known whether the TLH gene alone has a limited distribution, or whether the entire subtelomere region is specific to rice-infecting isolates. To address this question, we used the complete subtelomere sequence to search genome assemblies of *M. oryzae* isolates from wheat (2 isolates), perennial ryegrass (2 isolates) and annual ryegrass (1 isolate), and a single isolate of *M. grisea* (crabgrass). The proximal region of subtelomere region II (II-A) was not found in any of the genomes analyzed, nor was the distal portion of region I (I-B) (Fig. 4.8). In contrast, various portions of the sequence between the RETRO6-1 insertion and the TLH gene were identified, but they were not contiguous in the other genomes. For example, sequences present in II-A and II-B were identified in most of the genomes that were examined but were never found adjacent to one another (Fig. 4.8). Region A occurred at a copy number of up to three in any given genome, and in all instances, the RETRO6-1 insertion was missing. In contrast, the homologous loci did possess an Mg-SINE insertion at the expected location, but the element was intact, whereas the one in the 70-15 subtelomere is 5′-truncated. All the "non-rice pathogen" genomes contained sequences matching region II-B, but these sequences were not bounded by Mg-SINE or HARs repeats. Furthermore, the newly identified sequences appear to represent an ancestral II-B locus structure because all of them could be aligned well beyond the points where the Mg-SINE and HARs are located in the 70-15 subtelomeres (Fig. 4.8).

Most of the genomes analyzed lacked HAR sequences, except for the *M. grisea* isolate (from crabgrass) and two *M. oryzae* genomes. In these cases, a single abbreviated HAR-DEF array was identified upstream of a very short segment from the 5′ end of the TLH ORF. Similarly-truncated TLH sequences were identified in all but one of the remaining genomes, but, in these cases, they were unlinked to any other subtelomere-related motifs. Only one non-rice pathogen—an isolate

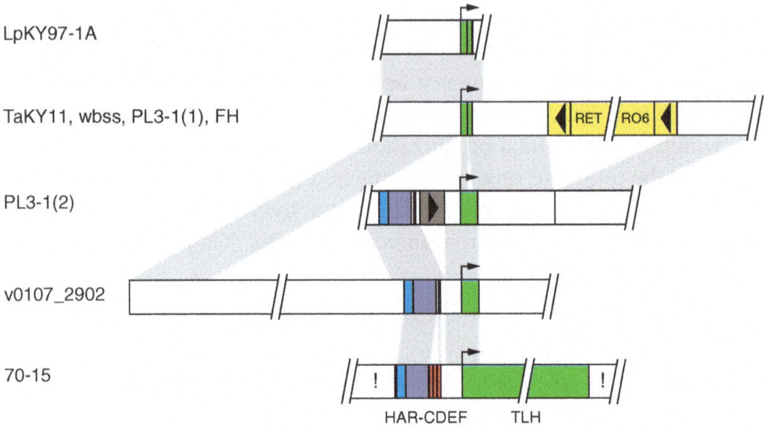


**Fig. 4.8** Distribution and organization of subtelomere sequences in Magnaporthe. The subtelomere sequence from strain 70-15 was used to search genome sequences of Magnaporthe isolates infecting other host species. These isolates were as follows: *M. oryzae*: TaKY11 and wbss—wheat; LpKY97-1A and FH—perennial ryegrass; and PL3-1—annual ryegrass. *M. grisea*: v0107—crabgrass. Notable sequence features are colored: *green* TLH sequences, *yellow* RETRO6 retrotransposon, *brown* RETRO7 LTR, *yellow*, *blue*, *lilac*, and *red* HAR-C, HAR-D, HAR-E, and HAR-F, respectively. Regions with significant sequence identity are connected with *gray shading*. Regions having no matches to any of the other chromosome segments in the figure are not connected with shading. Sequences that are unique to a specific isolate are marked with *exclamation marks*

from wheat—appeared to contain an almost intact TLH ORF (possibly lacking its 3′ end), but the corresponding sequences were contained on several very short contigs, and consequently, their positions relative to the sequences shown in Fig. 4.8 (and to one another) are unknown. Based on these data, it appears that the conserved subtelomere sequence found at multiple chromosome ends in the rice pathogens is actually a mosaic of much shorter sequences derived from several internal chromosome regions, with transposons and HARs repeats defining the fusion points between the different segments.

The different host-specific forms of Magnaporthe exhibited significant structural variation at the loci containing TLH gene relics (Fig. 4.8). The homologous loci differed in transposon insertion/excision patterns and showed evidence of having underdone a number of translocations. So, while these relics appear to be no longer located near telomeres, the abundance of rearrangements is probably a reflection of their past residence in the highly dynamic terminal regions. Interestingly, the distribution patterns of the TLH genes/relics are strikingly similar to those of the MoTeRs. Specifically, full-length genes are widely present and occur at high copy numbers only in certain pathogen populations (in this case, the rice pathogens). Intact genes can be found in other host-specialized forms but are extremely rare, whereas truncated forms (in this case, 3′ truncated copies) are almost ubiquitously present within Magnaporthe but are found at internal genome locations.

### 4.9.3 Subtelomeres as Regions of Genome Innovation

Although we were able to identify portions of TLH regions in other host-specific forms of Magnaporthe, these sequences did not appear to be linked to telomeres. Likewise, analysis of 39 telomeres in strains from crabgrass, foxtail, wheat, and perennial ryegrass (7 strains total) provided no evidence for the existence of TLH-related sequences in the associated subtelomere regions. This suggested that the subtelomeres of non-rice strains have an entirely different origin and structure. Further insight into the organization and evolution of TLH-lacking subtelomeres was gained first by utilizing genome data to expand the available information on the telomere-linked sequences, followed by the use of BLAST to identify the corresponding regions in the 70-15 genome. This revealed that only 23 of the 39 newly identified subtelomere regions had matches in 70-15, and among this subset, only 11 matched a genomic location that was within 100 kb of a 70-15 telomere. Indeed, many of the sequences that were subtelomeric in the non-rice pathogens were firmly embedded in the middle of a 70-15 chromosome (Table 4.2). Although it was not possible to perform a complete analysis in the reciprocal direction due to incomplete genome assemblies for the non-rice pathogens, a number of the sequences that are subterminal in 70-15 were located in the middle of large sequence contigs in the other strains (data not shown).

The discovery of so many apparent subterminus ←→ internal region translocations was surprising because genetic mapping studies show that, in general, the genomes of different host-specialized forms of *M. oryzae* tend to be co-linear. The chromosome regions in question were usually comprised of single-copy sequences, making it unlikely that asymmetric duplications are responsible for this discrepancy. Instead, it would appear that subterminal sequences frequently insert themselves at internal genome locations, with destroying regional organization. It is important to note, however, that we cannot rule out the existence of sequence duplications that were "collapsed" during sequence assembly.

For 29 of the newly identified subterminal sequences, the sequences immediately adjacent to the telomere repeats had no matches to the 70-15 genome. The lengths of the novel sequences ranged from 70 bp to more than 30 kb (average = 16 kb) (Table 4.2). Likewise, the distal portions of 12 subtelomeres in 70-15 lacked matches in the other strains, although, not surprisingly, most of the non-matching segments involved portions of the TLH region. Nevertheless, at least one subterminal region in 70-15 contains almost 14 kb of sequence that was not found in any of the other strains analyzed.

Gene finding programs predicted that many of the longer novel DNA segments have protein-coding potential. BLASTx and BLASTp searches of the NCBI nr protein database failed to identify any putative functions for the corresponding proteins. Interestingly, however, the top matches in the database were usually Magnaporthe proteins encoded by the 70-15 genome. So, although the novel subterminal sequence regions had no obvious similarity to the reference genome at the nucleotide level, protein similarity was retained. This suggests that the novel

**Table 4.2** Comparative analysis of telomere-adjacent sequences in different Magnaporthe strains

| Host strain | Matching scaffold in 70-15[a] | Distance from 70-15 telomere[b] | Length of unique sequence |
|---|---|---|---|
| *M. oryzae* | | | |
| *Foxtail* | | | |
| **Arcadia 2** | | | |
| 1 | – | – | >285 bp |
| 2 | 7.5e | 1.2 Mb | – |
| 3 | – | – | >1.3 kb |
| 4 | 7.4b | 135 kb | – |
| 5 | 7.4b | 139 kb | 1.85 kb |
| 6 | – | – | |
| 7 | 7.4e | 9 kb | – |
| 8 | – | – | |
| 9 | 7.7e | 759 kb | – |
| *Prg* | | | |
| **LpKY97-1** | | | |
| 1 | rDNA | – | – |
| 2 | 7.3b | 44.9 kb | – |
| 3 | 7.5b | 26.5 kb | 1.2 kb |
| 4 | – | – | 8 kb |
| 5 | 7.1b | 2.8 Mb | 2.6 kb |
| 6 | 7.4b | 30.5 kb | 2.6 kb |
| 7 | 7.4b | 135 kb | >21 kb |
| 8 | 7.2e | 153 kb | 9.6 kb |
| 9 | 7.8e | ? | 1.3 kb |
| **FH** | | | |
| 2 | – | 140 bp | 9.1 kb |
| 3 | rDNA | – | – |
| 5 | 7.3e | 17.9 kb | – |
| 8 | 7.6e | 19.6 kb | – |
| 11 | 7.5b | 26.5 kb | 1.2 kb |
| *Wheat* | | | |
| **WHTQ** | | | |
| 1 | – | | >3.2 kb |
| 2 | – | | >233 bp |
| **WBSS** | | | |
| 1 | – | – | 19.8 kb |
| **TaKY11** | | | |
| 1 | 7.1e | 2.5 Mb | 34 kb |
| 2 | 7.5b | 83 kb | 22.5 kb |
| *M. grisea* | | | |
| Crabgrass | | | |
| **Dc217** | | | |
| 1 | – | – | 70 bp |
| 2 | 7.5b | 70 kb | >300 bp |
| 3 | 7.1b | 2.75 Mb | >1.4 kb |

(continued)

Table 4.2 (continued)

| Host strain | Matching scaffold in 70-15[a] | Distance from 70-15 telomere[b] | Length of unique sequence |
|---|---|---|---|
| 4 | – | – | ~10.7 kb |
| 5 | – | – | ~8 kb |
| 6 | 7.6e | 20.9 kb | 1.4 kb |
| 7 | – | – | 2.6 kb |
| 8 | 7.5 | 1.3 Mb | – |
| 9 | – | – | >8.3 kb |
| 10 | – | – | |
| 11 | – | – | 2.6 kb |

[a] *b* match is nearest to the beginning of scaffold; *e* match is nearest to the end; "–" no match; poor match or repeated sequence
[b] Indicates position of query sequence relative to the nearest telomere in 70-15. "?" distance to telomere unknown

subterminal sequences are actually highly diverged copies of internal genome regions. Closer examination of these presumed duplications points to a complex origin because separate genes that are closely linked in the novel chromosomal segments appear to have been derived from multiple, dispersed genome locations. The same properties were exhibited by the telomere-adjacent sequences at the chromosome ends of the related fungus *N. crassa* (Wu et al. 2009). Thus, it may prove to be a common phenomenon that fungal chromosome ends are repositories for novel sequences generated via terminal duplication and subsequent divergence of internal chromosome regions.

The above comparative studies highlight the dynamic nature of the subtelomeric/subterminal regions. Specifically, we find that sequences found near to the telomeres in one strain of the fungus are rarely subterminal/subtelomeric in another. This implies the frequent shuttling of sequences back and forth between the chromosome ends and the rest of the genome. Such movement is often coupled with the duplication of the sequences involved. The net effect is to generate subterminal sequences that are often mosaics of highly diverged copies of dispersed regions from the genome interior.

## 4.10 Conclusion

Here, we show that the chromosome ends of certain *M. oryzae* strains experience alterations at an extremely high frequency. Indeed, in some cases, it appears that occasional failure of the telomere maintenance may be genetically programmed as a means to generate terminal sequence diversity. As noted above, the subterminal regions of the *M. oryzae* chromosomes harbor avirulence genes that tend to trigger host defense responses—a situation paralleling the variant surface glycoproteins of *Trypanosoma brucei* (Berriman et al. 2005) and the variant surface antigens of

*Plasmodium falciparum* (Hernandez-Rivas et al. 1997). In the latter pathogens, telomere proximity is believed to be instrumental in antigenic switching, wherein the expression of the variant surface proteins is continually changed (Horn and Barry 2005), allowing the pathogens to evade the host's immune system. Clearly, the events we have documented at the *M. oryzae* telomeres have the potential to alter gene expression, be it through gene deletion, translocation, or repression/activation. Consequently, the observed telomere dynamics may serve to switch avirulence gene expression in *M. oryzae*. This would be beneficial to the fungus because it would allow *M. oryzae* to alter its secreted profile to avoid recognition by the plant's surveillance system, in turn expanding the potential host range.

It is worth pointing out that the majority of avirulence genes identified to date have been discovered in rice blast pathogens. One might question the advantage of these genes residing near to telomeres that are relatively stable. However, while we observed stability over the short term, the structures of the rice pathogen subtelomeres point to histories replete with terminal truncations, and the subterminal regions bear the hallmarks of numerous deletions of chromosome segments between transposon insertions (Rehmeyer et al. 2006). Thus, a potential for avirulence gene deletion remains, albeit at a slower rate than what would be expected for strains containing MoTeRs, or interstitial telomeres.

The *M. oryzae* genome codes for upward of 500 small, secreted proteins (SSPs), many of which are translocated into the host cell along with the effectors that confer avirulence. Only a very small proportion of the SSP genes map near to telomeres (Rehmeyer et al. 2006). It is, therefore, highly significant that the ones with avirulence activity (i.e., host defense triggering capability) and, hence, experiencing the strongest negative selection, are found so often in subterminal locations. This situation implies that SSP genes with avirulence activity are somehow able to migrate to the chromosome ends. Our studies of telomere dynamics have uncovered plausible mechanisms by which this could be accomplished. Specifically, we have shown that telomeres are capable of capturing internal sequences, presumably in response to terminal damage. Conversely, mechanisms exist by which terminal sequences can be relegated to internal chromosome positions. Acting together, these processes could allow fungi to take maximal advantage of the dynamic telomere environment by accelerating the evolution of genes with critical ecological roles.

## References

Aksoy, S., Williams, S., Chang, S., & Richards, F. F. (1990). SLACS retrotransposon from *Trypanosoma brucei gambiense* is similar to mammalian LINEs. *Nucleic Acids Research, 18*, 785–792.

Arkhipova, I. R., & Morrison, H. G. (2001). Three retrotransposon families in the genome of Giardia lamblia: Two telomeric, one dead. *Proceedings of the National Academy of Sciences of the United States of America* (Vol. 98 pp. 14497–14502). National Academy of Sciences: USA.

Berriman, M., Ghedin, E., Hertz-Fowler, C., Blandin, G., Renauld, H., et al. (2005). The genome of the African trypanosome *Trypanosoma brucei*. *Science, 309*, 416–422.

Bonman, J. M. (1992). Durable resistance to rice blast disease—environmental influences. *Euphytica, 63*, 115–123.

Butler, D. K. (1992). Ribosomal DNA is a site of chromosome breakage in aneuploid strains of Neurospora. *Genetics, 131*, 581–592.

Chao, C. C. T., & Ellingboe, A. H. (1991). Selection for mating competence in Magnaporthe grisea pathogenic on rice. *Canadian Journal of Botany, 69*, 2130–2134.

Chuma, I., Hotta, Y., & Tosa, Y. (2011). Instability of subtelomeric regions during meiosis in *Magnaporthe oryzae*. *Journal of General Plant Pathology, 77*, 317–325.

Couch, B. C., & Kohn, L. M. (2002). A multilocus gene genealogy concordant with host preference indicates segregation of a new species, *Magnaporthe oryzae*, from *M. grisea*. *Mycologia, 94*, 683–693.

Couch, B. C., Fudal, I., Lebrun, M. H., Tharreau, D., Valent, B., et al. (2005). Origins of host-specific populations of the blast pathogen Magnaporthe oryzae in crop domestication with subsequent expansion of pandemic clones on rice and weeds of rice. *Genetics, 170*, 613–630.

Dean, R. A., Talbot, N. J., Ebbole, D. J., Farman, M. L., Mitchell, T. K., et al. (2005). The genome sequence of the rice blast fungus *Magnaporthe grisea*. *Nature, 434*, 980–986.

Farman, M. L. (2002). *Pyricularia grisea* isolates causing gray leaf spot on perennial ryegrass (*Lolium perenne*) in the United States: relationship to *P. grisea* isolates from other host plants. *Phytopathology, 92*, 245–254.

Farman, M. L. (2007). Telomeres in the rice blast fungus: The world of the end as we know it. *FEMS Microbiology Letters, 273*, 125–132.

Farman, M. L. (2011). Targeted cloning of fungal telomeres. In J.-R. Xu (Ed.), *Methods in Molecular Biology*.

Farman, M. L., & Kim, Y.-S. (2005). Telomere hypervariability in *Magnaporthe oryzae*. *Molecular Plant Pathology, 6*(3), 287–298.

Farman, M. L., & Leong, S. A. (1995). Genetic and physical mapping of telomeres in the rice blast fungus, *Magnaporthe grisea*. *Genetics, 140*, 479–492.

Farman, M. L., & Leong, S. A. (1998). Chromosome walking to the AVR1-CO39 avirulence gene of Magnaporthe grisea: discrepancy between the physical and genetic maps. *Genetics, 150*, 1049–1058.

Farman, M. L., Tosa, Y., Nitta, N., & Leong, S. A. (1996). MAGGY, a retrotransposon in the genome of the rice blast fungus *Magnaporthe grisea*. *Molecular and General Genetics, 251*, 665–674.

Gabriel, A., Yen, T. J., Schwartz, D. C., Smith, C. L., Boeke, J. D., et al. (1990). A rapidly rearranging retrotransposon within the miniexon gene locus of *Crithidia fasciculata*. *Molecular and Cellular Biology, 10*, 615–624.

Gangloff, S., Zou, H., & Rothstein, R. (1996). Gene conversion plays the major role in controlling the stability of large tandem repeats in yeast. *EMBO Journal, 15*, 1715–1725.

Gao, W., Khang, C. H., Park, S.-Y., Lee, Y.-H., & Kang, S. K. (2002). Evolution and organization of a highly dynamic, subtelomeric helicase gene family in the rice blast fungus *Magnaporthe grisea*. *Genetics, 162*, 103–112.

Gao, X., Hou, Y., Ebina, H., Levin, H. L., & Voytas, D. F. (2008). Chromodomains direct integration of retrotransposons to heterochromatin. *Genome Research, 18*, 359–369.

Gladyshev, E. A., & Arkhipova, I. R. (2007). Telomere-associated endonuclease-deficient Penelope-like retroelements in diverse eukaryotes. *Proceedings of the National Academy of Sciences of the United States of America* (Vol. 104 pp. 9352–9357). National Academy of Sciences: USA.

Goodwin, T. J., & Poulter, R. T. (2001). The diversity of retrotransposons in the yeast *Cryptococcus neoformans*. *Yeast, 18*, 865–880.

Hernandez-Rivas, R., Mattei, D., Sterkers, Y., Peterson, D. S., Wellems, T. E., et al. (1997). Expressed var genes are found in Plasmodium falciparum subtelomeric regions. *Molecular and Cellular Biology, 17*, 604–611.

Horn, D., & Barry, J. D. (2005). The central roles of telomeres and subtelomeres in antigenic variation in African trypanosomes. *Chromosome Research, 13*, 525–533.

Itzhaki, J. E., Barnett, M. A., Maccarthy, A. B., Buckle, V. J., Brown, W. R., et al. (1992). Targeted breakage of a human chromosome mediated by cloned human telomeric DNA. *Nature Genetics, 2*, 283–287.

Kang, S., Lebrun, M. H., Farrall, L., & Valent, B. (2001). Gain of virulence caused by insertion of a Pot3 transposon in a Magnaporthe grisea avirulence gene. *Molecular plant-microbe interactions: MPMI, 14*, 671–674.

Khang, C. H., Berruyer, R., Giraldo, M. C., Kankanala, P., Park, S. Y., et al. (2010). Translocation of Magnaporthe oryzae effectors into rice cells and their subsequent cell-to-cell movement. *Plant Cell, 22*, 1388–1403.

Levy, M., Romao, J., Marchetti, M. A., & Hamer, J. E. (1991). DNA fingerprinting with a dispersed repeated sequence resolves pathotype diversity in the rice blast fungus. *Plant Cell, 3*, 95–102.

Li, W., Rehmeyer, C. J., Staben, C., & Farman, M. L. (2005). TERMINUS—telomeric end-read mining in unassembled sequences. *Bioinformatics, 21*, 1695–1698.

Mcclure, M. A., Donaldson, E., & Corro, S. (2002). Potential multiple endonuclease functions and a ribonuclease H domain in retroposon genomes. *Virology, 296*, 147–158.

McEachern, M. J. (2007). Telomeres: Guardians of genomic integrity or double agents of evolution? In J. Noseck & L. Tomaska (Eds.), *Origins and evolution of telomeres*. Georgetown, TX: Landes Bioscience, Eureka Press.

Murakami, J., Tosa, Y., Kataoka, T., Tomita, R., Kawasaki, J., et al. (2000). Analysis of host species specificity of *Magnaporthe grisea* toward wheat using a genetic cross between isolates from wheat and foxtail millet. *Phytopathology, 90*, 1060–1067.

Murakami, J., Tomita, R., Kataoka, T., Nakayashiki, H., Tosa, Y., et al. (2003). Analysis of host species specificity of *Magnaporthe grisea* toward foxtail millet using a genetic cross between isolates from wheat and foxtail millet. *Phytopathology, 93*, 42–45.

Nakayashiki, H., Awa, T., Tosa, Y., & Mayama, S. (2005). The C-terminal chromodomain-like module in the integrase domain is crucial for high transposition efficiency of the retrotransposon MAGGY. *FEBS Letters, 579*, 488–492.

Nelson, A. D., Lamb, J. C., Kobrossly, P. S., & Shippen, D. E. (2011). Parameters affecting telomere-mediated chromosomal truncation in Arabidopsis. *Plant Cell, 23*, 2263–2272.

Okazaki, S., Ishikawa, H., & Fujiwara, H. (1995). Structural analysis of TRAS1, a novel family of telomeric repeat-associated retrotransposons in the silkworm, *Bombyx mori Molecular and Cellular Biology, 15*, 4545–4552.

Orbach, M. J., Farrall, L., Sweigard, J. A., Chumley, F. G., & Valent, B. (2000). A telomeric avirulence gene determines efficacy for the rice blast resistance gene *Pi-ta. The Plant Cell, 12*, 2019–2032.

Peacock, C. S., Seeger, K., Harris, D., Murphy, L., Ruiz, J. C., et al. (2007). Comparative genomic analysis of three Leishmania species that cause diverse human disease. *Nature Genetics, 39*, 839–847.

Petes, T. D. (1980). Unequal meiotic recombination within tandem arrays of yeast ribosomal DNA genes. *Cell, 19*, 765–774.

Peyyala, R., & Farman, M. L. (2006). Magnaporthe oryzae isolates causing gray leaf spot of perennial ryegrass possess a functional copy of the AVR1-CO39 avirulence gene. *Molecular Plant Pathology, 7*, 157–165.

Pryde, F. E., Gorham, H. C., & Louis, E. J. (1997). Chromosome ends: All the same under their caps? *Current Opinion in Genetics and Development, 7*, 822–828.

Rehmeyer, C., Li, W., Kusaba, M., Kim, Y.-S., Brown, D., et al. (2006). Organization of chromosome ends in the rice blast fungus *Magnaporthe oryzae. Nucleic Acids Research, 34*, 4685–4701.

Rehmeyer, C. J., Li, W., Kusaba, M., & Farman, M. L. (2009). The telomere-linked helicase (TLH) gene family in *Magnaporthe oryzae*: revised gene structure reveals a novel TLH-specific protein motif. *Current Genetics, 55*, 253–262.

Schwartz, S. L., & Farman, M. L. (2010). Systematic overrepresentation of DNA termini and underrepresentation of subterminal regions among sequencing templates prepared from hydrodynamically sheared linear DNA molecules. *BMC Genomics, 11*, 87.

Starnes, J. H., Thornbury, D. W., Novikova, O.S., Rehmeyer, C. J., & Farman, M. L., (2012). Telomere-targeted retrotransposons in the rice blast fungus Magnaporthe oryzae: agents of telomere instability. *Genetics, 191*, 389–406.

Sweigard, J. A., Carroll, A. M., Kang, S., Farrall, L., Chumley, F. G., et al. (1995). Identification, cloning, and characterization of PWL2, a gene for host species specificity in the rice blast fungus. *The Plant Cell, 7*, 1221–1233.

Takahashi, H., Okazaki, S., & Fujiwara, H. (1997). A new family of site-specific retrotransposons, SART1, is inserted into telomeric repeats of the silkworm, *Bombyx mori*. *Nucleic Acids Research, 25*, 1578–1584.

Teng, S. C., Wang, S. X., & Gabriel, A. (1995). A new non-LTR retrotransposon provides evidence for multiple distinct site-specific elements in *Crithidia fasciculata* miniexon arrays. *Nucleic Acids Research, 23*, 2929–2936.

Tosa, Y., Tamba, H., Tanaka, K., & Mayama, S. (2006). Genetic analysis of host species specificity of *Magnaporthe oryzae* isolates from rice and wheat. *Phytopathology, 96*, 480–484.

Villanueva, M. S., Williams, S. P., Beard, C. B., Richards, F. F., & Aksoy, S. (1991). A new member of a family of site-specific retrotransposons is present in the spliced leader RNA genes of *Trypanosoma cruzi*. *Molecular and Cellular Biology, 11*, 6139–6148.

Wu, C., Kim, Y.-S., Smith, K. M., Li, W., Hood, H. M., et al. (2009). Characterization of chromosome ends in the filamentous fungus *Neurospora crassa*. *Genetics, 181*, 1129–1145.

Yu, W., Lamb, J. C., Han, F., & Birchler, J. A. (2006). Telomere-mediated chromosomal truncation in maize. *Proceedings of the National Academy of Science* (Vol. 103 pp. 17331–17336). USA.

Zolan, M. E. (1995). Chromosome-length polymorphism in fungi. *Microbiological Reviews, 59*, 686–698.

# Chapter 5
# *Pneumocystis carinii* Subtelomeres

**James R. Stringer**

**Abstract** *Pneumocystis carinii* is a yeast-like fungus that dwells exclusively in rats, where it is subjected to the host immune response. Immune pressure is countered by three subtelomeric gene families that generate antigenic diversity in populations of *P. carinii*. Members of the three gene families are grouped together in tandem arrays that appear to have been generated by recombination and modified by further recombination events after array formation. One of these gene families, MSG, encodes various forms of a major surface glycoprotein. In a given *P. carinii* organism, all but one of the members of the MSG gene family appear to be transcriptionally silent. The expressed MSG gene is adjacent to a unique locus (the expression site). Different MSG genes can occupy the expression site, suggesting that recombination can take a gene from a pool of silent donors and install it at the expression site, thereby extinguishing transcription of the previous expression site resident, activating transcription of the newly installed MSG gene, and changing the surface of the microbe. Switching at the expression site is probably facilitated by the subtelomeric locations of expressed and silent MSG genes. A second subtelomeric gene family, MSR, is not strictly regulated at the transcriptional level, but may contribute to phenotypic diversity via high-frequency frameshifting caused by coding poly(G) tracts in some MSR genes.

## 5.1 *P. carinii* and Other Species in the Genus *Pneumocystis*

Pneumocystis carinii is a yeast-like member of the fungal phylum ascomycota, which includes the well-studied yeasts *Candida albicans*, *Saccharomyces cerevisiae,* and *Schizosaccharomyces pombe* (Eriksson 1994; Redhead et al. 2006;

J. R. Stringer (✉)
Department of Molecular Genetics, Biochemistry and Microbiology,
University of Cincinnati, Cincinnati, OH 45220-0524, USA
e-mail: stringjr@ucmail.uc.edu

E. J. Louis and M. M. Becker (eds.), *Subtelomeres*, DOI: 10.1007/978-3-642-41566-1_5, 

Liu et al. 2009). However, unlike these three yeasts, which are ubiquitous in the environment, proliferate robustly in culture, and are amenable to study using both genetics and biochemistry, *P. carinii* has been found only in the lungs of Norwegian rats and proliferates little in culture (Stringer 2002; Keely et al. 2003; Cushion and Stringer 2010). Consequently, *P. carinii* is relatively difficult to study.

Members of the genus *Pneumocystis* have been seen in numerous other mammalian species. DNA sequences suggest that each species of mammal carries at least one species of *Pneumocystis*. Four additional species have been named so far. *P. jirovecii* is found in humans, laboratory mice can carry *P. murina*, and rabbits are routinely colonized by *P. oryctolagi* (Stringer et al. 2002, 2009; Keely et al. 2004; Cushion and Stringer 2005; Dei-Cas et al. 2006; Brubaker et al. 2009). Another species, *P. wakefieldiae*, occurs in Norwegian rats, sometimes concurrently with *P. carinii* (Cushion et al. 2004, 2005). Although some early studies reported that *Pneumocystis* from humans can proliferate in mice, these reports appeared before it was possible to discriminate one species of *Pneumocystis* from another (Sethi 1992). It has since been shown that *Pneumocystis* species are host specific (Gigliotti et al. 1993; Durand-Joly et al. 2002; Aliouat-Denis et al. 2008).

*Pneumocystis* species are famous for causing pneumonia in immunodeficient/immunocompromised mammals. For example, *P. jirovecii* causes lethal pneumonia in immunocompromised individuals, such as victims of AIDS (Smulian et al. 1994; Keely et al. 1996; Miller 1999; Huang et al. 2011). However, the natural habitat of *Pneumocystis* species is the airway of immunocompetent mammals, where they are subjected to immune attack (Cushion et al. 2010). Such attacks can lead to the elimination of the fungus from an individual host. Though limited, the survival time in an immunocompetent individual tends to be sufficient to allow transmission to a new individual. This situation appears to have developed in rats millions of years ago, and similar relationships were formed between other *Pneumocystis* and mammalian species (Keely et al. 2004; Keely and Stringer 2005; Fischer et al. 2006).

## 5.2 Surface Variation Via Restricted Expression of a Subtelomeric Gene Family

Complete dependence on the Norwegian rat makes *P. carinii* an obligate parasitic fungus. Microbes that depend upon proliferation in immunocompetent hosts tend to have mechanisms that work to delay destruction by the immune response, thereby raising the probability of transmission to a new host (Deitsch et al. 2009). One such mechanism is rapid surface variation caused by differential expression of one or more gene families encoding different forms of proteins found on the surface of the microbe. These gene families are often located subtelomerically, a location that can facilitate DNA recombination, which is a mechanism that can be used to control the expression of gene families in a digital fashion, whereby activating the expression of one gene or genes automatically inactivates the expression of others.

In *P. carinii*, the MSG gene family resides at subtelomeres and encodes different isoforms of a major surface glycoprotein (Stringer et al. 1991; Sunkin and Stringer 1996; Stringer and Keely 2001; Stringer 2003, 2007; Keely et al. 2005; Keely and Stringer 2009). Different cells in a population of *P. carinii* can express different *MSG* isoforms (Wada et al. 1995; Sunkin and Stringer 1996, 1997). Expression of the MSG gene family appears to be controlled by recombination (Fig. 5.1). The MSG gene that is adjacent to a unique expression site is expressed, while all other MSG genes remain untranscribed, but can be moved to the expression site (Sunkin and Stringer 1996, 1997; Stringer and Keely 2001). Installation of a new MSG gene at the expression stops transcription of the formerly expressed gene and simultaneously activates transcription of the newly installed gene. The model would allow the expression of only one MSG gene at a time in a haploid cell, and the vast majority of the cells in a population of *P. carinii* are haploid.

The lack of a method to produce clonally derived populations of *P. carinii* has precluded direct tests of the hypothesis that MSG expression is completely

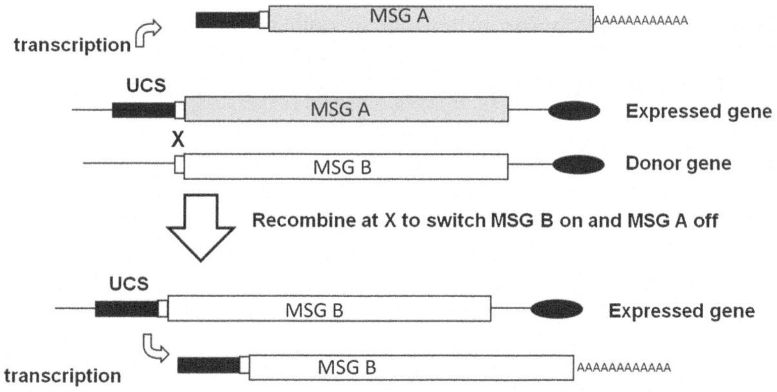


**Fig. 5. 1** The expression site model of MSG gene transcription. The expression site is a unique locus at the end of one of the 17 chromosomes in the *P. carinii* genome. This site contains the only genomic copy of the sequence encoding the upstream conserved sequence (*UCS*), which is found at the beginning of mRNAs encoding diverse MSGs, because the UCS is transcribed along with the adjacent MSG gene. When the expression site is occupied by the MSG A gene, MSG A mRNA is produced. (Translation starts in the UCS, and the UCS-encoded peptide functions to send the protein into the ER, where it is processed and modified for deposition on the cell surface.) The MSG gene that resides at the expression site can be changed via recombination. In the example shown, recombination occurs between the expression site and a donor MSG gene called MSG B. Thus, recombination would turn MSG A transcription off and turn MSG B transcription on (the genome of *P. carinii* contains approximately 73 donor MSG genes, most of which have been seen at the expression site). The recombination event shown occurs between copies of a 25-bp sequence known as the conserved recombination junction element (*CRJE*). Every MSG gene begins with a copy of the CRJE, and a copy of the CRJE is located between the UCS and adjacent MSG gene. However, there is no direct evidence for CRJE × CRJE recombination, and recombination could occur within MSG coding sequences. *Filled ovals* represent telomeres

dependent on linkage to the expression site. Nevertheless, three lines of evidence strongly support the expression site model depicted in Fig. 5.1. (1) The expression site contains the upstream conserved sequence (UCS), which encodes the RNA sequence found at the 5′ ends of transcripts encoding diverse MSGs (Wada et al. 1995; Edman et al. 1996). No transcript encoding an MSG and lacking the UCS has been described. (2) If *P. carinii* were to make UCS-less MSG transcripts, these would probably not be efficiently translated because MSG genes lack canonical translational initiation sites. By contrast, the UCS encodes such a site, and the first amino acids in MSG precursor proteins are UCS-encoded. In addition, the UCS leader peptide appears to be required to send an MSG preprotein into the endoplasmic reticulum, where it is glycosylated, trimmed off its leader peptide, and then deposited on the cell surface (Sunkin et al. 1998). (3) Studies employing monoclonal antibodies have supported the hypothesis that linkage of an MSG gene to the expression site occurs in cells that have the protein encoded by the UCS-linked gene on their surface. In these studies, antibodies were used to identify and quantify the fraction of *P. carinii* cells that expressed the MSG isoform that contained an epitope unique to that isoform (the C11 epitope), and PCR was used to determine the fraction of *P. carinii* cells that had a UCS-linked MSG gene encoding the C11 epitope. Populations of *P. carinii* from different rats were observed to differ by up to tenfold in both respects. The fraction of cells that expressed the C11 epitope expression correlated with the fraction of cells where the UCS was linked to the MSG gene encoding this epitope (Schaffzin and Stringer 2004).

The MSG gene family has been seen in six *Pneumocystis* species (Stringer et al. 1993; Garbe and Stringer 1994; Wright et al. 1994, 1995a, b; Haidaris et al. 1998; Schaffzin et al. 1999a; Schaffzin and Stringer 2000; Keely et al. 2007). However, *P. carinii* is the only species in which subtelomeres have been studied in detail. These data offer insights into the origin and function of MSG genes and other gene families in this species.

## 5.3 *P. carinii* Subtelomeres

Underwood et al. (1996) were the first to describe a cloned *P. carinii* subtelomere, which they obtained by using a *S. cerevisiae* artificial chromosome vector in *S. cerevisiae*. One clone contained an 8.3-kb segment that ended with tandem copies of TTAGGG, the telomere repeat usually found in organisms from the animal and fungal kingdoms, but not at *S. cerevisiae* telomeres. Copies of the TTAGGG repeat were shown to be present in all *P. carinii* chromosomes and to be hypersensitive to *Bal*31 exonuclease digestion, indicating that they are located at chromosome ends (Underwood et al. 1996).

The subtelomeric region in the DNA segment studied by Underwood et al. contained part of a member of the MSG gene family. It had been shown that the expression of the MSG gene family involves a unique expression site that contains a sequence present in the 5′ ends of messenger RNA molecules encoding different isoforms of MSG (Fig. 5.1). This sequence came to be known as the UCS. The UCS is located near a telomere because this locus was rapidly degraded when chromosomes were treated with *Bal*31 exonuclease (Sunkin and Stringer 1996; Wada and Nakamura 1996). In addition, a second cloned *P. carinii* subtelomere contained the UCS, which was adjacent to an MSG gene, which was followed by subtelomeric repetitive sequences and telomere-specific tandem repeats of TTAGGG (Wada et al. 1996).

Later studies employed two-dimensional pulsed-field gel electrophoresis to show that many MSG genes lie within 60 kb of telomeres (Cornillot et al. 2002). These data suggested that it might be possible to clone complete subtelomeric gene arrays by using a cosmid vector and screening for cosmids that contain both an MSG gene and a telomere repeat. In addition, other data had shown that *P. carinii* MSG genes tend to be linked to members of two other gene families, PRT1 and MSR (Wada and Nakamura 1994; Lugli et al. 1997; Russian et al. 1999; Schaffzin et al. 1999b; Ambrose et al. 2004).

A pWEB cosmid library was screened for hybridization to probes for MSG, MSR, PRT1, and a conserved sequence known to reside between protein-coding genes and the telomere repeat. This approach yielded 60 candidate clones. Analysis of these clones identified three clones that contained the UCS, at least one MSG gene and copies of the telomere repeat and seven that contained DNA segments that lacked the UCS, but included at least one MSG gene, and at least one MSR genes. PRT1 genes were present in six of these clones (Keely et al. 2005).

The inserts in the UCS clones were partially sequenced, and the inserts in the seven cosmids carrying MSGs not linked to the UCS were completely sequenced. The inserts in two cosmids that lacked the UCS, 11H12, and 1B2 came from the same subtelomere, but started and ended in different locations. These data were combined to produce contig 11H12 + 1B2. Therefore, cosmid cloning produced six different subtelomeres containing MSG genes that were not at the UCS (i.e., donor MSG genes, see Fig. 5.2).

### 5.3.1 UCS Subtelomere Structures

The three clones that contained the UCS resembled a lambda clone previously described (Wada and Nakamura 1996) and contained a single UCS-linked MSG gene, followed by subtelomeric and telomeric repeats. Three previously characterized UCS clones had the same structure (Sunkin and Stringer 1996). However, previous studies had described three clones that contained the UCS followed by more than one MSG gene, showing that the region between the UCS and the telomere can contain more than one gene (Sunkin and Stringer 1996). The number and types of genes downstream of the UCS-linked MSG gene are of interest when considering possible recombination events that could be used to move a gene from a donor subtelomere to the UCS locus, especially in light of the organization of donor MSG genes.

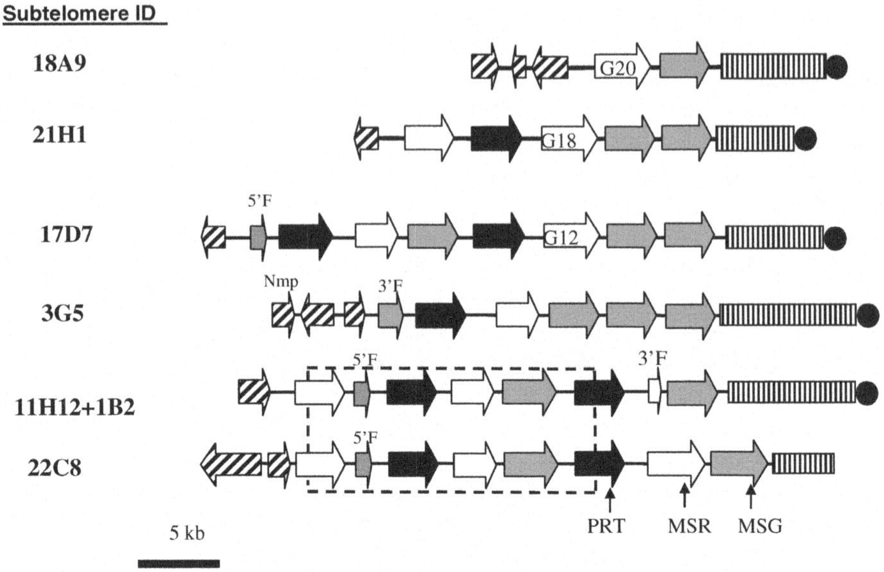


**Fig. 5.2** Structures of six subtelomeric gene arrays. *Arrows* are open reading frames (*ORFs*) and point in the direction of transcription. *Crosshatched arrows* are ORFs that are present in one copy in the *P. carinii* genome. *Solid arrows* are members of the PRT1 (*black*), MSR (*white*), and MSG (*gray*), gene families. *Rectangles with vertical lines* represent noncoding DNA. *Filled circles* represent telomeres. All features except the telomere are drawn to scale. 5′F and 3′F indicate ORFs corresponding to the 5′ and 3′ ends, respectively, of either an MSG (*gray arrow*) or an MSR (*white arrow*) gene. G12, G18, and G20 indicate MSR genes that contain a coding poly G mononucleotide tract containing 12, 18, and 20 Gs, respectively. The *dashed-line boxes* enclose regions that were identical

### *5.3.2 Structures of Subtelomeres Carrying Donor MSG Genes*

Figure 5.2 shows maps produced from sequencing cloned subtelomeric gene arrays (Keely et al. 2005). The maps were produced by assembling shotgun sequence reads from each insert. The accuracy of the maps produced from assembled sequence reads was confirmed by the analysis of cosmid DNA with restriction endonucleases.

One end of each subtelomeric segment begins with a sequence that is not repeated in the *P. carinii* genome. The uniqueness of these sequences was shown by Southern blot hybridization to *P. carinii* chromosomes that had been separated by pulsed-field gel electrophoresis. These experiments also mapped these unique sequences to five of the 17 *P. carinii* chromosomes, thereby allowing each repeated gene array to be assigned to a chromosome (Keely et al. 2005).

In all but one case, 22C8, the other end of the sequence contains copies of the telomere repeat. There is a complex repeat between the last MSG gene and the telomere repeat. The complex repeat regions in the six subtelomeres are all different, but resemble each other and those described previously (Underwood et al. 1996; Wada and Nakamura 1996).

The cloned donor gene arrays contain genes from different gene families, and all such genes point in the same direction. This situation has implications for evolution and expression of all three gene families.

### 5.3.3 Evolution of Gene Families in *P. carinii*

Often, the mechanism of gene family expansion can be inferred from the locations and orientations of family members. For example, it is known that when there are two adjacent copies of a sequence, this allows unequal reciprocal homologous recombination, usually involving sister chromatids, to produce additional copies of that sequence. Expansion via this mechanism produces identical copies that are located next to one another and pointed in the same direction (a tandem array).

The *P. carinii* subtelomeres shown in Fig. 5.2 feature tandem arrays of genes that are all pointed in the same direction, consistent with the hypothesis that unequal reciprocal homologous recombination contributed to the formation of these arrays. The structure PRT1-MSR-MSG occurs seven times. In addition to the seven perfect copies of PRT1-MSR-MSG, the 11H12 + 1B2 subtelomeric gene array has what appears to be a degenerate PRT1-MSR-MSG repeat because there is a fragment of an MSR gene (Fig. 5.2, 3′F) between the terminal PRT1 and MSG genes. Apparent fragments of MSG/MSR genes also occur in other arrays (17D7 and 3G5). Tandem arrays can be imperfect because mutational events can occur after duplication. The prevalence of the PRT1-MSR-MSG motif suggests that this three-gene unit expanded in number to generate all three gene families.

All of the PRT genes were followed by a 600-bp element that is 90 % identical to the 5′end of the *P. carinii* thioredoxin reductase gene, which is also located downstream of one PRT1 gene (Kutty et al. 2003). These data indicate that the evolution of the PRT1 gene family involved co-amplification of a part of the thioredoxin reductase gene along with the upstream PRT1 gene. It is not known whether the PRT1 gene that resides upstream of the complete *P. carinii* thioredoxin reductase gene is located in a subtelomere, or wheather there are MSG or MSR genes nearby. In any event, it is clear that the gene amplification events that generated the *P. carinii* PRT1 gene family included part of the thioredoxin gene.

It is interesting to note that other species of *Pneumocystis* do not contain multiple copies of the PRT1 gene. There is single copy of this gene (called Kex1) in both *P. murina* and *P. jirovecii* (Kutty and Kovacs 2003). Given the evidence that the PRT1 family formed in concert with the formation of the MSG gene family in *P. carinii*, the lack of a PRT1 family in other species suggests that the *P. carinii* gene families formed after *P. carinii* diverged from the last common ancestor of the

*Pneumocystis* species that have been studied. The alternative view that other species have lost the additional copies of the PRT1 gene is very unlikely given that PRT1 genes are interdigitated with MSG genes in *P. carinii*. It is not known whether MSR genes exist in other species, but if they do not, this would further illustrate the dramatic genomic difference between *P. carinii* and other members of the genus. Such genomic differences are difficult to reconcile with the view that *Pneumocystis* contains a single species (Cushion and Stringer 2005; Stringer et al. 2006).

The absence of the PRT1 gene family from other *Pneumocystis* species seems even more remarkable given that other species have multiple MSG genes that occur in clusters and rely on the attachment to a unique transcriptional expression site. It seems possible that each species has developed its own MSG gene family independently. In keeping with this speculation, each species has a UCS locus, but UCS loci vary considerably among species (Keely et al. 2007).

While the prevalence of the PRT-MSR-MSG motif in *P. carinii* is striking, and suggests a role for this motif in the amplification of all three gene families, other data indicate that other types of recombination events must have contributed extensively to forming the gene arrays seen today. When a gene array expands via unequal reciprocal homologous recombination, the new copy or copies of the gene are identical to the older copies. Mutation can occur thereafter, but still, one might expect to find the MSG genes within an array to be more similar to one another than to MSG gene in other arrays. However, this is not the case (Keely et al. 2005). The MSG genes within an array are not more closely related. It is possible that this situation reflects rapid change post duplication, which would be possible if selection for variation among MSG gene sequences was occurring, as seems to the case. However, variation also occurred in ways that did not influence the sequence of the encoded protein, suggesting that divergence of adjacent MSG gene was not solely driven by selection at the protein level. Nevertheless, analysis of intergenic regions, which would not be expected to be subjected to such selection, showed that these regions were less diverged than the MSG genes, supporting the hypothesis that the arrays were formed by duplication events, followed by diversification due to homologous recombination events that cause MSG genes to move to locations remote from their birthplace (Keely et al. 2005). The conservation in intergenic regions would facilitate such movement.

Recombination presumably created the situation seen when the arrays 22C8 and 11H12 + 1B2 are compared. These two subtelomeres share a 15.4-kb central segment, but regions flanking this shared segment are different, and the unique genes in the two inserts mapped to different chromosomes. The presence of the same 15.4-kb segment of DNA in two different gene arrays can be explained as being the result of homologous recombination, because the junctions that define the boundaries of the two 15.4-kb segments lie in gene family members.

Homologous recombination can also cause deletions, and it is possible that MSG gene arrays that are uninterrupted by PRT1 and MSR genes were formed by such events. In addition, recombination between MSG and MSR genes would be expected to occur because MSR genes are similar to MSG genes. One such event appears to have produced the MSR gene in array 18A9. The 5′ half of this gene has a sequence closely related to that of MSR genes and contains the intron characteristic of MSR genes. However, the sequence of the 3′ half of this gene is more similar to an MSG gene than an MSR gene. Such a gene could have been formed by homologous recombination between an MSR and an MSG gene (Keely et al. 2005).

### 5.3.4 *P. carinii Subtelomeric Tandem Gene Array Organization Holds Implications for the Expression Site Model*

All of the gene arrays end with an MSG gene, and in three arrays, all MSG genes are terminal. The fact that there are no other genes between the telomere and an MSG gene is of interest because such an arrangement would allow the terminal MSG gene to move via a single reciprocal recombination event (telomere swapping), without moving any other gene in the process. Seven of ten cloned copies of the UCS subtelomere contain a single MSG gene between the UCS and the telomere. However, not all MSG genes in the subtelomeric gene arrays are terminal, suggesting that telomere swapping is probably not the only mechanism that installs new MSG genes at the expression site (Sunkin and Stringer 1996).

Nevertheless, telomere swapping might contribute to the large number of different karyotypes exhibited by *P. carinii*. Pulsed-field gel studies have identified more than a dozen different *P. carinii* karyotypes, each of which features between 13 and 15 bands (Rebholz and Cushion 2001). Band intensities suggest that two bands in the 15-band karyotype contain a pair of co-migrating chromosomes, suggesting that there are 17 chromosomes, and hence 34 subtelomeres in the *P. carinii* genome. The different karyotypes appeared in different rat colonies, suggesting that different strains of *P. carinii* exist in different populations of rats (Cushion 1998). Whereas karyotypic variation is common among *P. carinii* populations, DNA sequence variation is rare, suggesting that events that change the length of one or more chromosomes occur much more frequently than does point mutation. Given the evidence for recombination involving MSG genes, it seems reasonable to suggest that such events have contributed to karyotypic diversity.

The role of subtelomeric gene array length variation due to either telomere swapping or homologous recombination events that expand or shrink a gene array has not been thoroughly explored, but the identification of unique genes adjacent to subtelomeric gene arrays makes it possible to begin looking into this issue. To illustrate, my laboratory used one of the unique genes (NMP in Fig. 5.2) at the beginning of array 3G5 to determine whether variation in the length occurs at the NMP end of the chromosome, which is the smallest in the genome and varies by as much as 30 kb in karyotypes exhibited by nine independent *P. carinii* populations. The sequence of the 3G5 array insert revealed that the *ApaL*1 restriction enzyme would cut in the NMP gene, but not cut the DNA between the end of that gene and the telomere. Therefore, *ApaL*1 was employed in two-dimensional pulsed-field gel electrophoresis analysis, where the DNA fragment released from the end of the chromosome by ApaL1 cleavage in the NMP gene was detected by preparing a Southern blot and probing it with DNA made from the NMP gene region that is telomeric to the *ApaL*1 site. Surprisingly, all nine karyotypes produced a band that migrated at approximately 35 kb. All of these bands co-migrated under conditions that would have separated fragments that differed by as little as 5 kb. These data showed that most if not all of the chromosome size variation was not due to changes in the size of the gene array located at the NMP end of the chromosome. Additional studies are needed to determine whether variation in the gene array at the other end of this chromosome might account for some or all of the variation in the size of this chromosome. However, the probability of a change occurring at the same end in all nine populations of *P. carinii* studied would be 0.002, unless the NMP-linked gene array is much more stable than the gene array that is presumably at the other end of this chromosome.

All of the gene family members in the subtelomeric gene arrays are pointed in the same direction, with 3′ ends pointed toward the telomere. This arrangement raises a challenge to the expression site model (Fig. 5.1) because there are PRT and MSR genes upstream of MSG genes. Both of these gene families are known to express independently of the UCS, and multiple family members are transcribed in a given *P. carinii* nucleus (Schaffzin et al. 1999b; Keely and Stringer 2003; Ambrose et al. 2004). It would seem possible that an MSG gene located downstream of an MSR gene could be transcribed by the same RNA polymerase that initiated transcription of the upstream MSR gene. Nevertheless, no read-through transcripts have been reported, and the role of the UCS in producing an MSG protein and sending it to the cell surface suggests that read-through transcription is not an important mechanism, if it occurs at all. In addition, read-through transcription of protein coding genes is the exception in eukaryotes, where transcriptional termination is tightly tied to polyadenylation, a nearly universal modification of eukaryotic messenger RNAs. PRT1 and MSR genes contain polyadenylation signal sequences, and mRNAs encoding PRT1 and MSR are polyadenylated (Huang et al. 1999; Keely et al. 1999; Schaffzin et al. 1999b).

The presence of MSR and PRT1 genes near MSG genes argues against the idea that donor MSG gene arrays might be silenced by regional chromatin modifications, because many MSR and PRT1 genes are simultaneously expressed, suggesting that expression of these gene families occurs in subtelomeres that are not expressing MSG genes.

## 5.4 MSR Genes Exhibit Structural Variation that Could Contribute to Surface Diversity in the Absence of Restricted Expression of the Gene Family

MSR genes resemble MSG genes at the sequence level, but the two gene families differ with respect to both structure and function (Schaffzin et al. 1999b; Keely et al. 2003; Ambrose et al. 2004). A major structural difference is that while MSG genes contain no introns, all MSR genes contain an intron near their 5′ ends. The lack of introns in the MSG gene family deviates from the norm in this species. The vast majority of *P. carinii* genes contain introns (Stringer 1994; Stringer et al. 1998; Thomas et al. 1999; Smulian et al. 2001; Cushion et al. 2007). The two exons that flank the MSR intron are very different in size. Exon 1 encodes only about 28 amino acids, while exon 2 encodes more than 300. MSR and MSG genes differ even more profoundly when it comes to transcription. Unlike MSG genes, MSR genes are not found adjacent to the UCS, transcripts from many MSR genes have been detected in populations of *P. carinii* that express very few MSG gene family members, and none of these transcripts carry a copy of the UCS, which is present on all MSG mRNAs (Schaffzin et al. 1999b; Ambrose et al. 2004). Therefore, it appears that the entire MSR gene family may be transcribed at the same time, rather than expressed one at a time.

Despite the promiscuous expression of the MSR gene family, MSR genes vary in a manner that suggests that they can contribute to surface variation without being subjected to strict transcriptional control. Three classes of MSR genes (L, G, and S) have been identified (Keely et al. 2005). Class L genes have a second exon of ~2.4 kb that includes a 1-kb segment lacking in Class S genes. Class G genes are similar to Class L but have a poly(G) tract in the middle of the second exon.

The poly(G) tracts in class G MSR genes do not appear to be a chance occurrence. The *P. carinii* genome is rich in adenosine and thymidine. Poly(G) tracts may play a role in antigenic variation, because as the simplest of simple sequence repeats, these tracts are inherently unstable and spontaneously change in length, most often by either adding or deleting one G residue (DePrimo et al. 1998; Hersh et al. 2002). Such events cause frameshifts. Other microbes exploit this property of simple sequence repeats to generate stochastic variation in the production of a surface protein (Deitsch et al. 2009).

All three of the class G genes in the arrays shown in Fig. 5.2 would produce mRNAs that cannot be translated beyond the stop codon located 13 codons downstream of the poly(G) tract. However, the sequence downstream of this stop codon contains a reading frame that encodes a peptide corresponding to the last half of the peptide encoded by the single ORF in second exons of class L MSR genes. Therefore, a frameshift mutation in the poly(G) tract would lead to the production of a full-length class L MSR protein. Although all three of the MSR genes found in these gene arrays were in the wrong frame to allow production of an L MSR protein, three other MSR genes with poly(G) tracts can be inferred to exist from cDNA data, and all of these have a 2.4-kb exon 2 ORF (Huang et al. 1999).

## 5.5 Summary

*P. carinii* subtelomeric gene families appear to have evolved in order to cope with attacks by the rat immune system. Recombination played a central role in forming the subtelomeric gene families, and recombination also appears to create antigenic variation by exerting strict control over MSG gene family transcription.

Subtelomeric gene arrays include MSR genes that contain coding poly(G) tracts, which undergo high-frequency length changes, causing frame shifting, which would generate phenotypic diversity from MSR genes, even though this family is promiscuously transcribed.

Similar subtelomeric gene arrays occur in bacteria and protozoa that must face the full onslaught of the host immune response (Barry et al. 2003; Deitsch et al. 2009). These gene arrays allow a population of microbes to produce antigenic variants that survive this onslaught, thereby postponing clearance. The positioning of gene family members in telomeric clusters facilitates antigenic variation because this location fosters recombination, which has two advantageous effects. Recombination both provides a mechanism to change antigen expression and fosters expansion and evolution of gene copies.

*P. carinii* appears to be completely dependent on Norwegian rats, where it eventually encounters the host immune response. In the laboratory, this response is capable of completely eliminating the fungus from a rat. However, there is reason to suspect that *P. carinii* is more or less a constant resident in individual rats, kept at low levels, but typically not eliminated by the immune response, thanks to phenotypic variation conferred by the gene families that reside at *P. carinii* subtelomeres. These families are complex and structurally prone to change. The combination of selection for phenotypic change and structural fluidity of genes that confer variation presumably drove the evolution of subtelomeres in *Pneumocystis* species and produced the complicated gene arrays and expression systems we see today.

**Acknowledgments** I would like to thank Saundra Stringer, Ann Wakefield, and Scott Keely for their help and encouragement as the work on *Pneumocystis* telomeres began, developed, and matured.

## References

Aliouat-Denis, C. M., Chabe, M., Demanche, C., Aliouat, E. M., Viscogliosi, E., Guillot, J., et al. (2008). *Pneumocystis* species, co-evolution and pathogenic power. *Infection, Genetics and Evolution, 8*, 708–726.

Ambrose, H. E., Keely, S. P., Aliouat, E. M., Dei-Cas, E., Wakefield, A. E., Miller, R. F., et al. (2004). Expression and complexity of the *PRT1* multigene family of *Pneumocystis carinii*. *Microbiology, 150*, 293–300.

Barry, J. D., Ginger, M. L., Burton, P., & McCulloch, R. (2003). Why are parasite contingency genes often associated with telomeres? *International Journal for Parasitology, 33*, 29–45.

Brubaker, R., Redhead, S. A., Stringer, J. R., Keely, S. P., & Cushion, M. T. (2009). Misinformation about *Pneumocystis*. *Clinical and Experimental Dermatology, 34*, e426–e427.

Cornillot, E., Keller, B., Cushion, M. T., Metenier, G., & Vivares, C. P. (2002). Fine analysis of the *Pneumocystis carinii* f. sp. *carinii* genome by two- dimensional pulsed-field gel electrophoresis. *Gene, 293*, 87–95.

Cushion, M. T. (1998). Genetic heterogeneity of rat-derived *Pneumocystis*. *FEMS Immunology and Medical Microbiology, 22*, 51–58.

Cushion, M. T., & Stringer, J. R. (2005). Has the name really been changed? it has for most researchers. *Clinical Infectious Diseases, 41*, 1756–1758.

Cushion, M. T., & Stringer, J. R. (2010). Stealth and opportunism: Alternative lifestyles of species in the fungal genus *Pneumocystis*. *Annual Review of Microbiology, 64*, 431–452.
Cushion, M. T., Keely, S. P., & Stringer, J. R. (2004). Molecular and phenotypic description of *Pneumocystis wakefieldiae* sp. nov., a new species in rats. *Mycologia, 96*, 429–438.
Cushion, M. T., Keely, S. P., & Stringer, J. R. (2005). Validation of the name *Pneumocystis wakefieldiae*. *Mycologia, 97*, 268.
Cushion, M. T., Smulian, A. G., Slaven, B. E., Sesterhenn, T., Arnold, J., Staben, C., et al. (2007). Transcriptome of *Pneumocystis carinii* during fulminate infection: Carbohydrate metabolism and the concept of a compatible parasite. *PLoS ONE, 2*, e423.
Dei-Cas, E., Chabe, M., Moukhlis, R., Durand-Joly, I., Aliouat, E. M., Stringer, J. R., et al. (2006). *Pneumocystis oryctolagi* sp. nov., an uncultured fungus causing pneumonia in rabbits at weaning: Review of current knowledge, and description of a new taxon on genotypic, phylogenetic and phenotypic bases. *FEMS Microbiology Reviews, 30*, 853–871.
Deitsch, K. W., Lukehart, S. A., & Stringer, J. R. (2009). Common strategies for antigenic variation by bacterial, fungal and protozoan pathogens. *Nature Reviews Microbiology, 7*, 493–503.
DePrimo, S. E., Cao, J., Hersh, M. N., & Stringer, J. R. (1998). Use of human placental alkaline phosphatase transgenes to detect somatic mutation in mice in situ. *Methods, 16*, 49–61.
Durand-Joly, I., Aliouat, E. M., Recourt, C., Guyot, K., Francois, N., Wauquier, M., et al. (2002). *Pneumocystis carinii* f. sp. hominis is not infectious for SCID mice. *Journal of Clinical Microbiology, 40*, 1862–1865.
Edman, J. C., Hatton, T. W., Nam, M., Turner, R., Mei, Q., Angus, C. W., et al. (1996). A single expression site with a conserved leader sequence regulates variation of expression of the *Pneumocystis carinii* family of major surface glycoprotein genes. *DNA and Cell Biology, 15*, 989–999.
Eriksson, O. E. (1994). *Pneumocystis carinii*, a parasite in lungs of mammals, referred to a new family and order (*Pneumocystidaceae, Pneumocystidales, Ascomycota*). *Systema Ascomycetum, 13*, 165–180.
Fischer, J. M., Keely, S. P., & Stringer, J. R. (2006). Evolutionary rate of Ribosomal DNA in *Pneumocystis* species is normal despite the extraordinarily low copy-number of rDNA genes. *Journal of Eukaryotic Microbiology, 53*(Suppl 1), S156–S158.
Garbe, T. R., & Stringer, J. R. (1994). Molecular characterization of clustered variants of genes encoding major surface antigens of human *Pneumocystis carinii*. *Infection and Immunity, 62*, 3092–3101.
Gigliotti, F., Harmsen, A. G., Haidaris, C. G., & Haidaris, P. J. (1993). *Pneumocystis carinii* is not universally transmissible between mammalian species. *Infection and Immunity, 61*, 2886–2890.
Haidaris, C. G., Medzihradsky, O. F., Gigliotti, F., & Simpson-Haidaris, P. J. (1998). Molecular characterization of mouse *Pneumocystis carinii* surface glycoprotein A. *DNA Research, 5*, 77–85.
Hersh, M. N., Stambrook, P. J., & Stringer, J. R. (2002). Visualization of mosaicism in tissues of normal and mismatch-repair-deficient mice carrying a microsatellite-containing transgene. *Mutation Research, 505*, 51–62.
Huang, S. N., Angus, C. W., Turner, R. E., Sorial, V., & Kovacs, J. A. (1999). Identification and characterization of novel variant major surface glycoprotein gene families in rat *Pneumocystis carinii*. *Journal of Infectious Diseases, 179*, 192–200.
Huang, L., Cattamanchi, A., Davis, J. L., den, B. S., Kovacs, J., Meshnick, S., et al. (2011). HIV-associated *Pneumocystis* pneumonia. *Proceedings of the American Thoracic Society, 8*, 294–300.
Keely, S. P., & Stringer, J. R. (2003). Sequence diversity of transcripts from *Pneumocystis carinii* gene families MSR and PRT1. *Journal of Eukaryotic Microbiology, 50*(Suppl), 627–628.
Keely, S. P., & Stringer, J. R. (2005). Nomenclature and genetic variation of Pneumocystis. *Pneumocystis Pneumonia, 3*, 39–59.
Keely, S. P., & Stringer, J. R. (2009). Complexity of the MSG gene family of *Pneumocystis carinii*. *BMC Genomics, 10*, 367.

Keely, S. P., Baughman, R. P., Smulian, A. G., Dohn, M. N., & Stringer, J. R. (1996). Source of *Pneumocystis carinii* in recurrent episodes of pneumonia in AIDS patients. *AIDS, 10*, 881–888.

Keely, S. P., Cushion, M. T., & Stringer, J. R. (1999). Determination of the maximum frequency of genetic rearrangements associated with *Pneumocystis carinii* surface antigen variation (In Process Citation). *Journal of Eukaryotic Microbiology, 46*, 128S.

Keely, S. P., Fischer, J. M., & Stringer, J. R. (2003). Evolution and speciation of *Pneumocystis*. *Journal of Eukaryotic Microbiology, 50*(Suppl), 624–626.

Keely, S. P., Fischer, J. M., Cushion, M. T., & Stringer, J. R. (2004). Phylogenetic identification of *Pneumocystis murina* sp. nov., a new species in laboratory mice. *Microbiology, 150*, 1153–1165.

Keely, S. P., Renauld, H., Wakefield, A. E., Cushion, M. T., Smulian, A. G., Fosker, N., et al. (2005). Gene arrays at *Pneumocystis carinii* telomeres. *Genetics, 170*, 1589–1600.

Keely, S. P., Linke, M. J., Cushion, M. T., & Stringer, J. R. (2007). *Pneumocystis murina* MSG gene family and the structure of the locus associated with its transcription. *Fungal Genetics and Biology, 44*, 905–919.

Kutty, G., & Kovacs, J. A. (2003). A single-copy gene encodes Kex1, a serine endoprotease of *Pneumocystis jiroveci*. *Infection and Immunity, 71*, 571–574.

Kutty, G., Huang, S. N., & Kovacs, J. A. (2003). Characterization of thioredoxin reductase genes (trr1) from *Pneumocystis carinii* and *Pneumocystis jiroveci*. *Gene, 310*, 175–183.

Liu, Y., Leigh, J. W., Brinkmann, H., Cushion, M. T., Rodriguez-Ezpeleta, N., Philippe, H., et al. (2009). Phylogenomic analyses support the monophyly of *Taphrinomycotina*, including *Schizosaccharomyces* fission yeasts. *Molecular Biology and Evolution, 26*, 27–34.

Lugli, E. B., Allen, A. G., & Wakefield, A. E. (1997). A *Pneumocystis carinii* multi-gene family with homology to subtilisin-like serine proteases. *Microbiology, 143*, 2223–2236.

Miller, R. F. (1999). *Pneumocystis carinii* infection in non-AIDS patients. *Current opinion in infectious diseases, 12*, 371–377.

Rebholz, S. L., & Cushion, M. T. (2001). Three new karyotype forms of *Pneumocystis carinii* f. sp. carinii identified by contoured clamped homogeneous electrical field (CHEF) electrophoresis. *Journal of Eukaryotic Microbiology, 48*(Suppl), 109S–110S.

Redhead, S. A., Cushion, M. T., Frenkel, J. K., & Stringer, J. R. (2006). *Pneumocystis* and *Trypanosoma cruzi*: Nomenclature and Typifications. *Journal of Eukaryotic Microbiology, 53*, 2–11.

Russian, D. A., Andrawis-Sorial, V., Goheen, M. P., Edman, J. C., Vogel, P., Turner, R. E., et al. (1999). Characterization of a multicopy family of genes encoding a surface- expressed serine endoprotease in rat *Pneumocystis carinii*. *Proceedings of the Association of American Physicians, 111*, 347–356.

Schaffzin, J. K., & Stringer, J. R. (2000). The major surface glycoprotein expression sites of two special forms of rat *Pneumocystis carinii* differ in structure. *Journal of Infectious Diseases, 181*, 1729–1739.

Schaffzin, J. K., & Stringer, J. R. (2004). Expression of the *Pneumocystis carinii* major surface glycoprotein epitope is correlated with linkage of the cognate gene to the upstream conserved sequence locus. *Microbiology, 150*, 677–686.

Schaffzin, J. K., Garbe, T. R., & Stringer, J. R. (1999a). Major surface glycoprotein genes from *pneumocystis carinii* f. sp. ratti. *Fungal Genetics and Biology, 28*, 214–226.

Schaffzin, J. K., Sunkin, S. M., & Stringer, J. R. (1999b). A new family of *Pneumocystis carinii* genes related to those encoding the major surface glycoprotein. *Current Genetics, 35*, 134–143.

Sethi, K. K. (1992). Multiplication of human-derived *Pneumocystis carinii* in severe combined immunodeficient (SCID) mice. *Experientia, 48*, 63–66.

Smulian, A. G., Linke, M. J., Cushion, M. T., Baughman, R. P., Frame, P. T., Dohn, M. N., et al. (1994). Analysis of *Pneumocystis carinii* organism burden, viability and antigens in bronchoalveolar lavage fluid in AIDS patients with pneumocystosis: Correlation with disease severity. *AIDS, 8*, 1555–1562.

Smulian, A. G., Sesterhenn, T., Tanaka, R., & Cushion, M. T. (2001). The ste3 pheromone receptor gene of *Pneumocystis carinii* Is surrounded by a cluster of signal transduction genes. *Genetics, 157*, 991–1002.

Stringer, J. R. (1994). Molecular Genetics of *Pneumocystis carinii*. In P. D. Walzer (Ed.), *Pneumocystis carinii pneumonia* (2nd ed., pp.73–90). Inc., New York: Marcel Dekker.

Stringer, J. R. (2002). Pneumocystis. *International Journal of Medical Microbiology, 292*, 391–404.

Stringer, J. R. (2003). The MSG gene family and antigenic variation in the fungus *P. carinii*. In A. Craig, A. Scherf (Eds.), *Antigenic Variation* (pp. 201–223). Amsterdam: Academic Press.

Stringer, J. R. (2007). Antigenic variation in *Pneumocystis*. *Journal of Eukaryotic Microbiology, 54*, 8–13.

Stringer, J. R., & Cushion, M. T. (1998). The genome of *Pneumocystis carinii*. *FEMS Immunology and Medical Microbiology, 22*, 15–26.

Stringer, J. R., & Keely, S. P. (2001). Genetics of surface antigen expression in *Pneumocystis carinii*. *Infection and Immunity, 69*, 627–639.

Stringer, S. L., Hong, S. T., Giuntoli, D., & Stringer, J. R. (1991). Repeated DNA in *Pneumocystis carinii*. *Journal of Clinical Microbiology, 29*, 1194–1201.

Stringer, S. L., Garbe, T., Sunkin, S. M., & Stringer, J. R. (1993). Genes encoding antigenic surface glycoproteins in *Pneumocystis* from humans. *Journal of Eukaryotic Microbiology, 40*, 821–826.

Stringer, J. R., Beard, C. B., Miller, R. F., & Wakefield, A. E. (2002). A new name (*Pneumocystis jirovecii*) for *Pneumocystis* from humans. *Emerging Infectious Diseases, 8*, 891–896.

Stringer, J. R., Cushion, M. T., & Redhead, S. A. (2006). Reply to weinberg, to hughes, and to limper. *Clinical Infectious Diseases, 42*, 1212–1214.

Stringer, J. R., Beard, C. B., & Miller, R. F. (2009). Spelling *Pneumocystis jirovecii*. *Emerging Infectious Diseases, 15*, 506.

Sunkin, S. M., & Stringer, J. R. (1996). Translocation of surface antigen genes to a unique telomeric expression site in *Pneumocystis carinii*. *Molecular Microbiology, 19*, 283–295.

Sunkin, S. M., & Stringer, J. R. (1997). Residence at the expression site is necessary and sufficient for the transcription of surface antigen genes of *Pneumocystis carinii*. *Molecular Microbiology, 25*, 147–160.

Sunkin, S. M., Linke, M. J., McCormack, F. X., Walzer, P. D., & Stringer, J. R. (1998). Identification of a putative precursor to the major surface glycoprotein of *Pneumocystis carinii*. *Infection and Immunity, 66*, 741–746.

Thomas, C. F, Jr, Leof, E. B., & Limper, A. H. (1999). Analysis of *Pneumocystis carinii* introns. *Infection and Immunity, 67*, 6157–6160.

Underwood, A. P., Louis, E. J., Borts, R. H., Stringer, J. R., & Wakefield, A. E. (1996). *Pneumocystis carinii* telomere repeats are composed of TTAGGG and the subtelomeric sequence contains a gene encoding the major surface glycoprotein. *Molecular Microbiology, 19*, 273–281.

Wada, M., & Nakamura, Y. (1994). MSG gene cluster encoding major cell surface glycoproteins of rat *Pneumocystis carinii*. *DNA Research, 1*, 163–168.

Wada, M., & Nakamura, Y. (1996). Unique telomeric expression site of major-surface-glycoprotein genes of *Pneumocystis carinii*. *DNA Research, 3*, 55–64.

Wada, M., Sunkin, S. M., Stringer, J. R., & Nakamura, Y. (1995). Antigenic variation by positional control of major surface glycoprotein gene expression in *Pneumocystis carinii*. *Journal of Infectious Diseases, 171*, 1563–1568.

Wright, T. W., Simpson Haidaris, P. J., Gigliotti, F., Harmsen, A. G., & Haidaris, C. G. (1994). Conserved sequence homology of cysteine-rich regions in genes encoding glycoprotein A in *Pneumocystis carinii* derived from different host species. *Infection and Immunity, 62*, 1513–1519.

Wright, T. W., Bissoondial, T. Y., Haidaris, C. G., Gigliotti, F., & Haidaris, P. J. (1995a). Isoform diversity and tandem duplication of the glycoprotein A gene in ferret *Pneumocystis carinii*. *DNA Research, 2*, 77–88.

Wright, T. W., Gigliotti, F., Haidaris, C. G., & Simpson-Haidaris, P. J. (1995b). Cloning and characterization of a conserved region of human and rhesus macaque *Pneumocystis carinii* gpA. *Gene, 167*, 185–189.

# Chapter 6
# Subtelomeres of *Aspergillus* Species

**Elaine M. Bignell**

**Abstract** Understanding of *Aspergillus* subtelomere biology remains in its infancy but has recently come under closer scrutiny due to possible involvement in plant, human and animal virulence, and natural product biosynthesis. Comparative genomics of pathogenic and non-pathogenic *Aspergillus* species has indicated that subtelomeres are hotbeds of genetic variation, thereby fuelling the search for virulence determinants in these regions. This chapter draws upon the recent availability of multiple *Aspergillus* genome sequences, and transcriptome analyses to summarise extant knowledge on *Aspergillus* subtelomeres, their organisation and gene content and regulatory influences upon subtelomeric gene expression. In particular, the entire subtelomeric gene inventory of the major mould pathogen of humans, *Aspergillus fumigatus*, is functionally categorised here by chromosome, and the evidence that *Aspergillus* subtelomeric genes mediate rapid responses to challenging environmental niches is assessed.

## 6.1 Introduction

The genus *Aspergillus* includes an estimated two hundred moulds. (Baker and Bennet 2008; Bennet 2010). Conidiophores are the morphological feature upon which *Aspergillus* taxonomy is defined, and constitute the stalk-like structures which bear asexual *Aspergillus* spores. *Aspergillus* species are ubiquitous in the natural environment and perform important roles in natural ecosystems as well as being highly significant to the human economy. As such the Aspergilli are the single most significant fungal genus to man.

The first domestication of *Aspergillus* species for food production is reported as occurring some 2,000 years ago when Chinese food fermentation processes,

E. M. Bignell (✉)
Institute of Inflammation and Repair, The University of Manchester, Oxford Road, Manchester M13 9PL, UK
e-mail: elaine.bignell@manchester.ac.uk

E. J. Louis and M. M. Becker (eds.), *Subtelomeres*, DOI: 10.1007/978-3-642-41566-1_6, 

collectively referred to as Koji processes, originated (Baker and Bennet 2008; Bennet 2010). Many different industrial processes use members of the genus, for example, in production of acidification agents (citric acid), hydrolytic enzymes (such as amylase used for hydrolysis of starch in bread and beer), invertase (used in manufacture of confectionary), and pectinases (for fruit juice and wine production). *Aspergillus oryzae* is widely used for the production of traditional fermented foods and beverages in Japan. As a toxin non-producing species, it is considered, alongside *Aspergillus niger*, to be an organism which is safe for use in human food production, in fact over half all bread production in the USA is thought to utilise *A. oryzae* proteases to liberate amino acids required for yeast growth and respiration (Bigelis 1992). *A. niger* has risen to prominence in industrial biotechnology as a highly efficient producer of polysaccharide-degrading enzymes (amylases, pectinases and xylanases) and organic acids (Andersen et al. 2011; Pel et al. 2007). The world market for such enzymes has an estimated worth of US$ 5 billion (in 2009) of which production in filamentous fungi accounts for one half of all production (Lubertozzi and Keasling 2009).

Statins, first derived as lovastatin from *Aspergillus terreus*, are a class of super drugs which reduce human cholesterol levels by competitively inhibiting 3-hydroxy 3-methylglutaryl CoA (HMG-CoA), the rate-limiting enzyme in cholesterol biosynthesis. Widespread use of such drugs has had a dramatic impact upon incidence of coronary artery diseases as well as having proven efficacy in prevention of stroke and peripheral vascular disease. Statins have also been implicated as beneficial for the treatment of Alzheimer's disease, cancer, dementia, Parkinson's disease, multiple sclerosis and rheumatoid arthritis (Ginter and Simko 2009).

Some *Aspergillus* species are harmful or pathogenic to man and animals. *Aspergillus clavatus* is thought to be the causative agent of extrinsic allergic alveolitis (EAA), also known as malt worker's lung. It has also been implicated as the cause of mycotoxin-mediated neurotoxicosis in farm animals, resulting from feeding upon infected grain (Kellerman et al. 1976). Similarly, aflatoxin production by the opportunistic plant pathogen *Aspergillus flavus* can result in human mycotoxicoses, since aflatoxin is the most potent of known carcinogenic natural compounds and is often found as a contaminant of corn and peanuts. Food spoilage due to aflatoxin contamination costs hundreds of millions of US dollars annually (Yu et al. 2005).

A range of human mycoses, of varying severity, result from inhalation of *Aspergillus* spores. The most frequent pathogen of the genus is *Aspergillus fumigatus* which accounts for more than 90 % of human infections (Latge 1999). Human aspergilloses include a range of allergic, chronic and life-threatening conditions, the most severe of which is invasive pulmonary aspergillosis (IPA). IPA results from fungal colonisation of the pulmonary cavity often leading to invasive and/or disseminated tissue involvement. Poor diagnostic capabilities and suboptimal efficacies of existing antifungal regimens contribute to the high mortality observed among IPA patients, which is typically over 50 % and can be up to 90 % among patients suffering from haematological malignancies. Immune dysfunction among organ and stem cell transplantees is the major risk factor for IPA, and the usefulness of immunosuppressive therapies has lead to measurably increased incidences of aspergillosis in recent decades. Reports of IPA among hospitalised critical care patients are on the rise but, paradoxically, diagnostic shortcomings prevent accurate predictions IPA incidence, which is therefore thought to be greater than epidemiological studies suggest.

As in most eukaryotes, *Aspergillus* telomeres consist of tandem arrays of direct nucleotide repeats, presumably, although not yet proven, to house telomere-capping proteins which ensure chromosomal end integrity. Early indications are that chromatin remodelling enzymes and gene-silencing mechanisms are operative in the Aspergilli and, moreover, have profound effects upon the expression of genes. Often, this affects genes and gene clusters implicit in the very pathways exploited by man for the harvest of biologically active compounds, and those exploited by the fungus to survive in punishing environments such as the human host. An exhaustive commentary on telomere biology of all, perhaps any, *Aspergillus* species is not possible at the current time, however, the available knowledge provides an intriguing backdrop against which to examine the organisation and gene content of some important species. Providing, as it does, compelling evidence for the importance of *Aspergillus* subtelomeres in supporting adaptation to challenging environments, the subtelomeric gene content of the major mould pathogen of humans, *A. fumigatus*, will be extensively considered in this chapter.

## *6.1.1 Aspergillus Telomere Structure and Organisation*

### 6.1.1.1 Information from Early Molecular Studies

As with most other eukaryotes, *Aspergillus telomeres*, due to their repetitive nature and tertiary structure, have proven difficult to map genetically and also to identify in genome-wide sequence programs. As defining components of chromosomal extremities, the earlier, descriptive, studies of *Aspergillus* telomere sequences have proven to be important for anchoring the sequence contigs obtained from more recent genome-scale sequencing. The model organism for the *Aspergillus* genus is *Aspergillus nidulans*. Due to genetic tractability of *A. nidulans,* a highly detailed physical map (Clutterbuck 1997) has been established over generations of classical genetic study and the association between sequenced and physical genome structures have become particularly well established for this species.

The cloning and sequencing of *A. nidulans* telomeres were reported in 1997 (Bhattacharyya and Blackburn 1997). The study identified three classes of chromosomal ends having a telomeric repeat, TTAGGG, identical to that of humans and multiple other eukaryotes. *A. nidulans* telomeric tracts were noted as being short (4–22 repeats) as well as remarkably stable in different cell types and at altered growth temperatures, suggesting a highly regulated mechanism for length

**Table 6.1** Telomeric repeat sequence units of selected *Aspergilli* and other fungal species

| Species | Telomeric repeat | Source |
|---|---|---|
| *Aspergillus clavatus* | TTAGGG | Broad Institute[a] |
| *Aspergillus flavus* | TTAGGGTCAACA | Chang et al. (2005) |
| *Aspergillus fumigatus* | TTAGGG | Broad Institute |
| *Aspergillus nidulans* | TTAGGG | Bhattacharyya and Blackburn (1997) |
| *Aspergillus oryzae* | TTAGGGTCAACA | Kusumoto et al. ( 2003) |
| *Fusarium oxysporum* | TTAGGG | Powell and Kistler (1990) |
| *Histoplasma capsulatum* | TTAGGG | Broad Institute |
| *Magnaporthe grisea* | TTAGGG | Farman and Leong (1995) |
| *Ustilago maydis* | TTAGGG | Guzman and Sanchez (1994) |
| *Neurospora crassa* | TTAGGG | Schechtman (1990) |

Modified after Chang et al. (2010). [a] From the fungal genome database at Broad Institute (http://www.broadinstitute.org/science/data#

control. In fact this finding, notwithstanding difficulties pertaining to accurate estimation of telomere length, appears to be true of all characterised *Aspergillus* species, although the telomeric repeat sequences (Table 6.1) have been found to vary slightly. *A. nidulans* telomeres were initially characterised by construction of a library of telomeric clones. This was constructed by treatment of total genomic DNA with T4 DNA polymerase to produce blunt ends, ligation to a linearised vector, restriction enzyme digestion and ligation. The library was screened by colony blotting using a radiolabelled $(TTAGGG)_4$ oligonucleotide probe and positive clones were sequenced. Among ten sequences hybridising to the probe, one was found to be 7–8 times higher in signal intensity leading to the conclusion that all 16 chromosome ends could be accounted for by the analysis. All sequenced clones contained a tract of 4–22 telomeric repeats composed of the basic TTAGGG repeat unit. The clones most frequently obtained contained 15–20 repeats, and the G-rich repeat strand was orientated as 5′ to 3′ towards the telomere sequence. Exploiting the preferential sensitivity of telomere sequences to Bal31 exonuclease a Bal31 time-series digestion experiment, coupled with Southern blot analysis detected sequential size reductions of detectable DNA fragments and eventual disappearance relative to an 18 s rDNA marker. At the time of near disappearance, there was little size difference from time zero suggesting that native telomeres are indeed short. The authors observed a striking uniformity of telomere fragment lengths, as evidenced by well-defined, non-smearing, DNA detection signals thereby suggesting a tight regulation of telomeric length. Assessment of differing developmental states using asexual spores and hyphae revealed no differences in telomere length in dormant versus vegetative cells; thus, in contrast to studies in *C. albicans,* which had demonstrated variance of telomere length in response to temperature shift, no effect on telomere length was observed for *A. nidulans* at temperatures of 30, 37 and 42 °C. The *A. nidulans* telomeres are therefore dramatically different from vertebrate telomeres in one respect, namely length of telomere tract. A similar approach to telomere sequencing in *A. oryzae* captured the unique dodeca-nucleotide sequence (TTAGGGTCAAACA) with a tract length of 114–136 bp (Kusumoto et al. 2003). The same telomeric repeat has been identified for *A. flavus* isolates.

#### 6.1.1.2 Information from Genetic and Physical Mapping

Despite the advances in genome sequencing technologies, the subtelomeric sequences of some important genus members remain to be validated at the time of writing and mark notable omissions from Table 6.1. The assembly and annotation of several sequenced *Aspergillus* genomes is still in progress and despite the existence of easily accessible sequence, telomere sequences remain to be appended. Physical maps of certain genomes have proven highly facilitative when it comes to anchoring contig sequences to telomere sequences, in particular for *A. nidulans* where the scorable phenotypes of hundreds of different mutants have been used to calculate recombination frequencies between markers (Clutterbuck 1997).

At the time of sequence release, the published Broad Institute *A. nidulans* genome was comprised of 173 contigs linked by BAC-end sequences and fosmid bridges into 16 scaffolds. 152 mapped genetic markers were used to anchor the genome sequence to the genetic linkage map, with 75 contigs remaining unassigned. Noting a paucity of telomeric sequences (only four of the scaffolds had typical telomeric sequence repeats, and a fifth was present but remained unanchored), Clutterbuck and Farman (2008) searched the NCBI sequence trace archive identifying a large number of candidate telomere reads which had not been used in the genome assembly. Using the TERMINUS (Li et al. 2005a) software to mine for telomere sequences in unassembled sequence databases, 11 contigs were constructed which ended in telomeric repeats and ranged in size from 973 to 1,055 bp. Five of these corresponded to the five already present in the sequence assembly.

A lack of overlap with established sequence assemblies prevented the appendation of these contigs to existing reads, and further effort was required to access the mate-pair sequences derived from the subtelomeric ends of the relevant clones. De novo BLAST and TRUMATCH post-processing analyses (Li et al. 2005b), combining the validation of matches with positional information, allowed ten new telomeres to be unequivocally positioned relative to the genome assembly and identified the genomic contigs that occupy terminal chromosomal locations. The authors have made the contigs used to map these telomeres available from a specified URL (Clutterbuck and Farman 2008).

One aspect of *A. nidulans* telomere architecture which was revealed by this specialised additional analysis was the widespread occurrence of distinct subtelomere domains consisting of sequences which are duplicated at multiple chromosome ends. The *A. nidulans* subtelomere domain was estimated by BLAST analyses to be 16.5 kb in length. Furthermore, the alignments between different subtelomere domains were observed to be discontinuous, the initial assumption being that insertion of transposable elements was responsible. However, although transposons were found to be present, and often immediately adjacent to the subtelomeres, there were no transposon insertions in the subtelomere domains. Divergences of up to 15 % at the level of nucleotide sequence typified by large numbers of G to A and C to T transition mutations are suggestive of the activity of a RIP-like process upon *A. nidulans* telomeres.

#### 6.1.1.3 Information from Whole-Genome Sequencing: Intra- and Interspecies Variation

Comparative analyses of the *A. nidulans* genome sequence with those of *A. fumigatus* and *A. oryzae*, performed by Galagan and co-workers, revealed the occurrence of large regions which lack long (evolutionary)-range synteny. The scrutiny of regions which differ between these evolutionarily distant species revealed multiple repeats and subtelomeric sequences. This finding is believed to have significant implications for *Aspergillus* biology, as subtelomeric regions in *Aspergillus* species are enriched for secondary metabolite gene clusters, which are thought to facilitate niche adaptation and virulence. The authors of this study proposed that rapid rearrangement of these regions might explain their species-specific evolution (Galagan et al. 2005).

*A. flavus* and *A. oryzae* are two highly related *Aspergillus* species, the genetic distinctions between which have been a source of much debate in recent years. Prior to availability of the genomes for these species, the distinctions were made predominantly upon the basis of aflatoxin productivity. In 2007, Rokas et al. concluded, on the basis of sequence similarities and relative numbers of species-specific genes, that *A. oryzae* is not a distinct species but more likely to be a distinct 'ecotype' Rokas et al. (2007). The divergence of *A. flavus* and *A. oryzae* isolates highlights the importance of telomeric regions as focal regions of intra- and interspecies diversity. Deletions of the subtelomeric aflatoxin gene cluster had been repeatedly reported among non-toxigenic *A. flavus* isolates harvested from various sources including food, feed and soils (Chang et al. 2005; Criseo et al. 2001). A subset of these isolates was found to harbour a breakpoint in homology to aflatoxigenic strains at the 5′ UTR of the *ver1* gene of the cluster. Sequence upstream and adjacent to the breakpoint is also found in *A. oryzae* SRRC2098 (ATCC11493), an isolate from soybean–wheat flour mixture in Japan and in SRRC2103 (ATCC10196). Closer scrutiny of the breakpoint by sequencing analyses (Tominaga et al. 2006) found approximately half of the 7.8-kb sequence to be of unknown origin, encoding a monooxygenase, while half was found to be identical to the subtelomeric region of *A. flavus* NRRL335. In the latter *A. flavus* isolate, an inverse telomeric repeat was hypothesised to be causative of chromosomal instability in the region. Carbone et al. assessing the ordering and clustering of genes in the aflatoxin biosynthetic gene cluster across multiple sequenced *Aspergillus* species (Carbone et al. 2007a, b) concluded that, during the course of evolution, recombination and balancing selection must have played a role in the organisation of the aflatoxin gene cluster.

These results support the case for subtelomeres as regions of high genetic diversity prone to recombination, inversions, deletions, translocations and other genomic rearrangements. As such, Chang and Ehrlich proposed them as ideal for the nurture of new synthetic machinery under the constraints of co-regulation (Chang and Ehrlich 2010). The same authors refer to telomeric repeat sequences and retrotransposons as a tool to aid the distinction between very closely related *Aspergillus* species. Such observations are upheld by our comparative analyses of LINE-1 retrotransposon insertions in the two sequenced *A. fumigatus* genomes (Huber and Bignell, unpublished). Strain-specific secondary metabolism genes have also been identified in distinct *A. niger* isolates during comparative genome analysis of citric acid and enzyme-producing isolates (Andersen et al. 2011) where the presence of telomere sequences in the analysed genome data confirmed an inversion of the complete right arm of chromosome VI.

Other comparative genome studies have aimed to maximise the resolution of genome comparisons using *Aspergillus* species and isolates having minimal evolutionary separation. Comparing the genomes of two *A. fumigatus* clinical isolates with those of the closely related but rarely pathogenic species *Neosartorya fisheri* and *Aspergillus clavatus* (Fedorova et al. 2008) identified the extent of species-specific gene content for each of *A. fumigatus, N. fisheri* and *A. clavatus* to be 8.5, 13.5, and 12.6 % of total genome content, respectively. Species-specific genes were found to be smaller than conserved genes, to contain fewer exons and to be non-randomly distributed, with a precedent for biased subtelomeric location. Most of them were found to cluster, comprising 13 groups, or 'gene islands' enriched for pseudogenes, transposons and other repetitive elements.

### 6.1.2 Subtelomeres of *A. fumigatus*: Gene Content and Organisation

*A. fumigatus* accounts for the majority of *Aspergillus*-related human disease. Currently, the molecular basis of such predominance over other *Aspergillus* species is unknown. Physical properties of *A. fumigatus* spores might contribute to the observed incidence, as *A. fumigatus* spores are smaller and lighter than those of closely related species. This promotes the aerosolisation of spores, and subsequent entry into human pulmonary cavities, but the abundance of *Aspergillus* spores in the airborne microflora would not indicate that a quantitative difference in airborne spore abundance is responsible (Latge 1999). The upheld view is that virulence is multifactorial, and that *A. fumigatus* possesses specific genetic traits which facilitate pathogenesis. Recently therefore, genome cohorts of lineage-specific genes have been catalogued and are under further scrutiny since they might disclose the basis of phenotypic differences among species, including virulence and niche occupation.

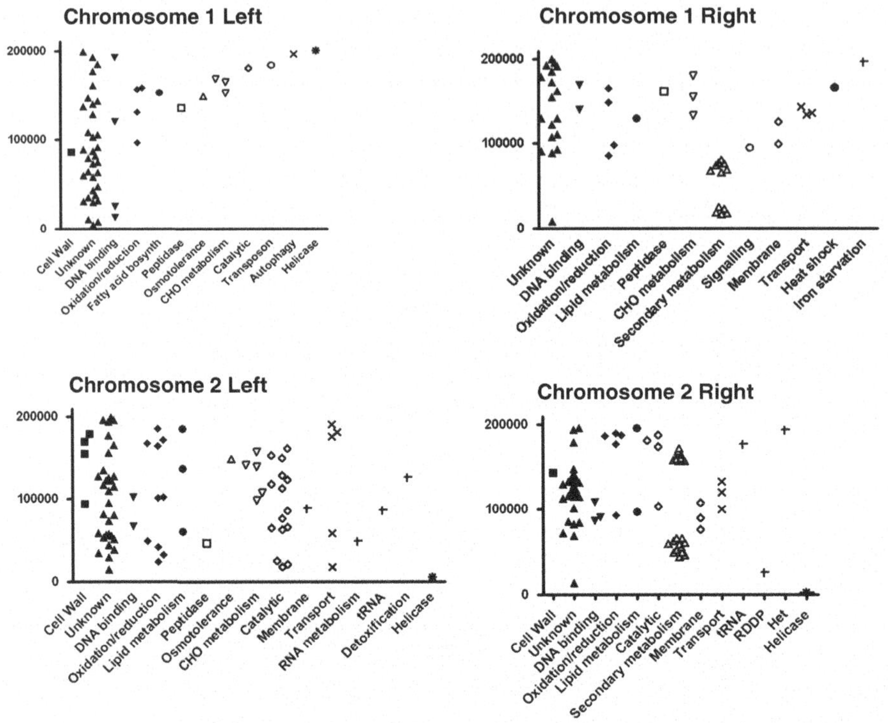


**Fig. 6.1** Functional inventory of *A. fumigatus* subtelomeric genes, chromosomes 1 and 2 (Af293) derived from the AspGD (http://www.aspgd.org/) database and plotted as a function of distance (kb) from sequenced chromosome ends

During analysis of the isolate-specific genomic islands identified in the *A. fumigatus* genome, Fedorova et al. noted that *A. fumigatus* chromosomes encode gene functions which have been commonly associated with other fungal telomeres, including transposons, telomere-linked helicases, clusters of secondary metabolite genes, cytochrome oxidases, hydrolases and molecular transporters (Fedorova et al. 2008). Isolate-specific islands were found to be composed of clustered blocks ranging in size from 10 to 400 kb. They appear to contain numerous pseudogenes and repeat elements. Moreover, supporting a model of gene duplication and diversification, 46 % of *A. fumigatus* genes with paralogs were reported as being telomere-proximal.

For the purposes of more thoroughly cataloguing the genes housed in *A. fumigatus* subtelomeres, genes identified by interrogation of the *A. fumigatus* Af293 genome (accessed via AspGD: http://www.aspgd.org/) are plotted as a function of distance from sequenced chromosome ends in Figs. 6.1, 6.2, 6.3, 6.4. This accounts for a total of 1,015 predicted protein-encoding genes housed within

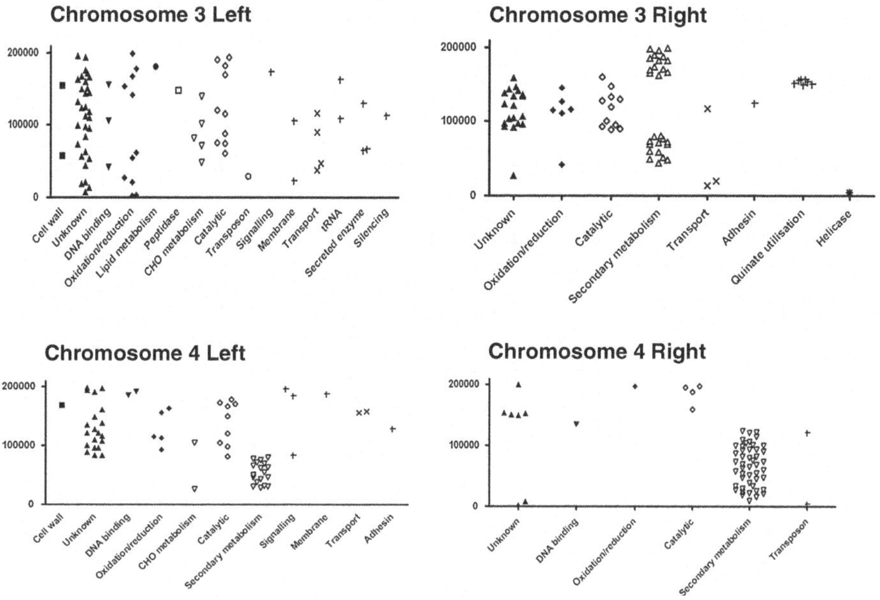


**Fig. 6.2** Functional inventory of *A. fumigatus* subtelomeric genes, chromosomes 3 and 4 (Af293) derived from the AspGD (http://www.aspgd.org/) database and plotted as a function of distance (kb) from sequenced chromosome ends

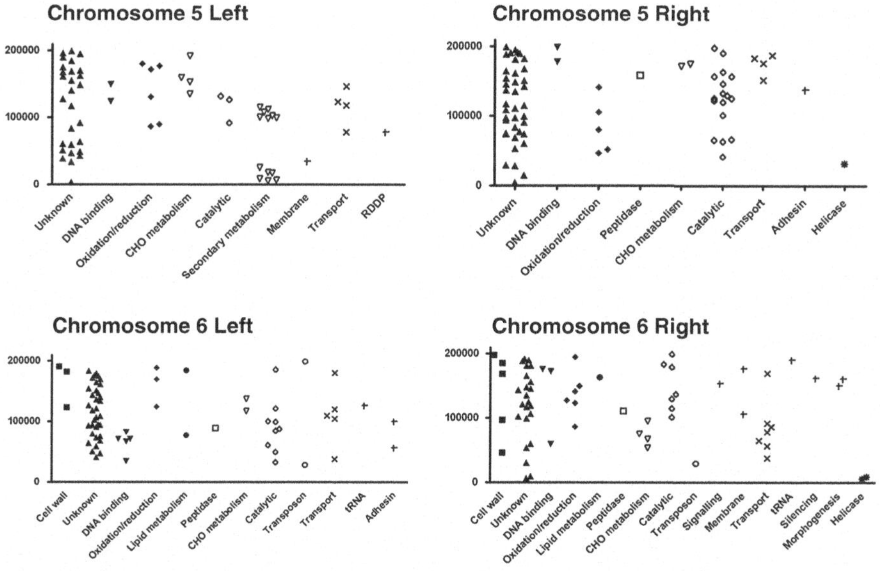


**Fig. 6.3** Functional inventory of *A. fumigatus* subtelomeric genes, chromosomes 5 and 6 (Af293) derived from the AspGD (http://www.aspgd.org/) database and plotted as a function of distance (kb) from sequenced chromosome ends

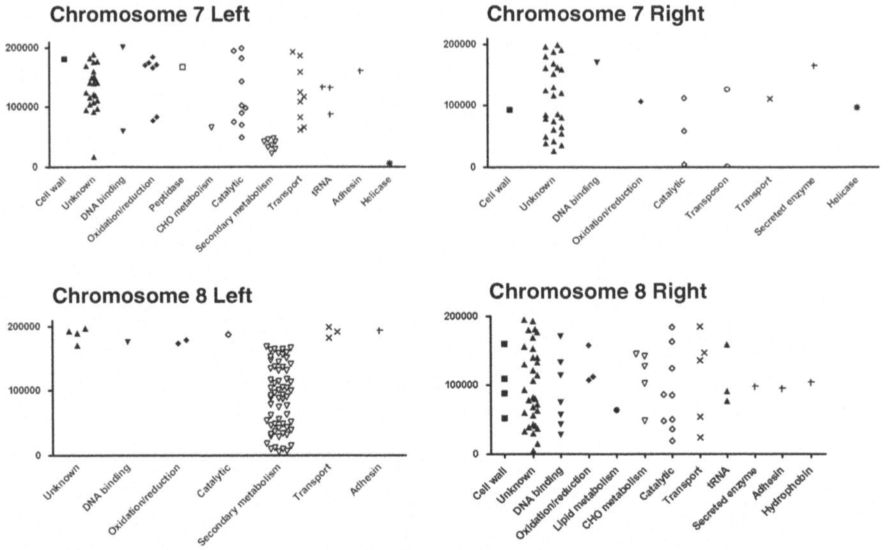


**Fig. 6.4** Functional inventory of *A. fumigatus* subtelomeric genes, chromosomes 7 and 8 (Af293) derived from the AspGD (http://www.aspgd.org/) database and plotted as a function of distance (kb) from sequenced chromosome ends

200 kb of the 16 sequenced *A. fumigatus* chromosomal termini. The most densely populated subtelomere is the left arm of chromosome 2 (78 predicted genes) and the least populated is the left arm of chromosome 7 (37 predicted genes). 11 out of 22 predicted *A. fumigatus* secondary metabolism gene clusters as predicted from in silico analyses are represented among 7 subtelomeric regions (Table 6.2).

**Telomere-linked helicase (TLH) genes**

Numerous cooperatively acting factors maintain the fidelity of chromosome replication, repair and segregation. These are critical house-keeping functions and have been highly conserved throughout evolution. The notable similarity between a chromosomal end and a double-stranded DNA break, for example, must be distinguishable at a cellular level to prevent catastrophic non-homologous end rejoining. Helicases of the RecQ family which unwind dsDNA, and promote strand annealing and fork regression, are crucial for chromosomal stability, and mutations in a human RecQ gene cause premature ageing (Brosh and Bohr 2007). Fungal RecQ helicases have been commonly found as associated with chromosome ends which would support a role for them in telomere maintenance.

*A. fumigatus* subtelomeres 1L, 2L, 2R, 3R, 5R, 6R ($n = 2$), 7L and 7R house a total of 9 predicted telomere helicase-encoding genes positioned between 2,025 and 199,230 kb from sequenced chromosome ends (Figs. 6.1–6.4). BLAST analysis reveals significant identity between predicted open reading frames of those which are most closely situated near telomeres (Afu2g18100, Afu3g15395,

**Table 6.2** Subtelomeric gene clusters of *A. fumigatus* with predicted or demonstrated secondary metabolic activities

| Subtelomere[a] | Location | Number of ST genes | Product | Pathogenesis | Reference |
|---|---|---|---|---|---|
| 1R | Afu1g17640–Afu1g17740 | 12[b] | Unknown | Unknown | |
| 2R | Afu2g17960–Afu2g18070 | 12 | Ergot alkaloids | Unknown | Bigelis (1992), Brosh and Bohr (2007), Carbone et al. (2007a), Lubertozzi and Keasling (2009) |
| 2R | Afu2g17510–Afu2g17600 | 10 | DHN melanin | Murine | Li et al. (2005b), Liang et al. (1997), Maiya et al. (2006) |
| 3R | Afu3g15200–Afu3g15340 | 15 | Pes3 | Murine | Pes3 |
| 3R | Afu3g14560–Afu3g14760 | 14[b] | Unknown | Unknown | |
| 4L | Afu4g00110–Afu4g00280 | 18 | Unknown | Unknown | |
| 4R | Afu4g14380–Afu4g14850 | 48 | Unknown | Unknown | |
| 5L | Afu5g00110–Afu5g00160 | 5 | Unknown | Unknown | |
| 5L | Afu5g00340–Afu5g00400 | 7 | Unknown | Unknown | |
| 7L | Afu7g00110–Afu7g00190 | 8 | Unknown | Unknown | |
| 8L | Afu8g00100–Afu8g00720 | 61 | Fumitremorgin B, Pseurotin A | Partial | Maiya et al. (2006), O'Hanlon et al. (2011) (Bignell unpublished) |

Modified after Perrin et al. (2008)

[a] L and R indicate left and right, respectively

[b] Predicted gene cluster extends beyond 200,000 kb from chromosome end

Afu5g15150 (395), Afu7g00090 for Afu2g0090). Only one of these, Afu2g18100 (851 amino acid residues) appears to be full-length.

**Secondary metabolism gene clusters**

During interstrain comparative genomic analysis, Fedorova et al. (2008) noted that only 10 % of secondary metabolism genes have orthologs in all sequenced *Aspergilli.* In *A. fumigatus,* only 30 % of secondary metabolism genes are shared with *A. fisheri* and *A. clavatus*. The three species also vary with respect to the numbers of enzymes which catalyse first steps in secondary metabolite biosynthesis such as non-ribosomal peptide synthases (NRPS), polyketide synthases (PKS) and dimethylallyl tryptophan synthases (DMAT). The *A. fisheri* genome actually contains the highest number of secondary metabolism genes.

*A. fumigatus* subtelomere 2R (chromosome 2, right arm) contains two secondary metabolite gene clusters, which encode the biosynthetic apparatus required for DHN melanin synthesis (Afu2g17510–Afu2g17600) and ergot alkaloid synthesis (Afu2g17960–Afu2g18070). At the time of its discovery, the dihydroxynapthalene (DHN) melanin gene cluster was the largest cluster of fungal biosynthetic genes to be reported as required for pigment synthesis. Gene deletion mutants lacking any of the six open reading frames including those encoding a polyketide synthase, scytalone dehydratase and HN reductase lacked the typical blue-green spore pigmentation characteristic of *A. fumigatus,* and regulation of gene expression was observed to be developmentally linked to sporulation (Tsai et al. 1999). Disruption of the scytalone dehydrogenase-encoding *arp1* gene results in the production of reddish-pink conidia and significantly increases the ability of the human complement component C3 to bind to these conidia thereby likely impacting upon the efficacy of phagocytosis of mutant spores (Tsai et al. 1997). The alb1 polyketide synthase is required for full virulence in a murine model of aspergillosis (Tsai et al. 1997). Thus, conidial pigment biosynthesis in *A. fumigatus* appears to be an important virulence factor in the establishment of infection.

The ergot alkaloid biosynthetic gene cluster encodes the genetic machinery for at least four ergot alkaloids which share a variously modified four-member ergoline ring (Panaccione and Coyle 2005). Ergot alkaloids are mycotoxins which have been shown to negatively impact multiple physiological systems of exposed humans and animals and are an abundant component of *A. fumigatus* spores. Currently, the role of this gene cluster in virulence of *A. fumigatus* is unknown.

Subtelomere 3R contains two predicted (Perrin et al. 2007) secondary metabolism gene clusters, Afu3g15200–Afu3g15340 and Afu3g14560–Afu3g14760, for neither of which the biosynthetic products are known. The former of the two gene clusters contains a polyketide synthase-encoding gene designated *pes3* which has recently been shown to affect murine virulence and insect virulence (O'Hanlon et al. 2011). A *Pes*3 null mutant was found to be increased for fungal burden in corticosteroid-treated mice at 5 days post-infection, a phenotype which was accompanied by a more rapid germination of spores within murine lung tissues and a reduction in proinflammatory cytokine release from macrophages exposed to Δ*pes3* mutants, relative to wild type, in vitro.

Subtelomeres 4L, 4R, 5L and 7L collectively accommodate five gene-predicted (Perrin et al. 2007) secondary metabolism gene clusters, the biosynthetic products of which are currently unknown.

Subtelomere 8R is exceptional, as it encodes genes for predominantly secondary metabolism ($n = 61$), including 2 polyketide synthases (Afu8g00370 and Afu8g00890), a hybrid polyketide synthase/NRPS encoding gene (Afu8g00540) and a further NRPS (Afu8g00170). Sheppard et al. (2005) noted the presence of genes which are known to be required for sterigmatocystin/aflatoxin biosynthesis, as well as genes required for ergot alkaloid synthesis in this region. Intriguingly, this combination of secondary metabolism genes is not found in other *Aspergillus* species. Previous gene deletion studies by Turner and colleagues have defined the biosynthetic products of some sub clusters within this supercluster on chromosome 8, including the pseurotin A gene cluster (Maiya et al. 2007) and the fumitremorgin gene cluster (Maiya et al. 2006), and work in my laboratory is currently addressing the role of such metabolites in murine virulence. Indications to date are that loss of at least one of the metabolites whose biosynthesis is directed by genes resident in the chromosome 8 supercluster can impact virulence at whole animal level (Bignell, unpublished). However, our recent analyses indicate only a partial requirement for these metabolites during infection, as strain-dependent variability is observed when similar mutants are constructed in different genetic backgrounds. Thus, further characterisation is required before firm conclusions on the role of the metabolites encoded by genes in this region can be conclusively ascertained. A further crucial point is that the combinatorial activity of *A. fumigatus* secondary metabolites is a possibility thus far remaining unexplored. It is highly feasible that certain secondary metabolites act in concert to disable residual host immune defences. The construction of mutants lacking multiple biosynthetic properties is therefore required to address this important question.

With respect to the importance of chromosomal context and regulation of secondary metabolism genes, Keller and Palmer offer the view that SM clusters are located in regions of facultative heterochromatin, described as genomic regions which can be silenced and activated by both canonical and novel chromatin-mediated mechanisms (Palmer and Keller 2010). Returning to the aforementioned *ver1* aflatoxin biosynthesis gene, this time in *A. parasiticus*, localisation of the gene was found to be fundamentally important for gene expression, whereby misplacement of the gene outside of the cluster boundaries resulted in a 500-fold reduction in expression (Liang et al. 1997).

#### 6.1.2.1 Expression of Subtelomeric Genes in *A. fumigatus*

Novel views on the roles and regulation of subtelomeric genes are beginning to emerge from whole-genome transcriptome analyses. With respect to *A. fumigatus* gene expression, genomic context has been afforded considerable scrutiny with intriguing results.

Sheppard et al. (2005) while studying the regulon of the *A. fumigatus* APSES transcription factor StuA during acquisition of developmental competence noted the StuA-dependent developmentally regulated expression of multiple genes of the subtelomere 8L supercluster (Fig. 6.4; Table 6.2). *Aspergillus* species are multicellular organisms and undergo distinct developmental cycles according to nutrient availability and external environmental cues. Following an initial period of isotropic (non-responsive) growth hyphae are considered to become developmentally competent and responsive to external stimuli governing morphological transitions. To assess the role of *Stu*A in regulation of morphogenetic genes, Sheppard et al. compared the transcriptome of wild-type and Δ*stuA* isolates in pre- and post-competent hyphae. Time-series analysis of the wild-type transcriptome (8, 24, and 30 h) revealed upregulation of multiple genes located in subtelomere 8L at 24 and 30 h of submerged culture, suggesting the developmental regulation of genes in this region. Direct comparisons of wild-type and Δ*stuA* gene expression profiles revealed extensive derepression of gene expression in this region in the mutant isolate. No effect on murine virulence was evident when infecting with a wild-type versus mutant isolate. The coordinate regulation of genes in this region during developmental growth, and the dependence of such regulation upon a developmental regulator provide some insights into the temporal and developmental programming of this large cluster of secondary metabolism genes.

A more recent analysis of gene expression during colony growth in vitro used RNAseq to assess the differences between *A. fumigatus* biofilm and liquid planktonic growth. An interesting finding from this study was that genes which were upregulated during biofilm growth were significantly over-represented in subtelomere regions. 165 out of 383 postulated (Perrin et al. 2007) *A. fumigatus* secondary metabolism genes were found to be upregulated during biofilm growth, including 45 genes of the subtelomere 8L supercluster.

The global regulator of *Aspergillus* chemical diversity, LaeA, is a putative methyltransferase, originally identified as a regulator of secondary metabolism genes in *A. fumigatus* and *A. nidulans* (see Sect. 6.3.3) and required for murine virulence. Given the dependency of biosynthesis of a known immunotoxin, gliotoxin, upon *A. fumigatus* LaeA, Perrin et al. hypothesised that a comparative assessment of *A. fumigatus* wild-type and Δ*laeA* transcriptomes would provide insight into the role of LaeA in mammalian pathogenicity. Strains were cultured in vitro at ambient temperature (25 °C) for 60 h. The study revealed almost global suppression of secondary metabolite gene expression whereby 97 % of secondary metabolite gene clusters were impacted to some extent by laeA gene deletion. A biased distribution of LaeA-dependent genes was reported whereby 54 % of the LaeA-regulated gene clusters were located within 300 kb of the telomeres (Perrin et al. 2007). This pattern of gene expression gained additional significance when subsequent reports on gene expression during mammalian infection became available. To identify genes upregulated during mammalian infection McDonagh et al. developed a means to obtain whole-genome transcriptional data from minute samplings of mouse-infecting *A. fumigatus* germlings. This first and, to date, only report of the infectious *A. fumigatus* transcriptome revealed a significant predominance of subtelomeric genes among genes upregulated during murine infection, relative to laboratory culture. While only 16 % of the predicted *A. fumigatus* gene repertoire is housed within 300 kb of chromosome ends, 29 % of transcripts upregulated during murine infection are located in subtelomeric areas, compared to just 11 % of downregulated transcripts. Thus, 28 % of the entire subtelomeric gene repertoire was found to be represented among upregulated genes compared to only 8 % among downregulated functions. Coordinate expression of physically clustered genes was noted to be a feature of the induced, but not repressed, gene set and 40 % of upregulated physically clustered genes were found to occur within 300 kb of chromosome ends. By analogy to the earlier study of Perrin et al., it was found that among 415 genes downregulated in the absence of LaeA, 99 genes had increased abundance during initiation of murine infection, further substantiating the importance of LaeA as a regulator of virulence-associated genes. Determining the proportions of subtelomeric and secondary metabolism cluster genes shared between the two datasets, 49 and 40 genes were identified, having subtelomeric locations and secondary metabolite biosynthetic functions, respectively.

## 6.1.3 Regulation of *Aspergillus* Subtelomeric Gene Expression

Studies reviewed for the writing of this chapter collectively portray an important role for *Aspergillus* subtelomeric gene repertoires during adaptation to niche-specific stresses and also during development and morphogenesis of these organisms. Given the important role of certain subtelomeric gene functions, including secondary metabolism, and the relevance of concerted regulation of gene neighbourhoods in fungal pathogenesis and natural product biosynthesis, the rationale for further investigating the regulation of subtelomeric gene expression is clear.

### 6.1.3.1 Telomere Position Effect

Telomere position effect (TPE) is a eukaryotic phenomenon resulting in gene repression in areas immediately adjacent to telomere caps. Palmer et al. (2010) have demonstrated the occurrence of such a phenomenon in *A. nidulans,* specifically associated with linkage groups II and VI (left and right arms, respectively). The phenomenon was uncovered during attempts to characterise a telomere-linked helicase gene (AN5092) in which transgene repression of the gene replacement marker (encoding pyrimidine auxotrophy), and an associated reduction in radial growth and sexual spore production, were shrewdly identified as silencing phenomena by the investigators. This provided the researchers with a set of simple, scorable phenotypes with which to probe the phenomenon further and also facilitated the interrogation of the roles of certain heterochromatin-associated proteins

HepA, ClrD, HdaA and NkuA in transgene silencing, all of which were implicated to some extent. Repression was found to be independent of the transgene or its orientation, as evidenced by a similar gene replacement using a pyridoxine auxotrophy marker. Clutterbuck and Farman's *A. nidulans* sequence gap-closing exercise, described earlier in this section (Sect. 6.1.2), proved instrumental in determining the estimated distance of AN5092, placing the gene, and therefore the reach of TPE, at approximately 20 kb from the telomere cap. These results obtained in *A. nidulans* suggest considerable mechanistic conservation of TPE between fungal species, including TPE regulation by core heterochromatin-modulatory proteins.

#### 6.1.3.2 Chromatin and Silencing

The compaction of transcriptionally silent chromatin is a plausible mechanism by which chromosomal regions, rather than single genes, might be rendered accessible or recalcitrant to transcriptional regulation. Histone modifications, such as acetylation and methylation, exert well-documented effects upon the structure and function of chromatin, and both types of modification impact production of secondary metabolites in *Aspergillus* species, as fungi treated with methyltransferease or histone deactylase HDAC inhibitors are found to display altered patterns of metabolite expression. In *A. nidulans*, the HdaA class 2 histone deacetylase is involved in regulation of telomere-proximal secondary clusters whereby the production of sterigmatocystin and penicillin is found to increase substantially in a *ΔHdaA* isolate (Lee et al. 2009; Shwab et al. 2007).

LaeA-mediated regulation of gene clusters in *A. nidulans* was found to be locationally biased as placement of *afl*R outside of the sterigmatocystin gene cluster removes it from LaeA regulation, and conversely, the placement of an irrelevant gene into the correct locational context renders it subject to LaeA regulatory control. Several studies have linked LaeA mode of action to chromatin modification. Mutations in *Aspergillus* chromatin-modifying enzymes can activate silent or poorly expressed gene clusters and can partially rescue loss of metabolite production in LaeA null strains. *In A. nidulans* HDAC, HepA (heterochromatin protein 1) and ClrD (H3K9 methyltransferase) null mutants, all of which lead to elevated levels of secondary metabolites, target the H3K9 residue. Currently, the only metabolite with significantly supportive evidence for a role in virulence is gliotoxin while other metabolites might, in the future be proven able to damage mammalian cells. Deletion of the *A. fumigatus hdaA* gene did not impact virulence of *Aspergillus fumigatus* in neutropenic mice, but an assay of mammalian cell toxicity did suggest a role for HdaA in host cell damage (Lee et al. 2009).

#### 6.1.3.3 Repetitive Elements

A conserved feature of subtelomeric DNA sequences is the presence of repetitive elements, either active transposable elements or transposon relics. A possible role for transposon-mediated regulation of a subtelomeric gene cluster was recently reported for the *A. nidulans* penicillin gene cluster. The cluster consists of only 3 genes and is located around 30 kb from the telomere of chromosome VI. Disruption of large areas of repetitive DNA sequences resulted in mutants producing significantly less penicillin. One area, a 3.7-kb repeat termed PbIa was required for full production of penicillin while control strains harbouring marker gene insertions to either side of PbIa had no effect on production. Subsequent transcomplementation experiments were unable to restore PN production, suggesting that a transposon-mediated mechanism of SM expression could involve localised chromatin modifications (Shaaban et al. 2010).

### *6.1.4 Conclusions and Perspective*

The precedent for telomeres and subtelomeric gene reservoirs to support rapid adaptation to new ecological niches would appear also to apply to members of the *Aspergillus* genus. More, perhaps, than any other group of any microbial eukaryotes, the Aspergilli provide us with a means to explore speciation concepts and the role of subtelomere biology as a driving force in niche adaptation and differentiation. Such resolution is afforded by both the number of available, well-annotated genomes as well as the rich legacy of natural product biology which is continually resulting from genome scrutiny. The study and manipulation of gene neighbourhoods, particularly in the subtelomeric regions of *Aspergillus* chromosomes, with a view to maximising the harvest of natural products and minimising the harmful effects of secondary metabolites will be the focus of telomere biology in this genus for a while to come.

## References

Andersen, M. R., Salazar, M. P., Schaap, P. J., van de Vondervoort, P. J ., Culley, D., Thykaer, J, et al. (2011). Comparative genomics of citric-acid-producing *Aspergillus niger* ATCC 1015 versus enzyme-producing CBS 513.88. *Genome Research, 21*(6), 885–897 (available from: PM:21543515).

Baker, S. E. & Bennet, J. W. (2008). An overview of the genus *Aspergillus*. In Goldman, G. H & Osmani, S. A,(Eds.) *The Aspergilli: Genomics, medical aspects and research methods* (Vol. 26 pp. 3–13). Boca Raton: Taylor & Francis Group.

Bennet, J. W. (2010). An overview of the genus *Aspergillus. In:* M. Machida & K. Gomi, (Eds.), *Aspergillus. molecular biology and genomics* (pp. 1–18). Norfolk: Caister Academic Press.

Bhattacharyya, A. & Blackburn, E. H. (1997). *Aspergillus nidulans* maintains short telomeres throughout development. *Nucleic Acids Research, 25*(7), 1426–1431 (available from: PM:9060439).

Bigelis, R. (1992). Food enzymes. In D. B. Finkelstein & C. Ball (Eds.), *Biotechnology of filamentous fungi: Technology and product* (pp. 361–415). Butterworth-Heinemann: Boston.

Brosh, R. M., Jr. & Bohr, V. A. (2007). Human premature aging, DNA repair and RecQ helicases. *Nucleic Acids Research, 35*(22), 7527–7544 (available from: PM:18006573).

Carbone, I., Jakobek, J. L., Ramirez-Prado, J. H., & Horn, B. W. (2007a). Recombination, balancing selection and adaptive evolution in the aflatoxin gene cluster of *Aspergillus parasiticus*. *Molecular Ecological*, *16*(20), 4401–4417 (available from: PM:17725568).

Carbone, I., Ramirez-Prado, J. H., Jakobek, J. L., & Horn, B. W. (2007b). Gene duplication, modularity and adaptation in the evolution of the aflatoxin gene cluster. *BMC Evolution Biology*, *7*(1), (111) (available from: PM:17620135).

Chang, P. K. & Ehrlich, K. C. (2010). What does genetic diversity of *Aspergillus flavus* tell us about *Aspergillus oryzae*? *International Journal of Food Microbiology*, *138*(3), 189–199 (available from: PM:20163884).

Chang, P. K., Horn, B. W., & Dorner, J. W. (2005). Sequence breakpoints in the aflatoxin biosynthesis gene cluster and flanking regions in nonaflatoxigenic *Aspergillus flavus* isolates. *Fungal Genetics and Biology*, *42*(11), 914–923 (available from: PM:16154781).

Clutterbuck, A. J. (1997). The validity of the *Aspergillus nidulans* linkage map. *Fungal Genetics and Biology*, *21*(3), 267–277 (available from: PM:9299197).

Clutterbuck, A. J., & Farman, M. L. (2008). *Aspergillus nidulans* linkage map and genome sequence: Closing gaps and adding telomeres. *In* Goldman, G. H., & Osmani,S. A (Eds.), *The Aspergilli: Genomics, medical aspects and research methods*, (Vol. 26 pp. 43–73). Boca Raton: Taylor & Francis Group.

Criseo, G., Bagnara, A., & Bisignano, G. (2001). Differentiation of aflatoxin-producing and non-producing strains of *Aspergillus flavus* group. *Letters in Applied Microbiology*, *33*(4), 291–295 (available from: PM:11559403).

Farman, M. L., & Leong, S. A. (1995). Genetic and physical mapping of telomeres in the rice blast fungus, Magnaporthe grisea. *Genetics*, *140*, 479–492.

Fedorova, N. D., Khaldi, N., Joardar, V. S., Maiti, R., Amedeo, P., Anderson, M. J., et al. (2008). Genomic islands in the pathogenic filamentous fungus *Aspergillus fumigatus*. *PLoS Genetics*, *4*(4), e1000046 (available from: PM:18404212).

Galagan, J. E., Calvo, S. E., Cuomo, C., Ma, L. J., Wortman, J. R., Batzoglou, S., Lee, S. I., et al. (2005). Sequencing of *Aspergillus nidulans* and comparative analysis with *A. fumigatus* and *A. oryzae*. *Nature*, *438*(7071), 1105–1115 (available from: PM:16372000).

Ginter, E., & Simko, V. (2009). Statins: the drugs for the 21st century? *Bratislavske Lekarske Listy, 110*(10) 664–668 (available from: PM:20017462).

Guzman, P. A., & Sanchez, J. G. (1994). Characterization of telomeric regions from Ustilago maydis. *Microbiology*, *140*(3), 551–557.

Kellerman, T. S., Pienaar, J. G., van der Westhuizen, G. C., Anderson, G. C., & Naude, T. W. (1976). A highly fatal tremorgenic mycotoxicosis of cattle caused by *Aspergillus clavatus*. *Onderstepoort Journal of Veterinary Research*, *43*(3), 147–154 (available from: PM:1012654).

Kusumoto, K. I., Suzuki, S., & Kashiwagi, Y. (2003). Telomeric repeat sequence of *Aspergillus oryzae* consists of dodeca-nucleotides. *Applied Microbiology and Biotechnology*, *61*(3), 247–251 (available from: PM:12698283).

Latge, J. P. (1999). *Aspergillus fumigatus* and aspergillosis. *Clinical Microbiology Reviews*, *12*(2), 310–350 (available from: PM:10194462).

Lee, I., Oh, J. H., Shwab, E. K., Dagenais, T. R., Andes, D., & Keller, N. P. (2009). HdaA, a class 2 histone deacetylase of *Aspergillus fumigatus*, affects germination and secondary metabolite production. *Fungal Genetics and Biology*, *46*(10) 782–790 (available from: PM:19563902).

Li, W., Rehmeyer, C. J., Staben, C., & Farman, M. L. (2005a). TERMINUS—telomeric end-read mining IN unassembled sequences. *Bioinformatics.*, *21*(8) 1695–1698 (available from: PM:15585532).

Li, W., Rehmeyer, C. J., Staben, C., & Farman, M. L. (2005b). TruMatch—a BLAST post-processor that identifies bona fide sequence matches to genome assemblies. *Bioinformatics.*, *21*(9) 2097–2098 (available from: PM:15671115).

Liang, S. H., Wu, T. S., Lee, R., Chu, F. S., & Linz, J. E. (1997). Analysis of mechanisms regulating expression of the ver-1 gene, involved in aflatoxin biosynthesis. *Applied and Environment Microbiology*, *63*(3), 1058–1065 (available from: PM:9055421).

Lubertozzi, D., & Keasling, J.D. (2009). Developing *Aspergillus* as a host for heterologous expression. *Biotechnology Advances*, *27*(1), 53–75 (available from: PM:18840517).

Maiya, S., Grundmann, A., Li, S. M., & Turner, G. (2006). The fumitremorgin gene cluster of *Aspergillus fumigatus*: Identification of a gene encoding brevianamide F synthetase. *Chembiochem*, *7*(7), 1062–1069 (available from: PM:16755625).

Maiya, S., Grundmann, A., Li, X., Li, S. M., & Turner, G. (2007). Identification of a hybrid PKS/NRPS required for pseurotin A biosynthesis in the human pathogen *Aspergillus fumigatus*. *Chembiochem*, *8*(14), 1736–1743 (available from: PM:17722120).

O'Hanlon, K. A., Cairns, T., Stack, D., Schrettl, M., Bignell, E. M., Kavanagh, K., Miggin, S.M. (2011). Targeted disruption of nonribosomal peptide synthetase pes3 augments the virulence of *Aspergillus fumigatus*. *Infection and Immunity*, *79*(10), 3978–3992 (available from: PM:21746855).

Palmer, J. M., & Keller, N. P. (2010). Secondary metabolism in fungi: Does chromosomal location matter? *Current Opinion in Microbiology*, *13*(4), 431–436 (available from: PM:20627806).

Palmer, J. M., Mallaredy, S., Perry, D. W., Sanchez, J. F., Theisen, J. M., Szewczyk, E., et al. (2010). Telomere position effect is regulated by heterochromatin-associated proteins and NkuA in *Aspergillus nidulans*. *Microbiology*, *156*(Pt 12) 3522–3531 (available from: PM:20724388).

Panaccione, D. G., & Coyle, C. M. (2005). Abundant respirable ergot alkaloids from the common airborne fungus *Aspergillus fumigatus*. *Applied Environmental Microbiology*, *71*(6), 3106–3111 (available from: PM:15933008).

Pel, H. J., de Winde, J. H., Archer, D. B., Dyer, P. S., Hofmann, G., Schaap, P. J., et al. (2007). Genome sequencing and analysis of the versatile cell factory *Aspergillus niger* CBS 513.88. *Nature Biotechnology*, *25* (2), 221–231 (available from: PM:17259976).

Powell, W. A., & Kistler, H. C. (1990) In vivo rearrangement of foreign DNA by Fusarium oxysporum produces linear self-replicating plasmids. *Journal of Bacteriology*, *172*, 3163–3171.

Perrin, R. M., Fedorova, N. D., Bok, J. W., Cramer, R. A., Wortman, J. R., Kim, H. S., et al. (2007). Transcriptional regulation of chemical diversity in *Aspergillus fumigatus* by LaeA. *PLoS Pathogens*, *3* (4), e50 (available from: PM:17432932).

Rokas, A., Payne, G., Fedorova, N. D., Baker, S. E., Machida, M., Yu, J., et al. (2007). What can comparative genomics tell us about species concepts in the genus *Aspergillus*? *Studies in Mycology*, 59, 11–17 (available from: PM:18490942).

Schechtman, M. G., (1990). Characterization of telomere DNA from Neurospora crassa. *Gene*, *88*, 159–165.

Shaaban, M., Palmer, J. M., El-Naggar, W. A., El-Sokkary, M. A., Habib, E. S. E., & Keller, N. P. (2010). Involvement of transposon-like elements in penicillin gene cluster regulation. *Fungal Genetics and Biology*, *47*(5), 423–432 (available from: PM:20219692).

Sheppard, D. C., Doedt, T., Chiang, L. Y., Kim, H. S., Chen, D., Nierman, W. C., & Filler, S. G. (2005). The *Aspergillus fumigatus* StuA protein governs the up-regulation of a discrete transcriptional program during the acquisition of developmental competence. *Molecular Biology of the Cell*, *16*(12), 5866–5879 (available from: PM:16207816).

Shwab, E. K., Bok, J. W., Tribus, M., Galehr, J., Graessle, S., & Keller, N. P. (2007). Histone deacetylase activity regulates chemical diversity in *Aspergillus*. *Eukaryotic Cell*, *6*(9), 1656–1664 (available from: PM:17616629).

Tominaga, M., Lee, Y. H., Hayashi, R., Suzuki, Y., Yamada, O., Sakamoto, K., et al. (2006). Molecular analysis of an inactive aflatoxin biosynthesis gene cluster in *Aspergillus oryzae* RIB strains. *Applied and Environment Microbiology*, *72*(1), 484–490 (available from: PM:16391082).

Tsai, H. F., Washburn, R. G., Chang, Y. C., & Kwon-Chung, K. J. (1997). *Aspergillus fumigatus arp*1 modulates conidial pigmentation and complement deposition. *Molecular Microbiology*, *26*(1), 175–183 (available from: PM:9383199).

Tsai, H. F., Wheeler, M. H., Chang, Y. C., & Kwon-Chung, K. J. (1999). A developmentally regulated gene cluster involved in conidial pigment biosynthesis in *Aspergillus fumigatus*. *Journal of Bacteriology*, *181*(20), 6469–6477 (available from: PM:10515939).

Yu, J., Cleveland, T. E., Nierman, W. C., & Bennett, J. W. (2005). *Aspergillus flavus* genomics: Gateway to human and animal health, food safety, and crop resistance to diseases. *Revista iberoamericana de micología*, *22*(4) 194–202 (available from: PM:16499411).

# Chapter 7
# *Trypanosoma brucei* Subtelomeres: Monoallelic Expression and Antigenic Variation

**Luisa M. Figueiredo and David Horn**

**Abstract** Large families of Variant Surface Glycoprotein (VSG) contingency genes are found at African trypanosome subtelomeres. These parasites offer an important evolutionary perspective on subtelomeres since they diverged very early from the eukaryotic lineage. The *VSGs* are also key virulence determinants. *Trypanosoma brucei* subtelomeres display a remarkable transcription pattern that allows *VSGs* to enable an extremely effective form of antigenic variation and host immune evasion. This involves monotelomeric, polycistronic transcription, driven by extranucleolar RNA polymerase I. Here, we discuss this novel epigenetic system in terms of regulatory factors and mechanisms.

## 7.1 Introduction

As it will become clear during this chapter, *Trypanosoma brucei* is a somewhat unusual (hence, interesting!) eukaryote. Subtelomeres in eukaryotes often accumulate large families of contingency genes and African trypanosomes are no exception. Indeed, they offer an important evolutionary perspective on this topic since they are among the Excavates and diverged very early from the eukaryotic lineage (Sogin et al. 1989). The subtelomeric genes are also key virulence determinants in these important parasites of humans and animals. In contrast to most eukaryotes in which subtelomeres are underrepresented in the

L. M. Figueiredo (✉)
Instituto de Medicina Molecular (IMM), Universidade de Lisboa, Lisbon, Portugal
e-mail: lmf@fm.ul.pt

D. Horn
Division of Biological Chemistry and Drug Discovery, College of Life Sciences, University of Dundee, Dundee DD1 5EH, UK
e-mail: d.horn@dundee.ac.uk

E. J. Louis and M. M. Becker (eds.), *Subtelomeres*, DOI: 10.1007/978-3-642-41566-1_7, 

genome databases, in African trypanosomes, the first large genome locus ever sequenced was a subtelomere (Berriman et al. 2002). This is because subtelomeric regions have been a subject of interest and study since the 1980s, when variant surface glycoprotein (VSG) genes were found adjacent to telomeric repeats (de Lange and Borst 1982). *VSGs* encode for the surface coat of this extracellular parasite and periodic VSG exchange permits the parasite to escape the host antibody immune response, by a mechanism known as antigenic variation. Nowadays, we know that there are ~2,000 *VSGs,* mainly in telomeric locations, but only one is transcribed at a time from a specialized subtelomeric locus called the bloodstream expression site (BES). Each BES extends over ~35 kbp and consists of a polycistronic unit that contains one *VSG* and up to twelve expression site-sssociated genes (ESAGs) (Becker et al. 2004; Hertz-Fowler et al. 2008; Fig. 7.1). In the 35 Mbp genome of *T. brucei*, there are several BES (15 in the most widely studied 427 strain), but only one is actively transcribed while the others remain largely silenced. Strikingly, BESs are transcribed by RNA polymerase I (Pol I), which in most organisms is exclusively dedicated to transcribing ribosomal RNA genes.

*Trypanosoma brucei* is a unicellular eukaryote that infects humans, cattle, or wild animals. The parasite is transmitted by an insect vector commonly known as tsetse (*Glossina* genus). Because tsetse are found in Africa, the disease caused by *T. brucei* is restricted to this continent. This is in contrast to *Trypanosoma cruzi*, which is found mainly in South America and causes Chagas' disease. The human form of the disease caused by *T. brucei* is known as sleeping sickness, and the number of new cases is estimated to be around 30,000 people each year (Simarro

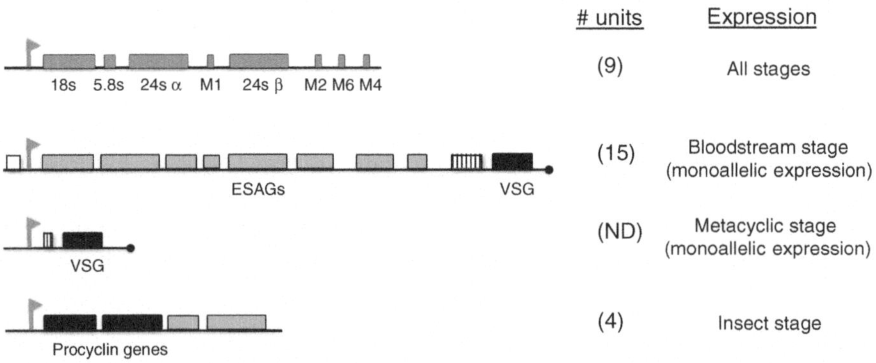


**Fig. 7.1** Genes transcribed by RNA Pol I in *T. brucei*. A typical *rDNA* transcription unit (White et al. 1986) and the other, more unusual, Pol I-transcribed units in *T. brucei* are depicted. The *rDNA* and procyclin units are found in diploid regions of the genome (haploid copy-number is indicated), while BESs and MESs are found in hemizygous subtelomeric domains (Callejas et al. 2006). All four Pol I promoters, depicted by flags, are distinct (see Fig. 7.2). *Darker boxes* indicate the *VSG* and procyclin genes. A 'consensus' arrangement of *ESAGs* is shown, excluding *T. b. rhodesiense SRA*. The *striped box* indicates the location of an array of 70-bp repeats that often serve as sites of recombination between BESs. *White box* depicts 50-bp repeats. The telomeric repeats are represented by the *black dot* downstream of the *VSG*. The transcription units are not to scale

et al. 2011). If left untreated, this disease is almost always fatal. sleeping sickness, coupled with nagana, the animal form of African trypanosomiasis, has been a major obstacle to sub-Saharan African rural development and a stumbling block to agricultural production. There is currently no vaccine and drug treatments are either toxic, difficult to administer or very expensive. Diagnosis is technologically outdated and complex (Wastling and Welburn 2011). It is therefore urgent to understand the biology of this parasite in order to identify new drug targets and to develop new diagnostic tools.

In the academic world, *T. brucei* became well known for the milestone discoveries that revealed new aspects of biology in eukaryotes. Indeed, it was studies in *T. brucei* that first identified that RNA can be edited (Benne et al. 1986), that proteins can be anchored to membranes via a glycophosphatidylinositol (GPI) glycolipid (Ferguson et al. 1985) and that two RNA molecules transcribed from two different genes can be ligated to produce a mature RNA via trans-splicing (Boothroyd and Cross 1982).

### *7.1.1 Organization of VSG Expression Sites*

It is a challenge to define the limits of subtelomeres in any organism and there is no strict definition in *T. brucei*. In fact, most of the sequences representing subtelomeric–non-subtelomeric junctions are not available. Here, we focus on the bloodstream *VSG* expression sites (BESs), but it should be noted that subtelomeres extend much further, maybe over 1 Mbp, in some cases. These extended regions are thought to comprise arrays of gene families, including *VSGs*, genes related to *ESAGs* (see below) and retrotransposons and retrotransposon hot spots. The organization, recombination, and evolution of the more extensive *VSG* archive are covered elsewhere and will not be covered in any detail here. Briefly, it is generally accepted that subtelomeric domains facilitate the evolution of large heterogeneous gene families in eukaryotes, and *VSG* organization in *T. brucei* supports this idea since *VSGs* occupy the subtelomeres of minichromosomes, intermediate-sized chromosomes, and the megabase-sized chromosomes, which encode the non-redundant 'housekeeping' genes. Thus, two of the three chromosome size-classes in *T. brucei* and all subtelomeres are dedicated to *VSG* gene archiving and evolution. The BESs themselves are only found on intermediate and megabase chromosomes.

## 7.2 Expression Site-Associated Genes

As noted above, *VSGs* are co-transcribed in the BES with up to twelve expression site-associated genes (*ESAG*s). The reason for multiple BES, and why *ESAG*s are resident at these loci, has remained a subject of debate. Pol I is thought to affords a high rate of transcription, which is thought to be important to make up the dense VSG coat but the ESAGs would not be expected to require a similarly high transcription rate; indeed, steady-state *ESAG* transcripts are much less abundant than *VSG* transcripts (Jensen et al. 2009). Yet, because the *VSG* is always found proximal to the telomere repeats, the *ESAGs* must surely be transcribed at the same, or a higher rate, than the *VSG*. Although this may sound wasteful, the apparent inefficiency may not represent a meaningful selective force in the context of a parasite living in a nutrient-rich bloodstream environment. A more likely selective force driving this association is for antigenic diversity, which in turn would be driven by host adaptive immunity or differential affinity for host nutrients (Hertz-Fowler et al. 2008; McCulloch and Horn 2009). In support of this idea, ESAGs display features consistent with cell-surface receptors, signaling molecules, and transporters. For example, ESAG6 and ESAG7 form a hetero-dimeric transferrin receptor (Pays et al. 2001) and ESAG4 is an adenylate cyclase (Paindavoine et al. 1992). Thus, BES transcription switching and recombination can switch *VSG* and *ESAG* expression simultaneously and generate new combinations of *ESAGs*, respectively.

One particular ESAG stands out because of its proven role in survival in primate hosts including humans. This serum resistance-associated gene (*SRA*) can convert *T. brucei brucei* from human serum-sensitive forms to human serum-resistant forms (Xong et al. 1998). *SRA*, unlike other *ESAGs*, is limited to one or few BESs and is only found in *T. b. rhodesiense*. In fact, *SRA* is now considered a diagnostic marker for *T. b. rhodesiense* (Picozzi et al. 2008). SRA appears to be a mutated *VSG* that fails to traffic to the cell surface. Sequestered in endosomes (Stephens and Hajduk 2011), *SRA* neutralizes the primate serum trypanolytic factor, ApoL1, and allows survival in primate hosts. A molecular 'arms race' is at play here, with evidence for selection of protective ApoL1 variants in African populations (Genovese et al. 2010). These variants fail to interact with SRA and are not neutralized. This SRA–lytic factor interplay is of great interest in terms of developing lytic factor-based therapeutics that are not neutralized by ESAGs or other parasite virulence factors.

## 7.3 Promoters

BESs share a conserved promoter, that is, on the other hand, different from all other promoters of Pol I-transcribed units. Other than rDNA and BES units, Pol I also transcribes surface-protein-encoding genes in other stages of the parasite life cycle. In procyclic forms, Pol I transcribes the EP/GPEET procyclin polycistronic units, while in metacyclic stages, Pol I mediates monoallelic transcription of metacyclic *VSG* expression sites (MES), which unlike BES are monocistronic units (Fig. 7.1). MES are expressed for a few days after trypanosomes are transmitted from the tsetse vector to the mammalian host (Pedram and Donelson 1999). Despite the sequence diversity between BES, MES, Procyclin, and rRNA promoters, there are some common structural features (Fig. 7.2). Procyclin and rRNA

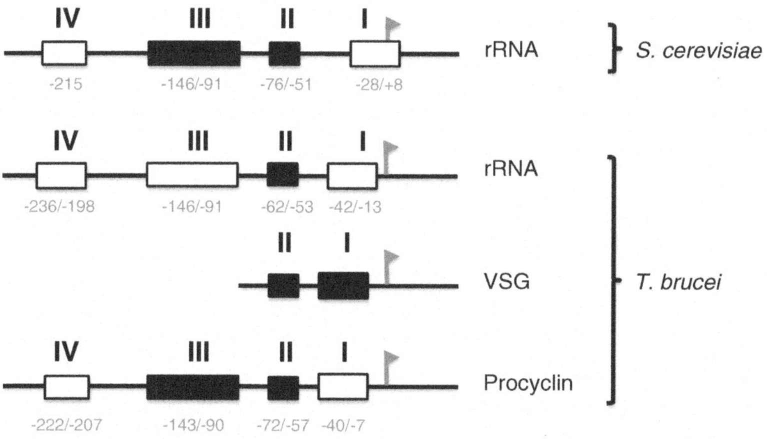


**Fig. 7.2** Structure of RNA Pol I promoters in *S. cerevisiae* and *T. brucei*. Promoter domains are indicated by *boxes*. *Numbers* indicate position relative to transcription start site. Domains shown in *black* are those important for stable binding of trans-activating factors in vitro. *VSG* BES promoter is the shortest Pol I promoter and it contains only two domains. *VSG* MES, although less studied, seems to have the same general organization (Kim and Donelson 1997)

promoters have a multi-domain structure resembling that of the yeast rDNA promoter. The BES promoter is shorter, with only two conserved elements that are involved in recruiting common Pol I transcription factors (Brandenburg et al. 2007). Metacyclic ES (MES) promoters are also short, but they are not conserved. Despite these differences, an rRNA promoter is interchangeable with a BES promoter (Rudenko et al. 1995), suggesting the involvement of epigenetic mechanisms in order to achieve stage-specific expression of surface proteins.

## 7.4 Other Conserved Non-Protein-Coding Sequences

There are a number of conserved non-protein-coding sequences found in BESs, and we now consider potential regulatory regions involved in controlling monoallelic *VSG* expression. We do not consider *ESAG*-associated sequences here since these sequences are not found in the MESs. Clearly, a competent BES must contain a Pol I promoter and BES promoters are only found at sub-telomeres. But within BESs, which other sequences might play a role in monotelomeric *VSG* expression control?

BESs are flanked by sequence repeats and also have internal repeats. The arrangement of these sequences is constant from one BES to the next, suggesting that location is important. Starting on the centromeric side of the BES, we see arrays of '50-bp repeats' upstream of the promoter, we then see arrays of '70-bp repeats' between the *ESAGs* and the *VSG* and, finally, the telomeric repeats beyond the *VSG* (Fig. 7.1). The 50-bp repeats may act as boundary elements that isolate silent and/or active domains from the rest of the chromosome (Sheader et al. 2003). The 70-bp repeats are strongly implicated in recombination and are dispensable at active (McCulloch et al. 1997) and silent (D.H., unpublished) BESs. The $T_2AG_3$ telomeric repeats are found 0.2–1.6 kbp downstream of the *VSG* and can extend for >15 kbp. Telomerase-dependent extension (Dreesen et al. 2005) is inversely related to telomere length (Horn et al. 2000) and is balanced by long-telomere instability. Small duplex t-loops have been seen at *T. brucei* telomeres (Munoz-Jordan et al. 2001).

The *VSG* 3′-UTR contains a short, but highly conserved motif (Majumder et al. 1981) involved in stage-specific *VSG* expression (Berberof et al. 1995). Finally, a conserved GC-rich element, with an 11-bp core that is complementary to, but inverted relative to, the telomeric repeats, is found between the *VSG* and the telomere (Horn and Barry 2005). It seems likely that sequences shared among BESs are important for *VSG* expression control, but the elements detailed above could equally be involved in other aspects of (sub) telomere function or even be more passive sequences that simply accumulate at these loci.

## 7.5 Basic Transcription Machinery and Nuclear Architecture

What is the basic transcription machinery involved in transcribing subtelomeric BES? Two major players have been identified and characterized to be involved in Pol I transcription: the RNA Pol I enzyme and a class I transcription factor complex (Brandenburg et al. 2007).

Like in other eukaryotes, *T. brucei* RNA Pol I is a multi-subunit enzyme. Some subunits are shared with all or some RNA polymerases, whereas others are Pol I specific. Due to the high degree of conservation, most subunits were identified by sequence homology and subsequently characterized biochemically. RPA1 is the largest subunit and is phosphorylated (Walgraffe et al. 2005). RPA2 is the second largest subunit and it carries a unique 50-kDa *N*-terminal extension whose function remains unknown (Schimanski et al. 2003). Another trypanosome peculiarity is that its genome contains two paralogues of the subunits RPB5, RPB6, and RPB10 (named RPB5z, RPB6z, and RPB10z). In contrast to RPB5 and RPB6, RPB5z and RPB6z are part of the Pol I complex and they localize at Pol I sites in the nucleus (Nguyen et al. 2006), suggesting that they are functionally different (Devaux et al. 2007). Why trypanosomes have two paralogues remains an intriguing evolutionary question (Kelly et al. 2005). The only study in which an active Pol I was purified revealed not only the above-mentioned subunits but three other proteins with apparent sizes of 31, 29, and 27 kDa. p31 is the only one that has been characterized and it consists of a trypanosome-specific subunit, absent in other eukaryotes (Nguyen et al. 2007). It was argued that the Pol II subunit, RPB7, is involved in Pol I transcription in *T. brucei* (Penate et al. 2009), but opposing results have also been reported (Park et al. 2011).

In eukaryotes, RNA polymerases are recruited to specific promoters because a set of transcription factors mark those loci. In trypanosomes, class I transcription factor A (C1TFA) has been biochemically purified and shown to be essential for Pol I transcription (BES, rRNA and Procyclin) (Brandenburg et al. 2007). It is composed of seven proteins that are trypanosome-specific. Apart from one of the proteins whose sequence suggested it was a dynein light chain, no conserved domains could be identified in the other six subunits. Future work should clarify the function of each of these subunits, which ones bind to the promoter, how they interact and recruit Pol I enzyme and if they interact with chromatin-modifying enzymes.

Textbooks say that Pol I transcription occurs in the nucleolus. For trypanosomes, this is not entirely true and that is because, as mentioned above, Pol I transcribes not only rRNA genes, but other protein-coding genes. In bloodstream forms, the life cycle stage in which one BES is monoallelically expressed and VSG is present at the surface, immunofluorescence analysis using an anti-Pol I antibody shows that there are two Pol I sites in the nucleus: the nucleolus and an extra-nucleolar expression site body (Navarro and Gull 2001). The actively transcribed BES is localized in this site throughout the cell cycle, and the sister chromatids display delayed dissociation at this locus during chromosome segregation (Landeira et al. 2009). Such enhanced cohesion is mediated by cohesin and, when this factor is diminished, the frequency of parasites that switch transcription to a new BES increases 10-fold. It is possible that the ESB may also restrict spatial access or contain limiting factor(s) required for BES transcription, but these hypotheses remain untested. When bloodstream forms differentiate into the insect stage, the active BES is silenced, and the ESB disappears. In this stage, transcription of procyclins, which is also Pol I mediated, does not seem to occur in any specialized extra-nucleolar compartment but rather at the nucleolus (Navarro et al. 2007). ELP3b controls rDNA transcription in the nucleolus but not at the ESB (Alsford and Horn 2011). This protein has homology to the catalytic component of Elongator, a complex that assists transcript elongation, acetylating nucleosomes in the path of the elongating polymerase (Svejstrup 2007).

## 7.6 Active and Silent BES Have Different Chromatin Structures

When a parasite divides, the daughter cell ‘remembers’ which *VSG* was actively transcribed by its mother cell and, in the majority of cases, it continues to use the same VSG coat. Stochastic switching to a different *VSG* occurs at a low frequency, which ranges between $10^{-2}$ and $10^{-6}$ switches per generation, depending on the strain. This means that the daughter cell inherited the information of which *VSG* should be expressed and switching happens without any changes in the DNA sequence. Therefore, it is currently accepted that monoallelic *VSG* expression is under epigenetic control. Based on this model, several studies have been published in the last six or seven years [reviewed by Horn and McCulloch 2010)] showing the role of candidate genes in *VSG* expression control (Table 7.1). There is now good evidence for the functional specialization of subtelomeric domains in terms of chromatin structure. However, given the early stages of these studies, we have little mechanistic information, including how the various factors are recruited to BESs, if they interact with each other, with a transcription factor or Pol I or if they affect chromatin condensation or histone modifications.

Most of the chromatin modifiers characterized so far are involved in transcription silencing (Table 7.1). Knockout or knockdown of chromatin modifiers can affect silencing in three distinct ways: (1) the whole BESs become partially derepressed (DOT1B, RAP1, ISWI); (2) only the BES promoter region becomes depressed, but no effect is detected in *VSG* transcripts (DAC3, FACT, NLP, ASF1A, CAF1), (3) only an exogenous promoter placed close to a telomere becomes derepressed (SIR2rp1, HAT1). The range and sites of derepression have been measured with different methods in different studies (Northern blotting, Western blotting, FACS GFP intensity, luciferase activity, quantitative real-time PCR), which may have introduced some variability in the interpretation ISWI depletion and DOT1B knockout were only linked to *VSG* derepression when sensitive quantitative PCR was used for the analysis, for example. Another issue worth considering here is the possibility of secondary effects, particularly when depleting factors that are required for continued growth; this is the case for all of the factors above, except for SIR2rp1 and DOT1B. When derepression was quantified, it ranged between 7- and 65-fold. Such fold increase in transcription is still far from a complete derepression, which would be a $10^4$- to $10^5$-fold increase. This suggests that we have either not found the key molecule(s) that regulate(s) *VSG* silencing, or that the system is redundant and multiple players would have to be simultaneously depleted in order to see a more pronounced phenotype. Only depletion of DAC1 resulted in repression rather than derepression, indicating that DAC1 antagonizes SIR2rp1-dependent telomeric silencing (Wang et al. 2010). Interestingly, this effect is developmentally regulated and fails to operate in insect-stage cells. It has become quite clear from these studies that two mechanistically distinct forms of transcription repression operate at *T. brucei* telomeres. SIR2rp1-dependent repression appears to be similar to the Sir2-dependent mechanism that has been documented in detail in the budding yeast, *Saccharomyces cerevisiae* (Grunstein 1997). The distinct mechanism that is considered to be of greater interest in *T. brucei* though is the one responsible for BES repression. Some evidence suggests that this form of repression extends over large arrays of *VSGs* (Horn and Cross 1997) and it is this mechanism that will likely dominate future research in this area.

Post-translation modifications can be important for positive or negative regulation of transcription. It was therefore puzzling not to obtain phenotypes in which transcription of the active BES was compromised when histone modifiers were depleted. A possible explanation for this observation came in 2010, when two groups showed that the chromatin of the active BES is drastically depleted of nucleosomes and, as a consequence, it has a more open chromatin structure (Figueiredo and Cross 2010; Stanne and Rudenko 2010). Silent BESs, on the other

**Table 7.1** Factors linked to transcription control at *T. brucei* telomeres

| Protein | Function | KO/KD phenotype | Reference |
|---|---|---|---|
| RAP1 | Telomeric protein | Entire silent BESs are derepressed: 7-fold close to the promoter and 25- to 50-fold close to the telomere in BSF | Yang et al. (2009) |
| ISWI | Chromatin remodeler | Entire silent BESs are derepressed: 30- to 60-fold in BSF and 10- to 17-fold in PF | Hughes et al. (2007), Stanne and Rudenko (2010) |
| DOT1B | Histone methyltransferase (H3K76me3) | Entire silent BESs are derepressed (10-fold in BSF); switching between BESs is slower | Figueiredo et al. (2008) |
| ASF1 | Histone chaperone | Silent BES promoter region is derepressed | (Alsford and Hom, unpublished data) |
| CAF1 | Histone chaperone | Silent BES promoter region is derepressed | (Alsford and Hom, unpublished data) |
| DAC3 | Histone deacetylase | Silent BES promoter region is derepressed | Wang et al. (2010) |
| FACT (Spt6) | Histone chaperone | Silent BES promoter region is derepressed (20- to 23-fold in BSF; 16- to 25-fold in PF); G2/M cell cycle arrest | Denninger et al. (2010) |
| NLP | Nucleoplasmin-like protein | Silent BES promoter region is derepressed: 45- to 65-fold in BSF | Narayanan et al. (2011) |
| SIR2rpl | Histone deacetylase | Telomere-proximal promoter is derepressed | Alsford et al. (2007) |
| HAT1 | Histone acetyltransferase | Telomere-proximal promoter is derepressed | Kawahara et al. (2008) |
| DAC1 | Histone deacetylase | Telomere-proximal promoter is repressed | Wang et al. (2010) |

hand, show a regular nucleosomal organization. Since there are more histones at silent BES, it is more likely that more post-translation modifications will play a role in silencing than in activating transcription.

Another discovery that emerged from studies on active and silent *VSG* BESs was of a novel DNA base modification that is also found in other trypanosomatids and some dinoflagellates but, intriguingly in African trypanosomes, is specific to the bloodstream life cycle stage (Borst and Sabatini 2008). This glucosylated version of thymidine, known as base J, has been found enriched at transcription initiation sites in the Pol II-transcribed regions in *T. brucei* and *Trypanosoma cruzi* (the South American trypanosome), where it plays a role in down-regulating transcription (Ekanayake et al. 2011). Although base J is enriched within silent BESs, no role has been demonstrated to date in the control of these subtelomeric Pol I transcription units.

## 7.7 Gene Expression Control

For antigenic variation to succeed, *T. brucei* subtelomeres display rather remarkable, and apparently unprecedented, transcription control. In order to express only one *VSG* at a time, all BESs but one must be silenced. Although monotelomeric *VSG* expression is an important and intensely studied topic, the primary mechanism underlying this phenomenon is not understood. We do know that the mechanism underlying BES control is developmentally regulated (Horn and Cross 1995), cross talk between the active telomere and silent telomeres is clearly important (Chaves et al. 1999) and it is transcription elongation rather than initiation that appears to be primarily regulated (Vanhamme et al. 2000). More recently, we have seen that SIR2rp1-dependent silencing is mechanistically distinct from BES silencing while depletion of the chromatin regulators linked to BES control fails to reveal major levels of derepression (see above). In light of these findings, it is worth reassessing the fundamental nature of the regulatory mechanism involved in monotelomeric *VSG* expression.

One may invoke a limiting 'competence' factor or single privileged domain to maintain monotelomeric *VSG* expression (Fig. 7.3a), but it is difficult to see how positive regulators would be limited or restricted within the nuclear environment and such models have proven difficult to test. An alternative '*trans*-acting negative control' model (Fig. 7.3b) that does not invoke a strictly limiting competence factor or restricted compartmentalization is tenable. The active *VSG* gene is always adjacent to a telomere, implicating the telomere as a possible positive controller. The telomere may allow sequences that can recruit Pol I (BES promoters and *rDNA* promoters) to do so in an efficient and productive manner. What if active BES-associated sequences were then to transmit a negative signal to other BESs? This would set up a competition between all telomeres with competent Pol I promoters. A single BES could emerge as the active one and this would cement the negative control signal that would silence all other BESs. At this stage, any

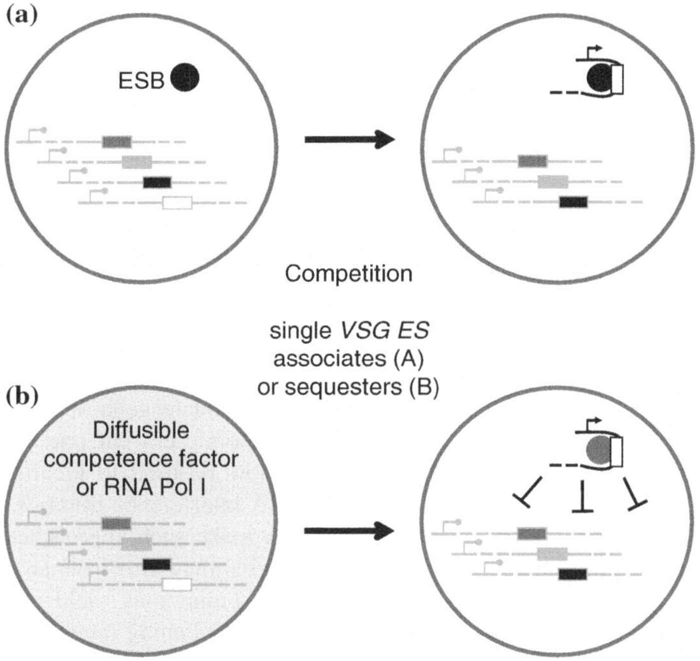


**Fig. 7.3** Models to explain monotelomeric *VSG* expression. The schematic shows nuclear compartments (*circles*) each containing four *VSG* expression sites. **a** The 'strictly limiting competence factor' or 'privileged domain' model. **b** The '*trans*-acting negative control' model. The active BES negatively controls other BESs in (**b**). Note that an expression site body (ESB), as defined by RNA Pol I accumulation, would be observed regardless of which mechanism operates

conserved BES sequence could be considered a candidate for transmitting and/or detecting the negative signal.

How would cells then switch transcription from one telomere to another? The negative signal would presumably need to be reduced or blocked for another BES to respond. Since DNA damage typically triggers major local chromatin remodeling, loss of the negative signal or failure to respond could be a result of DNA damage and repair at the active or a silent BES, respectively, which could 'reset' a positive or negative chromatin structure and initiate a new competition for activation. This could also involve synthesis of a sister BES that fails to inherit an epigenetic signal due to disrupted segregation timing, for example. Indeed, cohesion knockdown allows premature segregation of active BES loci and leads to increased BES transcription switching (Landeira et al. 2009). Another important finding that is consistent with this idea is that RAD51 is important for transcription switching (McCulloch and Barry 1999). RAD51-dependent recombination and repair could certainly occur within the telomeric repeats or between sister chromatids without leaving any detectable DNA sequence change and could disrupt the positive and/or negative controls that maintain monoallelic *VSG* expression as outlined above.

## 7.8 Concluding Remarks and Future Perspectives

African trypanosomes present an evolutionary divergent perspective on the organization and function of subtelomeres. They also present a remarkable gene expression phenomenon involving monotelomeric Pol I transcription. Studies on antigenic variation, the *VSGs* found at these loci, and their expression control have yielded several important discoveries of biological novelty. Studies on epigenetic control at these loci are more recent but are now yielding insights into this important problem. It has been, and will continue to be, a major challenge to distinguish between direct and indirect impacts on *VSG* expression in an allelic exclusion system involving cross talk between silent and active loci. Clearly, it will be important to endeavor to dissect the positive and negative controls that impact this remarkable monotelomeric expression system. Candidate gene approaches have prevailed in this area at the outset of the postgenomic era but one would hope that powerful forward-genetic approaches can now be used to identify positive and negative *VSG* regulators. Genome-scale RNA interference libraries in bloodstream-form *T. brucei* (Alsford et al. 2011) now make such approaches feasible. Different epigenetic states at *T. brucei* subtelomeres may also have an impact on recombination. In this regard, epigenetics-based studies may yield insights into the mechanism underlying the vastly different *VSG* switching rates ($10^{-2}$ to $10^{-6}$ switches per generation) reported for different African trypanosome isolates.

**Acknowledgments** Work in the authors' laboratories is funded by The Wellcome Trust (DH) and the Fundação Calouste Gulbenkian and EMBO (LMF).

## References

Alsford, S., & Horn, D. (2011). Elongator protein 3b negatively regulates ribosomal DNA transcription in African trypanosomes. *Molecular and Cellular Biology, 31*, 1822–1832. doi:10.1128/MCB.01026-10.

Alsford, S., Kawahara, T., Isamah, C., & Horn, D. (2007). A sirtuin in the African trypanosome is involved in both DNA repair and telomeric gene silencing but is not required for antigenic variation. *Molecular Microbiology, 63*, 724–736.

Alsford, S., Turner, D. J., Obado, S. O., Sanchez-Flores, A., Glover, L., Berriman, M. (2011). High-throughput phenotyping using parallel sequencing of RNA interference targets in the African trypanosome. *Genome Research, 21*, 915–924. doi:10.1101/gr.115089.110.

Becker, M., Aitcheson, N., Byles, E., Wickstead, B., Louis, E., & Rudenko, G. (2004). Isolation of the repertoire of VSG expression site containing telomeres of *Trypanosoma brucei* 427 using transformation-associated recombination in yeast. *Genome Research, 14*, 2319–2329. doi:10.1101/gr.2955304.

Benne, R., Van den Burg, J., Brakenhoff, J. P., Sloof, P., Van Boom, J. H., & Tromp, M. C. (1986). Major transcript of the frameshifted coxII gene from trypanosome mitochondria contains four nucleotides that are not encoded in the DNA. *Cell, 46*, 819–826. doi:10.1016/0092-8674(86)90063-2.

Berberof, M., Vanhamme, L., Tebabi, P., Pays, A., Jefferies, D., Welburn, S. (1995). The 3′-terminal region of the mRNAs for VSG and procyclin can confer stage specificity to gene expression in *Trypanosoma brucei*. *EMBO Journal, 14*, 2925–2934.

Berriman, M., Hall, N., Sheader, K., Bringaud, F., Tiwari, B., Isobe, T. (2002). The architecture of variant surface glycoprotein gene expression sites in *Trypanosoma brucei*. *Molecular and Biochemical Parasitology, 122*, 131–140. doi:10.1016/S0166-6851(02)00092-0.

Boothroyd, J. C., & Cross, G. A. (1982). Transcripts coding for variant surface glycoproteins of Trypanosoma brucei have a short, identical exon at their 5′ end. *Gene, 20*, 281–289. doi:10.1016/0378-1119(82)90046-4.

Borst, P., & Sabatini, R. (2008). Base J: Discovery, biosynthesis, and possible functions. *Annual Review of Microbiology,*. doi:10.1146/annurev.micro.62.081307.162750.

Brandenburg, J., Schimanski, B., Nogoceke, E., Nguyen, T. N., Padovan, J. C., Chait, B. T. (2007). Multifunctional class I transcription in *Trypanosoma brucei* depends on a novel protein complex. *EMBO Journal, 26*, 4856–4866. doi:10.1038/sj.emboj.7601905.

Callejas, S., Leech, V., Reitter, C., & Melville, S. (2006). Hemizygous subtelomeres of an African trypanosome chromosome may account for over 75 % of chromosome length. *Genome Research, 16*, 1109–1118. doi:10.1101/gr.5147406.

Chaves, I., Rudenko, G., Dirks-Mulder, A., Cross, M., & Borst, P. (1999). Control of variant surface glycoprotein gene-expression sites in *Trypanosoma brucei*. *EMBO Journal, 18*, 4846–4855. doi:10.1093/emboj/18.17.4846.

de Lange, T., & Borst, P. (1982). Genomic environment of the expression-linked extra copies of genes for surface antigens of *Trypanosoma brucei* resembles the end of a chromosome. *Nature, 299*, 451–453. doi:10.1038/299451a0.

Denninger, V., Full brook, A., Bessat, M., Ersfeld, K., & Rudenko, G. (2010). The FACT subunit Tbspt16 is involved in cell cyde specific control of VSG expression sites in *Trypanosoma brucei*. *Moi Microbiology, 76*, 459–474.

Devaux, S., Kelly, S., Lecordier, L., Wickstead, B., Perez-Morga, D., Pays, E. (2007). Diversification of function by different isoforms of conventionally shared RNA polymerase subunits. *Molecular Biology of the Cell, 18*, 1293–1301. doi:10.1091/mbc.E06-09-0841.

Dreesen, O., Li, B., & Cross, G. A. (2005). Telomere structure and shortening in telomerase-deficient Trypanosoma brucei. *Nucleic Acids Research, 33*, 4536–4543. doi:10.1093/nar/gki769.

Ekanayake, D. K., Minning, T., Weatherly, B., Gunasekera, K., Nilsson, D., Tarleton, R. (2011). Epigenetic regulation of transcription and virulence in *Trypanosoma cruzi* by O-linked thymine glucosylation of DNA. *Molecular Cell Biology,*. doi:10.1128/MCB.01277-10.

Ferguson, M. A., Low, M. G., & Cross, G. A. (1985). Glycosyl-sn-1,2-dimyristylphosphatidyli nositol is covalently linked to *Trypanosoma brucei* variant surface glycoprotein. *Journal of Biological Chemistry, 260*, 14547–14555.

Figueiredo, L. M., & Cross, G. A. M. (2010). Nucleosomes are depleted at the VSG expression site transcribed by RNA polymerase I in African trypanosomes. *Eukaryotic Cell, 9*, 148–154. doi:10.1128/EC.00282-09.

Figueiredo, L. M., Janzen, C. J., & Cross, G. A. (2008). A histone methyltransferase modulates antigenic variation in African trypanosomes. *PLoS Biology, 6*.

Genovese, G., Friedman, D. J., Ross, M. D., Lecordier, L., Uzureau, P., Freedman, B. I. (2010). Association of trypanolytic ApoL1 variants with kidney disease in African Americans. *Science, 329*, 841–845. doi:10.1126/science.1193032.

Grunstein, M. (1997). Molecular model for telomeric heterochromatin in yeast. *Current Opinion in Cell Biology, 9*, 383–387. doi:10.1016/S0955-0674(97)80011-7.

Hertz-Fowler, C., Figueiredo, L. M., Quail, M. A., Becker, M., Jackson, A., Bason, N. (2008). Telomeric expression sites are highly conserved in *Trypanosoma brucei*. *PLoS ONE, 3*, e3527. doi:10.1371/journal.pone.0003527.

Horn, D., & Barry, J. D. (2005). The central roles of telomeres and subtelomeres in antigenic variation in African trypanosomes. *Chromosome Research, 13*, 525–533. doi:10.1007/s10577-005-0991-8.

Horn, D., & Cross, G. A. M. (1995). A developmentally regulated position effect at a telomeric locus in *Trypanosoma brucei*. *Cell, 83*, 555–561. doi:10.1016/0092-8674(95)90095-0.

Horn, D., & Cross, G. A. M. (1997). Position-dependent and promoter-specific regulation of gene expression in *Trypanosoma brucei*. *EMBO Journal, 16*, 7422–7431. doi:10.1093/emboj/16.24.7422.

Horn, D., & McCulloch, R. (2010). Molecular mechanisms underlying the control of antigenic variation in African trypanosomes. *Current Opinion in Microbiology, 13*, 700–705. doi:10.1016/j.mib.2010.08.009.

Horn, D., Spence, C., & Ingram, A. K. (2000). Telomere maintenance and length regulation in *Trypanosoma brucei*. *EMBO Journal, 19*, 2332–2339. doi:10.1093/emboj/19.10.2332.

Hughes, K., Wand, M., Foulston, L., Young, R., Harley, K., Terry, S., et al. (2007). A novel ISWI is involved in VSG expression site downregulation in African trypanosomes. *EMBO Journal, 26*, 2400–2410.

Jensen, B. C., Sivam, D., Kifer, C. T., Myler, P. J., & Parsons, M. (2009). Widespread variation in transcript abundance within and across developmental stages of *Trypanosoma brucei*. *BMC Genomics, 10*, 482. doi:10.1186/1471-2164-10-482.

Kawahara, T., Siegel, T. N., Ingram, A. K., Alsford, S., Cross, G. A., & Horn, D. (2008). Two essential MYST-family proteins display distinct roles in histone H4K10 acetylation and telomeric silencing in trypanosomes. *Molecular Microbiology, 69*, 1054–1068.

Kelly, S., Wickstead, B., & Gull, K. (2005). An in silico analysis of trypanosomatid RNA polymerases: Insights into their unusual transcription. *Biochemical Society Transactions, 33*, 1435–1437. doi:10.1042/BST20051435.

Kim, K. S., & Donelson, J. E. (1997). Co-duplication of a variant surface glycoprotein gene and its promoter to an expression site in African trypanosomes. *Journal of Biological Chemistry, 272*, 24637–24645. doi:10.1074/jbc.272.39.24637.

Landeira, D., Bart, J. M., Van Tyne, D., & Navarro, M. (2009). Cohesin regulates VSG monoallelic expression in trypanosomes. *Journal of Cell Biology, 186*, 243–254. doi:10.1083/jcb.200902119.

Majumder, H. K., Boothroyd, J. C., & Weber, H. (1981). Homologous 3′-terminal regions of mRNAs for surface antigens of different antigenic variants of *Trypanosoma brucei*. *Nucleic Acids Research, 9*, 4745–4753. doi:10.1093/nar/9.18.4745.

McCulloch, R., & Barry, J. D. (1999). A role for RAD51 and homologous recombination in *Trypanosoma brucei* antigenic variation. *Genes and Development, 13*, 2875–2888. doi:10.1101/gad.13.21.2875.

McCulloch, R., & Horn, D. (2009). What has DNA sequencing revealed about the VSG expression sites of African trypanosomes? *Trends in Parasitology, 25*, 359–363. doi:10.1016/j.pt.2009.05.007.

McCulloch, R., Rudenko, G., & Borst, P. (1997). Gene conversions mediating antigenic variation in *Trypanosoma brucei* can occur in variant surface glycoprotein expression sites lacking 70-base-pair repeat sequences. *Molecular and Cellular Biology, 17*, 833–843.

Munoz-Jordan, J. L., Cross, G. A. M., de Lange, T., & Griffith, J. D. (2001). T-loops at trypanosome telomeres. *EMBO Journal, 20*, 579–588. doi:10.1093/emboj/20.3.579.

Narayanan, M. S., Kushwaha, M., Ersfeld, K., Full brook, A., Stanne, T. M., & Rudenko, G. (2011). NLP is a novel transcription regulator involved in VSG expression site control in *Trypanosoma brucei*. *Nucleic Acids Research, 39*, 2018–2031.

Navarro, M., & Gull, K. (2001). A pol I transcriptional body associated with VSG mono-allelic expression in *Trypanosoma brucei*. *Nature, 414*, 759–763. doi:10.1038/414759a.

Navarro, M., Penate, X., & Landeira, D. (2007). Nuclear architecture underlying gene expression in *Trypanosoma brucei*. *Trends in Microbiology, 15*, 263–270. doi:10.1016/j.tim.2007.04.004.

Nguyen, T. N., Schimanski, B., Zahn, A., Klumpp, B., & Gunzl, A. (2006). Purification of an eight subunit RNA polymerase I complex in *Trypanosoma brucei*. *Molecular and Biochemical Parasitology, 149*, 27–37. doi:10.1016/j.molbiopara.2006.02.023.

Nguyen, T. N., Schimanski, B., & Gunzl, A. (2007). Active RNA polymerase I of Trypanosoma brucei harbors a novel subunit essential for transcription. *Molecular and Cellular Biology, 27*, 6254–6263. doi:10.1128/MCB.00382-07.

Paindavoine, P., Rolin, S., Van Assel, S., Geuskens, M., Jauniaux, J. C., Dinsart, C. (1992). A gene from the variant surface glycoprotein expression site encodes one of several transmembrane adenylate cyclases located on the flagellum of *Trypanosoma brucei*. *Molecular and Cellular Biology, 12*, 1218–1225.

Park, S. H., Nguyen, T. N., Kirkham, J. K., Lee, J. H., & Gunzl, A. (2011). Transcription by the multifunctional RNA polymerase I in *Trypanosoma brucei* functions independently of RPB7. *Molecular and Biochemical Parasitology, 180*, 35–42. doi:10.1016/j.molbiopara.2011.06.008.

Pays, E., Lips, S., Nolan, D., Vanhamme, L., & Perez-Morga, D. (2001). The VSG expression sites of *Trypanosoma brucei*: Multipurpose tools for the adaptation of the parasite to mammalian hosts. *Molecular and Biochemical Parasitology, 114*, 1–16. doi:10.1016/S0166-6851(01)00242-0.

Pedram, M., & Donelson, J. E. (1999). The anatomy and transcription of a monocistronic expression site for a metacyclic variant surface glycoprotein gene in *Trypanosoma brucei*. *Journal of Biological Chemistry, 274*, 16876–16883. doi:10.1074/jbc.274.24.16876.

Penate, X., Lopez-Farfan, D., Landeira, D., Wentland, A., Vidal, I., & Navarro, M. (2009). RNA pol II subunit RPB7 is required for RNA pol I-mediated transcription in *Trypanosoma brucei*. *EMBO Reports, 10*, 252–257. doi:10.1038/embor.2008.244.

Picozzi, K., Carrington, M., & Welburn, S. C. (2008). A multiplex PCR that discriminates between *Trypanosoma brucei brucei* and zoonotic *T. b. rhodesiense*. *Experimental Parasitology, 118*, 41–46. doi:10.1016/j.exppara.2007.05.014.

Rudenko, G., Blundell, P. A., Dirks-Mulder, A., Kieft, R., & Borst, P. (1995). A ribosomal DNA promoter replacing the promoter of a telomeric VSG gene expression site can be efficiently switched on and off in *T. brucei*. *Cell, 83*, 547–553. doi:10.1016/0092-8674(95)90094-2.

Schimanski, B., Klumpp, B., Laufer, G., Marhofer, R. J., Selzer, P. M., & Gunzl, A. (2003). The second largest subunit of *Trypanosoma brucei*'s multifunctional RNA polymerase I has a unique N-terminal extension domain. *Molecular and Biochemical Parasitology, 126*, 193–200. doi:10.1016/S0166-6851(02)00273-6.

Sheader, K., Berberof, M., Isobe, T., Borst, P., & Rudenko, G. (2003). Delineation of the regulated variant surface glycoprotein gene expression site domain of *Trypanosoma brucei*. *Molecular and Biochemical Parasitology, 128*, 147–156. doi:10.1016/S0166-6851(03)00056-2.

Simarro, P. P., Diarra, A., Postigo, J. A. R., Franco, J. R., & Jannin, J. G. (2011). The human African trypanosomiasis control and surveillance programme of the World Health Organization 2000–2009: The way forward. *PLoS Neglected Tropical Diseases, 5*, e1007. doi:10.1371/journal.pntd.0001007.

Sogin, M. L., Gunderson, J. H., Elwood, H. J., Alonso, R. A., & Peattie, D. A. (1989). Phylogenetic meaning of the kingdom concept: An unusual ribosomal RNA from Giardia lamblia. *Science, 243*, 75–77. doi:10.1126/science.2911720.

Stanne, T. M., & Rudenko, G. (2010). Active VSG expression sites in Trypanosoma brucei are depleted of nucleosomes. *Eukaryotic Cell, 9*, 136–147. doi:10.1128/EC.00281-09.

Stephens, N. A., & Hajduk, S. L. (2011). Endosomal localization of the serum resistance-associated protein in African trypanosomes confers human infectivity. *Eukaryotic Cell, 10*, 1023–1033. doi:10.1128/EC.05112-11.

Svejstrup, J. Q. (2007). Elongator complex: How many roles does it play? *Current Opinion in Cell Biology, 19*, 331–336. doi:10.1016/j.ceb.2007.04.005.

Vanhamme, L., Poelvoorde, P., Pays, A., Tebabi, P., Van Xong, H., & Pays, E. (2000). Differential RNA elongation controls the variant surface glycoprotein gene expression sites of *Trypanosoma brucei*. *Molecular Microbiology, 36*, 328–340. doi:10.1046/j.1365-2958.2000.01844.x.

Walgraffe, D., Devaux, S., Lecordier, L., Dierick, J. F., Dieu, M., Van den Abbeele, J., et al. (2005). Characterization of subunits of the RNA polymerase I complex in *Trypanosoma brucei*. *Molecular and Biochemical Parasitology, 139*, 249–260. doi:10.1016/j.molbiopara.2004.11.014.

Wang, Q. P., Kawahara, T., & Horn, D. (2010). Histone deacetylases play distinct roles in telomeric VSG expression site silencing in African trypanosomes. *Molecular Microbiology,*. doi:10.1111/j.1365-2958.2010.07284.x.

Wastling, S. L., & Welburn, S. C. (2011). Diagnosis of human sleeping sickness: Sense and sensitivity. *Trends in Parasitology, 27*, 394–402. doi:10.1016/j.pt.2011.04.005.
White, T. C., Rudenko, G., & Borst, P. (1986). Three small RNAs within the 10 kb trypanosome rRNA transcription unit are analogous to domain VII of other eukaryotic 28S rRNAs. *Nucleic Acids Research, 14*, 9471–9489. doi:10.1093/nar/14.23.9471.
Xong, H. V., Vanhamme, L., Chamekh, M., Chimfwembe, C. E., Van Den Abbeele, J., Pays, A., et al. (1998). A VSG expression site-associated gene confers resistance to human serum in *Trypanosoma rhodesiense*. *Cell, 95*, 839–846. doi:10.1016/S0092-8674(00)81706-7.
Yang, X., Flgueiredo, L. M., Espinal, A., Okubo, E., & Li, B. (2009). RAP1 is essential for silencing telomeric variant surface glycoprotein genes in *Trypanosoma brucei*. *Cell, 137*, 99–109.

# Chapter 8
# Human and Primate Subtelomeres

**M. Katharine Rudd**

Subtelomeres are an unusual part of primate genomes, enriched in genes, repetitive DNA, structural polymorphisms, and chromosome rearrangements. As with subtelomeres of other orders, such genomic variation in primates can lead to genetic diversity, the birth of new genes, and an explosion of gene families. However, rearrangements in human subtelomeres can also alter developmentally critical genes, causing intellectual disability and birth defects. Analysis of subtelomeric breakpoints has revealed "hot spots" of chromosome breakage that may be initiated by specific types of repetitive DNA abundant in subtelomeres. In most cases, subtelomeric breaks are repaired by non-homologous end-joining and DNA replication processes, rather than homologous recombination. Comparative genomic studies of orthologous subtelomeres in closely related primates show even greater diversity between species, consistent with the rapid evolution of chromosome ends.

## 8.1 Primate Subtelomere Organization

Primate subtelomeres are enriched in repetitive elements, including segmental duplications (SDs), satellite DNA, tandem repeats, and degenerate telomere repeats (Riethman et al. 2004; Linardopoulou et al. 2005) (Fig. 8.1). Though this repetitive structure may be important for subtelomere biology and evolution, it has made assembling these parts of the genome a challenge. Although the human genome assembly is more "complete" than other primate genomes, in the most recent build (GRCh37/hg19), only 17 of 46 of chromosome ends have traversed

M. K. Rudd (✉)
Department of Human Genetics, Emory University School of Medicine,
Atlanta, GA 30322, USA
e-mail: katie.rudd@emory.edu

E. J. Louis and M. M. Becker (eds.), *Subtelomeres*, DOI: 10.1007/978-3-642-41566-1_8, 

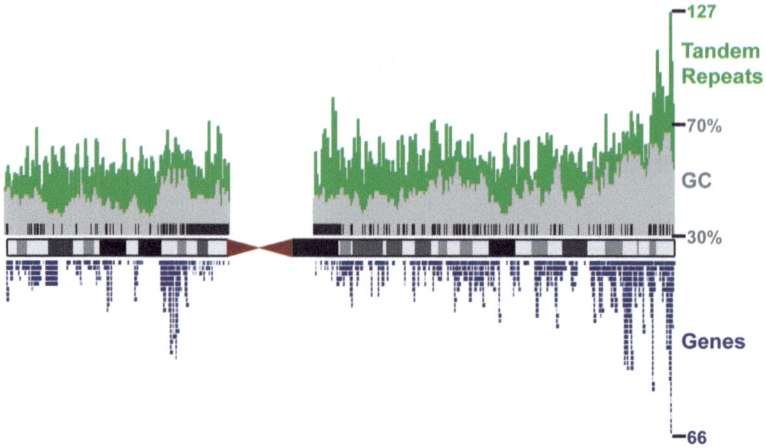


**Fig. 8.1** Repeat content and gene density of human chromosome 9. Tandem repeats (*green*), percent GC (*gray*), segmental duplications (*black*), and genes (*blue*) are shown along the length of chromosome 9 as represented in the UCSC Genome Browser (http://genome.ucsc.edu/)

subtelomeric sequences to reach the end of the chromosome, terminating in perfect telomeric repeats, (TTAGGG)n. Other primate genomes [chimpanzee (The Chimpanzee Sequencing and Analysis Consortium 2005), orangutan (Locke et al. 2011), and rhesus (Gibbs et al. 2007)] have been assembled using at least some comparisons to the human genome, so sequence gaps in the human reference genome as well as non-aligning regions between species are likely to remain as gaps in the assemblies of non-human primate genomes. In addition, subtelomeric sequences are incredibly polymorphic, and only a handful of subtelomeric alleles have been captured in the reference genome assembly (Trask et al. 1998; Linardopoulou et al. 2005). Thus, despite the successes assembling more and more primate genomes, the subtelomeric genome assemblies of human and non-human primates remain largely incomplete. Most subtelomeric genomic studies have focused on particular subtelomeres for the comparative analysis of primates. Here, we will describe human subtelomeric organization and discuss the limited non-human primate subtelomeric data for a subset of chromosome ends.

Human subtelomeres are made up of two major zones: a terminal region consisting of SDs and an adjacent region of chromosome-specific (non-duplicated) sequences (Fig. 8.2). SDs are operationally defined as DNA sequences 1 kb or larger that have another copy in the genome with ≥90 % identity. They make up more than 5 % of the human genome and are preferentially located at pericentromeres and subtelomeres (Bailey et al. 2002). In human subtelomeres, SDs occupy the terminal 5–300 kb of chromosomes. Each SD is shared between a subset of chromosome ends, and individual SDs range from 3 to 50 kb each (Linardopoulou et al. 2005). Copies of the same SD are 88–99.9 % identical and are occupy between 2 and 18 different chromosome ends, consistent with recent duplications that have rapidly spread to multiple chromosomes.

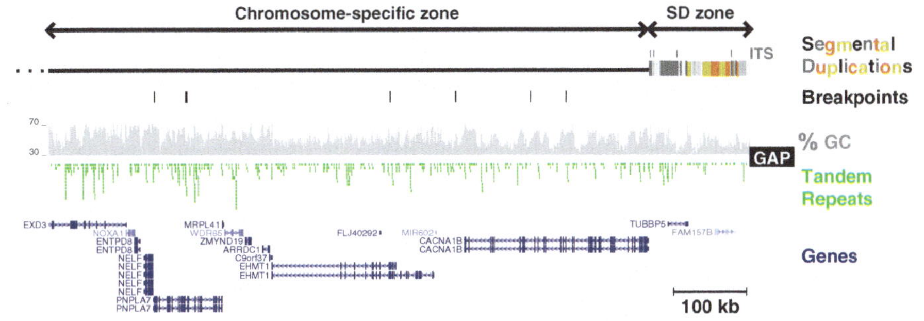


**Fig. 8.2** Human subtelomere organization. The terminal 1 Mb of chromosome 9q has a distal SD zone and an adjacent chromosome-specific zone. Segmental duplications (*orange*, *yellow*, and *gray*), percent GC (*gray*), assembly gaps (*black*), tandem repeats (*green*), and genes (*blue*) are shown as in the UCSC Genome Browser (http://genome.ucsc.edu/). Interstitial telomere sequences (ITS, *gray vertical lines*) were identified by RepeatMasker (http://www.repeatmasker.org). Breakpoints of subtelomeric rearrangements that cause intellectual disability (*black vertical lines*) were fine-mapped in (Luo et al. 2011)

Fluorescence in situ hybridization (FISH) analyses have shown that subtelomeric SDs are highly polymorphic, varying in copy number and chromosomal location from person to person (Trask et al. 1998; Linardopoulou et al. 2005). Given the number of subtelomeric SDs in the genome and the degree of polymorphism, it is likely that each human has a unique repertoire of subtelomeric SD sequence.

Many of the genes in subtelomeric SDs are part of gene families, such as odorant and cytokine receptors, tubulins, and transcription factors (Linardopoulou et al. 2005). The redundancy of duplicated subtelomeric genes may allow some copies to acquire new functions and some copies to mutate, while other copies retain their original function. Frequent interchromosomal exchanges can also juxtapose parts of different subtelomeric genes, potentially creating novel hybrid genes. The olfactory receptors (ORs) are a striking example of a gene family that expanded in primate subtelomeres. There are over 900 OR genes in the human genome, a subset of which are found at subtelomeric locations (Glusman et al. 2001). Some subtelomeric ORs are no longer functional and have become pseudogenes, whereas other ORs are transcribed in certain tissues, such as olfactory epithelium and testis (Linardopoulou et al. 2001).

Just proximal to subtelomeric SDs begins a region of chromosome-specific DNA (Fig. 8.2). Some deletions and duplications of this region have been detected in phenotypically normal individuals, suggesting that, like in the SD zones, some variation in the chromosome-specific regions is tolerated (Ballif et al. 2000; Ravnan et al. 2006; Redon et al. 2006; Balikova et al. 2007; Mills et al. 2011). Nevertheless, larger rearrangements of the chromosome-specific subtelomeric regions are associated with intellectual disability and birth defects (Ravnan et al. 2006; Ballif et al. 2007; Martin et al. 2007; Shao et al. 2008). Such rearrangements were originally identified by chromosome banding and FISH (National Institutes of Health and Institute of Molecular Medicine Collaboration 1996; Knight et al. 2000) and are now detected via genomic microarrays (Rudd 2011). Studies of clinically relevant copy number variation (CNV) have shown that subtelomeric rearrangements are overrepresented among CNVs that cause intellectual disability. For example, microarray analysis of 15,749 developmentally disabled individuals revealed that 16.3 % of pathogenic chromosome anomalies lie within the terminal 5 Mb of chromosome ends (Kaminsky et al. 2011), which accounts for only 7 % of the human genome. These chromosome rearrangements include deletions, duplications, and unbalanced translocations that are typically hundreds of kb to several Mb in size and include tens to hundreds of genes.

Loss, gain, and mutation of genes in the chromosome-specific zone of subtelomeres can cause a clinically recognized phenotype. Studies of patients with common phenotypic features and overlapping CNVs have pinpointed critical regions and genes associated with disease in a given subtelomere. The 9q subtelomeric deletion syndrome was first identified in patients with overlapping deletions, including the *EHMT1* gene, which is responsible for the phenotype, as *EHMT1* mutations cause a typical 9q deletion phenotype (Harada et al. 2004; Stewart et al. 2004; Kleefstra et al. 2006). Terminal deletions of chromosome 22q cause the 22q13 deletion syndrome, and mutations in the *SHANK3* gene in the critical region also cause those language disorders associated with the syndrome (Phelan et al. 2001; Wilson et al. 2003; Durand et al. 2007). Given the gene density at chromosome ends (Fig. 8.1), a host of candidate genes could be responsible for other "subtelomeric syndromes."

## 8.2 Subtelomeric Hot Spots and Rearrangement Mechanisms

Analysis of subtelomeric breakpoints has revealed recurrent sites of chromosome breakage. Given the enrichment of particular types of repeats in subtelomeres, such "hot spots" are likely related to DNA sequence and/or chromatin structure. Though not all types of repetitive DNA are linked to chromosome breakage, tandem repeats, trinucleotide repeats, satellite DNA, and G-rich sequences are known to underlie chromosomal fragility at other loci (Sutherland 2003; Bacolla et al. 2006; Zhao et al. 2010) and are strong candidates for DNA sequence-dependent causes of subtelomeric rearrangement. Uncovering how such sequences could form secondary structures that might interfere with cellular processes, including recombination and DNA replication, is crucial to untangling the molecular mechanisms that give rise to subtelomeric rearrangements.

One of the best examples of a subtelomeric hot spot lies in chromosome band 22q13.3. Rearrangements of this subtelomere have been independently identified in numerous studies, and fine-mapped breakpoints cluster between exons 8 and 9 of the *SHANK3* gene (Wong et al. 1997; Anderlid et al. 2002; Bonaglia et al. 2006, 2011; Durand et al. 2007; Philippe et al. 2008; Dhar et al. 2010; Luo et al. 2011; ). At least 13 published terminal deletion breakpoints lie in this 1.2-kb hot spot, which is made up of G-rich tandem repeats that are predicted to form G-quadruplexes. G-rich sequences that contain four tracts of at least three guanines separated by other bases can form G-quadruplexes by pairing between the four G-rich tracts (Huppert and Balasubramanian 2005; Burge et al. 2006). Such G-rich sequences can assemble highly stable G-quadruplexes in vitro (Neaves et al. 2009; Sanders 2010), and without specific helicases to unwind them, G-quadruplexes can cause chromosome breakage and genomic instability in vivo (Kruisselbrink et al. 2008; Ribeyre et al. 2009). Human subtelomeres are G-rich (Fig. 8.1), and there are many subtelomeric loci that contain the G-quadruplex consensus sequence, $G_{3-5}N_{1-7}G_{3-5}N_{1-7}G_{3-5}N_{1-7}G_{3-5}$ (Huppert and Balasubramanian 2005). Although functional studies of fragility at the 22q13.3 hot spot are still lacking, the recurrent breakpoints and predicted G-quadruplex motifs are suggestive of a region that is particularly susceptible to double-strand breaks (DSBs). It is likely that other subtelomeric rearrangement breakpoints are also caused by DSBs in G-rich sequences that assemble G-quadruplexes or other secondary structures.

Another indicator of elevated DSBs in subtelomeres comes from studies of sister chromatid exchange (SCE) in chromosome ends. The rate of SCE is significantly elevated in telomeres and subtelomeres, as demonstrated using a fluorescence method called chromosome orientation FISH (CO-FISH) (Cornforth and Eberle 2001; Londono-Vallejo et al. 2004; Rudd et al. 2007). Seventeen percent of all SCE occurs in the most terminal ~100 kb of chromosomes, translating to a 160-fold elevation of the rate of subtelomeric SCE compared with the rest of the genome (Rudd et al. 2007). More direct evidence of DSBs at chromosome ends comes from chromatin immunoprecipitation studies of the DSB-binding protein, γ-H2AX (d'Adda di Fagagna et al. 2003). In senescent primary cells, γ-H2AX is enriched 60 kb–1.5 Mb from the telomere, across different chromosome ends (Meier et al. 2007). These physical measurements of DSBs suggest that subtelomeres incur more breaks than other parts of the genome, consistent with the concentration of breakpoints in human subtelomeres.

DSBs in subtelomeres may be resolved via various DNA repair pathways, and analyses of breakpoint junctions in the chromosome-specific and SD zones provide insight into the rearrangement mechanisms that have shaped these regions. There are two major types of DNA repair, one that requires long stretches of sequence homology (homologous recombination) and one that does not (non-homologous repair). Comparing subtelomeric breakpoint junctions to the pre-rearrangement genomic state can distinguish the two types of DNA repair. A large-scale analysis of over 100 subtelomeric breakpoints in the chromosome-specific zone revealed that three of 21 sequenced breakpoint junctions were the product of homologous recombination between interspersed repeats, including LINE and *Alu* elements. The remaining 18 rearrangements did not involve significant sequence homology at the junctions and were formed via non-homologous end-joining (NHEJ) and DNA replication processes (Luo et al. 2011). Other studies of subtelomeric breakpoint junctions in chromosome-specific zones have also found a preponderance of NHEJ versus homologous recombination (Ballif et al. 2003, 2004; Gajecka et al. 2006, 2008; Bonaglia et al. 2006; Yatsenko et al. 2009).

A similar trend regarding homologous and non-homologous repair is evident in subtelomeric junctions in the SD zone. This part of the genome is organized as a patchwork of SDs shared between a subset of chromosome ends; however, subtelomeric SDs are not organized in a random manner. Instead, subtelomeric SDs are almost always in the same orientation and relative order, suggesting translocation, rather than transposition, as the mechanism of sequence transfer (Linardopoulou et al. 2005). The alignment of paralogous SDs in human subtelomeres highlights the interchromosomal sequence transfers responsible for the highly polymorphic organization of subtelomeric SDs. Forty-nine out of 53 SD breakpoint junctions are the product of NHEJ, while only four are mediated by homologous recombination (Linardopoulou et al. 2005). Thus, non-homologous DNA repair is the predominant mechanism underlying subtelomeric breakpoints in the chromosome-specific and SD zones.

## 8.3 Subtelomere Evolution

Investigations into the subtelomeric differences *between* species have also given us insight into the DNA breakage and repair processes involved in this rapidly evolving part of the genome. Comparative genomic analyses of the great apes have shown that although most orthologous sequences are highly conserved, chromosome ends are far more diverse. Since most primate subtelomeres are not sequenced, comparative studies have relied on a combination of FISH, PCR, chromosome flow-sorting, and BAC sequencing to generate syntenic maps of these regions (Monfouilloux et al. 1998; Trask et al. 1998; Martin et al. 2002; Fan et al. 2002; Ventura et al. 2003, 2011; Linardopoulou et al. 2005; Rudd et al. 2009). Detailed analyses of several chromosome ends have found that subtelomeric sequences vary dramatically in copy number and genomic location between closely related species; however, the reticulate nature of subtelomeric DNA exchanges complicates the interpretation of the DNA sequence transfers that have shaped modern-day primate chromosome ends. Chromosome fissions and fusions that give rise to the birth and death, respectively, of subtelomeres are ideal for teasing apart the steps involved in subtelomere evolution. Fissions and fusions punctuate subtelomeric events, making it possible to track a given subtelomere before and after a major chromosomal change.

Human chromosome 2, for example, is the product of a head-to-head fusion of two ancestral chromosomes that remained separate in the other great apes (Yunis and Prakash 1982; Ijdo et al. 1991; Fan et al. 2002). The fused chromosome 2 inactivated one centromere and two telomeres, and the human 2q13–2q14.1

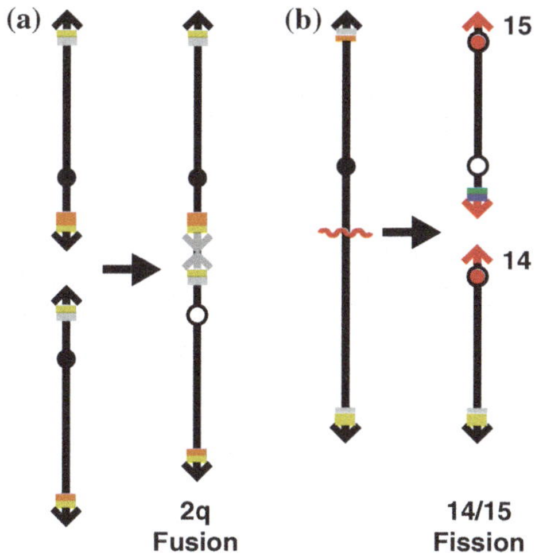


**Fig. 8.3** Chromosome fission and fusion in primates. Centromeres are represented as *circles*, telomeres are represented as *arrowheads*, and segmental duplications are represented as *colored rectangles*. **a** The chromosome fusion that gave rise to human chromosome 2 resulted in inactivation of one centromere (*open circle*) and the fusion of two telomeres (*gray*). **b** The chromosome fission (*red squiggly line*) in the ancestor of the great apes resulted in the birth of three new telomeres and two new centromeres (*red*) and the inactivation of one centromere (*open circle*). New segmental duplications (*green* and *blue*) were transferred to the 15q subtelomere post-fission

fusion site is marked by inverted telomere repeats and subtelomeric SDs that once resided at two independent chromosome ends (Fig. 8.3). These SDs are paralogous to several human subtelomeres, including 9p and 22q, consistent with multiple interchromosomal exchanges (Fan et al. 2002; Linardopoulou et al. 2005). The inverted telomere repeats at the fusion site are not perfect telomere arrays, but rather are 14 % diverged from the canonical telomere repeat, (TTAGGG)n. This could be due to the rapid divergence of perfect telomere repeats post-fusion, or it could indicate that the chromosomal fusion occurred at degenerate telomere repeats in the subtelomeres of the ancestral chromosomes, rather than as a fusion of the most terminal telomere sequences (Fan et al. 2002).

A chromosomal fission in the ancestor of great apes gave rise to human chromosomes 14 and 15 (Fig. 8.3). Rhesus macaque chromosome 7 represents the ancestral locus, in which regions orthologous to human chromosomes 15 and 14 are arranged in a head-to-tail configuration. After the fission of the ancestral chromosome, one new pericentromere (on chromosome 14) and one new subtelomere (on chromosome 15) were created at the fission site (Wienberg et al. 1992; Ventura et al. 2003; Rudd et al. 2009). In addition, the ancestral centromere inactivated, two new centromeres activated, and both chromosomes 14 and 15 acquired acrocentric short arms with new telomeres (Fig. 8.3). Since its birth at the chromosome fission, the 15q subtelomere has engaged in rampant sequence transfers. The orthologous regions of the 15q subtelomere in four great apes and an Old World monkey exist as completely different genomic structures in each species (Rudd et al. 2009). Terminal deletions, interstitial deletions, duplications, and interchromosomal exchanges have created a unique subtelomeric configuration in the genomes of rhesus macaque, orangutan, gorilla, chimpanzee, and human. The fission site was home to at least 21 olfactory receptor (OR) genes in the ancestral chromosome, and since the fission, ORs have been gained and lost in a lineage-specific manner in the genomes of all the great apes (Rudd et al. 2009).

Like human subtelomeres, non-human primate subtelomeres are also enriched in satellite DNA and SDs. However, different classes of repetitive DNA have expanded in different species, typical of concerted evolutionary processes. Heterochromatic "caps" of chromosome ends have been seen in chimpanzee and gorilla, but not in human (Yunis and Prakash 1982; Royle et al. 1994). Recent sequence analyses of chimpanzee and gorilla subtelomeres have revealed that both species have a 32-bp satellite at chromosome ends, but SDs that make up the chimpanzee caps are derived from the chromosome 2 fusion site, whereas the gorilla subtelomeric SDs are derived from a chromosome 10 sequence (Ventura et al. 2011).

Analysis of the SDs in the human genome assembly also provides information on the evolutionary timing of primate subtelomeres. Fifty percent of human subtelomeric SD sequence is >98.7 % identical to another chromosome end, indicating that the sequence transfer occurred since human and chimpanzee diverged (Linardopoulou et al. 2005). Further, FISH analysis of a subset of human subtelomeric SDs revealed variation in copy number and genomic location *between* individuals and heterozygosity for subtelomeric SDs *within* a single individual (Trask et al. 1998; Linardopoulou et al. 2005). Such data are consistent with subtelomeric SDs being one of the most rapidly evolving regions of the human genome.

Rearrangements in primate subtelomeres are a source of variation and disease. Although small rearrangements represent normal polymorphism, larger gains and losses involving dosage-sensitive genes can cause intellectual disabilities and birth defects, making these regions particularly relevant to studies of human disease and diversity. Though subtelomeric variation is recognized in the human genome, the causes of DSBs in chromosome ends are unknown. Functional studies of the DNA sequences underlying subtelomeric breakpoints are a crucial next step to discovering the risk factors and mechanisms of subtelomeric rearrangements.

**Acknowledgments** MKR is supported by a grant from the National Institute of Mental Health (1R01MH092902) and a grant from the March of Dimes (12-FY11-203).

## References

Anderlid, B. M., Schoumans, J., Anneren, G., Tapia-Paez, I., Dumanski, J., Blennow, E., et al. (2002). FISH-mapping of a 100-kb terminal 22q13 deletion. *Human Genetics, 110*(5), 439–443.

Bacolla, A., Wojciechowska, M., Kosmider, B., Larson, J. E., & Wells, R. D. (2006). The involvement of non-B DNA structures in gross chromosomal rearrangements. *DNA Repair (Amsterdam), 5*(9–10), 1161–1170.

Bailey, J. A., Gu, Z., Clark, R. A., Reinert, K., Samonte, R. V., Schwartz, S., et al. (2002). Recent segmental duplications in the human genome. *Science, 297*(5583), 1003–1007.

Balikova, I., Menten, B., de Ravel, T., Le Caignec, C., Thienpont, B., Urbina, M., et al. (2007). Subtelomeric imbalances in phenotypically normal individuals. *Human Mutation, 28*(10), 958–967.

Ballif, B. C., Kashork, C. D., & Shaffer, L. G. (2000). The promise and pitfalls of telomere region-specific probes. *American Journal of Human Genetics, 67*(5), 1356–1359.

Ballif, B. C., Sulpizio, S. G., Lloyd, R. M., Minier, S. L., Theisen, A., Bejjani, B. A., et al. (2007). The clinical utility of enhanced subtelomeric coverage in array CGH. *American Journal of Medical Genetics Part A, 143*(16), 1850–1857.

Ballif, B. C., Wakui, K., Gajecka, M., & Shaffer, L. G. (2004). Translocation breakpoint mapping and sequence analysis in three monosomy 1p36 subjects with der(1)t(1;1)(p36;q44) suggest mechanisms for telomere capture in stabilizing de novo terminal rearrangements. *Human Genetics, 114*(2), 198–206.

Ballif, B. C., Yu, W., Shaw, C. A., Kashork, C. D., & Shaffer, L. G. (2003). Monosomy 1p36 breakpoint junctions suggest pre-meiotic breakage-fusion-bridge cycles are involved in generating terminal deletions. *Human Molecular Genetics, 12*(17), 2153–2165.

Bonaglia, M. C., Giorda, R., Beri, S., De Agostini, C., Novara, F., Fichera, M., et al. (2011). Molecular mechanisms generating and stabilizing terminal 22q13 deletions in 44 subjects with Phelan/McDermid Syndrome. *PLoS Genetics, 7*(7), e1002173.

Bonaglia, M. C., Giorda, R., Mani, E., Aceti, G., Anderlid, B. M., Baroncini, A., et al. (2006). Identification of a recurrent breakpoint within the SHANK3 gene in the 22q13.3 deletion syndrome. *Journal of Medical Genetics, 43*(10), 822–828.

Burge, S., Parkinson, G. N., Hazel, P., Todd, A. K., & Neidle, S. (2006). Quadruplex DNA: Sequence, topology and structure. *Nucleic Acids Research, 34*(19), 5402–5415.

National Institutes of Health and Institute of Molecular Medicine Collaboration. (1996). A complete set of human telomeric probes and their clinical application. *Natural Genetics, 14*(1), 86–89.

The Chimpanzee Sequencing and Analysis Consortium. (2005). Initial sequence of the chimpanzee genome and comparison with the human genome. *Nature, 437*(7055), 69–87.

Cornforth, M. N., & Eberle, R. L. (2001). Termini of human chromosomes display elevated rates of mitotic recombination. *Mutagenesis, 16*(1), 85–89.

d'Adda di Fagagna, F., Reaper, P. M., Clay-Farrace, L., Fiegler, H., Carr, P., Von Zglinicki, T., et al. (2003). A DNA damage checkpoint response in telomere-initiated senescence. *Nature, 426*(6963), 194–198.

Dhar, S. U., del Gaudio, D., German, J. R., Peters, S. U., Ou, Z., Bader, P. I., et al. (2010). 22q13.3 deletion syndrome: Clinical and molecular analysis using array CGH. *American Journal of Medical Genetics Part A, 152A*(3), 573–581.

Durand, C. M., Betancur, C., Boeckers, T. M., Bockmann, J., Chaste, P., Fauchereau, F., et al. (2007). Mutations in the gene encoding the synaptic scaffolding protein SHANK3 are associated with autism spectrum disorders. *Nature Genetics, 39*(1), 25–27.

Fan, Y., Linardopoulou, E., Friedman, C., Williams, E., & Trask, B. J. (2002). Genomic structure and evolution of the ancestral chromosome fusion site in 2q13–2q14.1 and paralogous regions on other human chromosomes. *Genome Research, 12*(11), 1651–1662.

Gajecka, M., Gentles, A. J., Tsai, A., Chitayat, D., Mackay, K. L., Glotzbach, C. D., et al. (2008). Unexpected complexity at breakpoint junctions in phenotypically normal individuals and mechanisms involved in generating balanced translocations t(1;22)(p36;q13). *Genome Research, 18*(11), 1733–1742.

Gajecka, M., Glotzbach, C. D., Jarmuz, M., Ballif, B. C., & Shaffer, L. G. (2006). Identification of cryptic imbalance in phenotypically normal and abnormal translocation carriers. *European Journal of Human Genetics, 14*(12), 1255–1262.

Gibbs, R. A., Rogers, J., Katze, M. G., Bumgarner, R., Weinstock, G. M., Mardis, E. R., et al. (2007). Evolutionary and biomedical insights from the rhesus macaque genome. *Science, 316*(5822), 222–234.

Glusman, G., Yanai, I., Rubin, I., & Lancet, D. (2001). The complete human olfactory subgenome. *Genome Research, 11*(5), 685–702.
Harada, N., Visser, R., Dawson, A., Fukamachi, M., Iwakoshi, M., Okamoto, N., et al. (2004). A 1-Mb critical region in six patients with 9q34.3 terminal deletion syndrome. *Journal of Human Genetics, 49*(8), 440–444.
Huppert, J. L., & Balasubramanian, S. (2005). Prevalence of quadruplexes in the human genome. *Nucleic Acids Research, 33*(9), 2908–2916.
Ijdo, J. W., Baldini, A., Ward, D. C., Reeders, S. T., & Wells, R. A. (1991). Origin of human chromosome 2: An ancestral telomere-telomere fusion. *Proceedings of the National Academy of Sciences of the United States of America, 88*(20), 9051–9055.
Kaminsky, E. B., Kaul, V., Paschall, J., Church, D. M., Bunke, B., Kunig, D., et al. (2011). An evidence-based approach to establish the functional and clinical significance of copy number variants in intellectual and developmental disabilities. *Genetics in Medicine: Official Journal of the American College of Medical Genetics, 13*(9), 777–784.
Kleefstra, T., Brunner, H. G., Amiel, J., Oudakker, A. R., Nillesen, W. M., Magee, A., et al. (2006). Loss-of-function mutations in euchromatin histone methyl transferase 1 (EHMT1) cause the 9q34 subtelomeric deletion syndrome. *American Journal of Human Genetics, 79*(2), 370–377.
Knight, S. J., Lese, C. M., Precht, K. S., Kuc, J., Ning, Y., Lucas, S., et al. (2000). An optimized set of human telomere clones for studying telomere integrity and architecture. *American Journal of Human Genetics, 67*(2), 320–332.
Kruisselbrink, E., Guryev, V., Brouwer, K., Pontier, D. B., Cuppen, E., & Tijsterman, M. (2008). Mutagenic capacity of endogenous G4 DNA underlies genome instability in FANCJ-defective *C. elegans*. *Current Biology, 18*(12), 900–905.
Linardopoulou, E., Mefford, H. C., Nguyen, O., Friedman, C., van den Engh, G., Farwell, D. G., et al. (2001). Transcriptional activity of multiple copies of a subtelomerically located olfactory receptor gene that is polymorphic in number and location. *Human Molecular Genetics, 10*(21), 2373–2383.
Linardopoulou, E. V., Williams, E. M., Fan, Y., Friedman, C., Young, J. M., & Trask, B. J. (2005). Human subtelomeres are hot spots of interchromosomal recombination and segmental duplication. *Nature, 437*(7055), 94–100.
Locke, D. P., Hillier, L. W., Warren, W. C., Worley, K. C., Nazareth, L. V., Muzny, D. M., et al. (2011). Comparative and demographic analysis of orang-utan genomes. *Nature, 469*(7331), 529–533.
Londono-Vallejo, J. A., Der-Sarkissian, H., Cazes, L., Bacchetti, S., & Reddel, R. R. (2004). Alternative lengthening of telomeres is characterized by high rates of telomeric exchange. *Cancer Research, 64*(7), 2324–2327.
Luo, Y., Hermetz, K. E., Jackson, J. M., Mulle, J. G., Dodd, A., Tsuchiya, K. D., et al. (2011). Diverse mutational mechanisms cause pathogenic subtelomeric rearrangements. *Human Molecular Genetics, 20*(19), 3769–3778.
Martin, C. L., Nawaz, Z., Baldwin, E. L., Wallace, E. J., Justice, A. N., & Ledbetter, D. H. (2007). The evolution of molecular ruler analysis for characterizing telomere imbalances: From fluorescence in situ hybridization to array comparative genomic hybridization. *Genetic Medicine, 9*(9), 566–573.
Martin, C. L., Wong, A., Gross, A., Chung, J., Fantes, J. A., & Ledbetter, D. H. (2002). The evolutionary origin of human subtelomeric homologies–or where the ends begin. *American Journal of Human Genetics, 70*(4), 972–984.
Meier, A., Fiegler, H., Munoz, P., Ellis, P., Rigler, D., Langford, C., et al. (2007). Spreading of mammalian DNA-damage response factors studied by ChIP-chip at damaged telomeres. *EMBO Journal, 26*(11), 2707–2718.
Mills, R. E., Walter, K., Stewart, C., Handsaker, R. E., Chen, K., Alkan, C., et al. (2011). Mapping copy number variation by population-scale genome sequencing. *Nature, 470*(7332), 59–65.
Monfouilloux, S., Avet-Loiseau, H., Amarger, V., Balazs, I., Pourcel, C., & Vergnaud, G. (1998). Recent human-specific spreading of a subtelomeric domain. *Genomics, 51*(2), 165–176.
Neaves, K. J., Huppert, J. L., Henderson, R. M., & Edwardson, J. M. (2009). Direct visualization of G-quadruplexes in DNA using atomic force microscopy. *Nucleic Acids Research, 37*(18), 6269–6275.

Phelan, M. C., Rogers, R. C., Saul, R. A., Stapleton, G. A., Sweet, K., McDermid, H., et al. (2001). 22q13 deletion syndrome. *American Journal of Medical Genetics, 101*(2), 91–99.
Philippe, A., Boddaert, N., Vaivre-Douret, L., Robel, L., Danon-Boileau, L., Malan, V., et al. (2008). Neurobehavioral profile and brain imaging study of the 22q13.3 deletion syndrome in childhood. *Pediatrics, 122*(2), e376–e382.
Ravnan, J. B., Tepperberg, J. H., Papenhausen, P., Lamb, A. N., Hedrick, J., Eash, D., et al. (2006). Subtelomere FISH analysis of 11 688 cases: An evaluation of the frequency and pattern of subtelomere rearrangements in individuals with developmental disabilities. *Journal of Medical Genetics, 43*(6), 478–489.
Redon, R., Ishikawa, S., Fitch, K. R., Feuk, L., Perry, G. H., Andrews, T. D., et al. (2006). Global variation in copy number in the human genome. *Nature, 444*(7118), 444–454.
Ribeyre, C., Lopes, J., Boule, J. B., Piazza, A., Guedin, A., Zakian, V. A., et al. (2009). The yeast Pif1 helicase prevents genomic instability caused by G-quadruplex-forming CEB1 sequences in vivo. *PLoS Genetics, 5*(5), e1000475.
Riethman, H., Ambrosini, A., Castaneda, C., Finklestein, J., Hu, X. L., Mudunuri, U., et al. (2004). Mapping and initial analysis of human subtelomeric sequence assemblies. *Genome Research, 14*(1), 18–28.
Royle, N. J., Baird, D. M., & Jeffreys, A. J. (1994). A subterminal satellite located adjacent to telomeres in chimpanzees is absent from the human genome. *Nature Genetics, 6*(1), 52–56.
Rudd, M. K. (2011). Structural variation in subtelomeres. In L. Feuk (Ed.), *Genomic structural variants: Methods and protocols, methods in molecular biology* (Vol. 838). New York: Springer.
Rudd, M. K., Endicott, R. M., Friedman, C., Walker, M., Young, J. M., Osoegawa, K., et al. (2009). Comparative sequence analysis of primate subtelomeres originating from a chromosome fission event. *Genome Research, 19*(1), 33–41.
Rudd, M. K., Friedman, C., Parghi, S. S., Linardopoulou, E. V., Hsu, L., & Trask, B. J. (2007). Elevated rates of sister chromatid exchange at chromosome ends. *PLoS Genetics, 3*(2), e32.
Sanders, C. M. (2010). Human Pif1 helicase is a G-quadruplex DNA-binding protein with G-quadruplex DNA-unwinding activity. *The Biochemical Journal, 430*(1), 119–128.
Shao, L., Shaw, C. A., Lu, X. Y., Sahoo, T., Bacino, C. A., Lalani, S. R., et al. (2008). Identification of chromosome abnormalities in subtelomeric regions by microarray analysis: A study of 5,380 cases. *American Journal of Medical Genetics Part A, 146A*(17), 2242–2251.
Stewart, D. R., Huang, A., Faravelli, F., Anderlid, B. M., Medne, L., Ciprero, K., et al. (2004). Subtelomeric deletions of chromosome 9q: A novel microdeletion syndrome. *American Journal of Medical Genetics Part A, 128A*(4), 340–351.
Sutherland, G. R. (2003). Rare fragile sites. *Cytogenet Genome Research 100*(1–4), 77–84.
Trask, B. J., Friedman, C., Martin-Gallardo, A., Rowen, L., Akinbami, C., Blankenship, J., et al. (1998). Members of the olfactory receptor gene family are contained in large blocks of DNA duplicated polymorphically near the ends of human chromosomes. *Human Molecular Genetics, 7*(1), 13–26.
Ventura, M., Catacchio, C. R., Alkan, C., Marques-Bonet, T., Sajjadian, S., Graves, T. A., et al. (2011). Gorilla genome structural variation reveals evolutionary parallelisms with chimpanzee. *Genome Research, 21*(10), 1640–1649.
Ventura, M., Mudge, J. M., Palumbo, V., Burn, S., Blennow, E., Pierluigi, M., et al. (2003). Neocentromeres in 15q24-26 map to duplicons which flanked an ancestral centromere in 15q25. *Genome Research, 13*(9), 2059–2068.
Wienberg, J., Stanyon, R., Jauch, A., & Cremer, T. (1992). Homologies in human and macaca fuscata chromosomes revealed by in situ suppression hybridization with human chromosome specific DNA libraries. *Chromosoma, 101*(5–6), 265–270.
Wilson, H. L., Wong, A. C., Shaw, S. R., Tse, W. Y., Stapleton, G. A., Phelan, M. C., et al. (2003). Molecular characterisation of the 22q13 deletion syndrome supports the role of haploinsufficiency of SHANK3/PROSAP2 in the major neurological symptoms. *Journal of Medical Genetics, 40*(8), 575–584.
Wong, A. C., Ning, Y., Flint, J., Clark, K., Dumanski, J. P., Ledbetter, D. H., et al. (1997). Molecular characterization of a 130-kb terminal microdeletion at 22q in a child with mild mental retardation. *American Journal of Human Genetics, 60*(1), 113–120.

Yatsenko, S. A., Brundage, E. K., Roney, E. K., Cheung, S. W., Chinault, A. C., & Lupski, J. R. (2009). Molecular mechanisms for subtelomeric rearrangements associated with the 9q34.3 microdeletion syndrome. *Human Molecular Genetics, 18*(11), 1924–1936.

Yunis, J. J., & Prakash, O. (1982). The origin of man: A chromosomal pictorial legacy. *Science, 215*(4539), 1525–1530.

Zhao, J., Bacolla, A., Wang, G., & Vasquez, K. M. (2010). Non-B DNA structure-induced genetic instability and evolution. *Cellular and Molecular Life Sciences, 67*(1), 43–62.

# Chapter 9
# FSHD: A Subtelomere-Associated Disease

**Andreas Leidenroth and Jane E. Hewitt**

**Abstract** Facioscapulohumeral muscular dystrophy (FSHD) is an autosomal dominant genetic disorder caused by an unusual genetic mutation: the contraction of a macrosatellite repeat array on the chromosome 4 subtelomere. Due to the unusual location of this mutation, FSHD research has provided a wealth of data about the evolutionary history of this human telomere. In this chapter, we will cover both the early and the most recent disease models that have been proposed to explain the molecular pathogenesis of this disorder and highlight some of the most interesting genetic, epigenetic and evolutionary findings contributed by this field.

## 9.1 Introduction

Facioscapulohumeral muscular dystrophy (FSHD) is usually diagnosed in the later teenage years, with affected males tending to show earlier onset and faster decline than females (Padberg 2004). The first recognised symptom is most often a weakness of the shoulder girdle muscles, the scapula fixators. While in about a third of cases the disease does not progress beyond this stage, muscle wasting in most individuals subsequently extends to include muscles of the face, the foot extensor and the pelvic girdle with about 10 % of affected individuals requiring wheelchair aid in later life (Padberg 2004). FSHD patients who show symptoms in the first decade are more likely to become wheelchair dependent and suffer from additional non-muscular symptoms such as retinal vascular disease and high-frequency hearing loss, features less frequently seen in those with later onset of the disease (Padberg 2004; Padberg et al. 1995).

A. Leidenroth · J. E. Hewitt (✉)
Centre for Genetics and Genomics, Nottingham, UK
e-mail: jane.hewitt@nottingham.ac.uk

A. Leidenroth
e-mail: andreas.leidenroth@gmail.com

E. J. Louis and M. M. Becker (eds.), *Subtelomeres*, DOI: 10.1007/978-3-642-41566-1_9, 

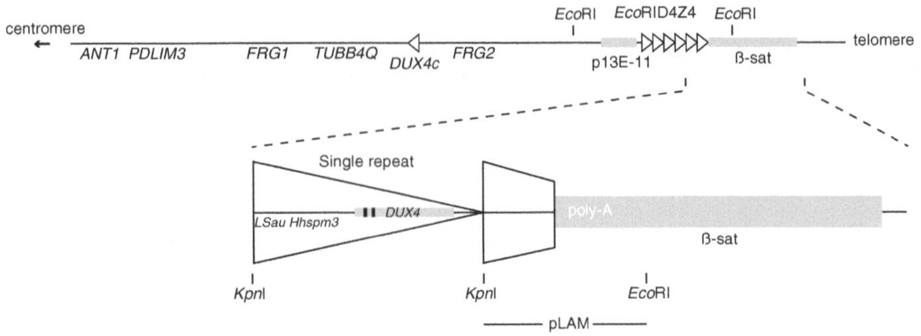


**Fig. 9.1** Genomic organisation of the human chromosome 4q35ter region and close-up of the distal end of the D4Z4 array. The distance from *ANT1* to the telomere is approximately 5 Mb. *Arrows* show orientation of *DUX4* ORF. *DUX4* homeoboxes are shaded *black*. The pLAM fragment includes the most distal, partial D4Z4 repeat and extends into the β-satellite array. Figure not to scale

In the early 1990s, classic linkage studies mapped the FSHD locus to chromosome 4q35-ter (Sarfarazi et al. 1989, 1992; Upadhyaya et al. 1992; Weiffenbach et al. 1992). When genomic DNA from affected families was digested with *Eco*RI and hybridised on a Southern blot with a probe (p13E-11) that had been sub-cloned from a cosmid (13E) mapping to this region, FSHD samples showed a notably smaller band than healthy controls. The short *Eco*RI fragment was found to co-segregate with the disease in families and was observed to occur *de novo* in sporadic cases (Wijmenga et al. 1992). Further analyses showed that the polymorphic fragment size was determined by the copy number of a subtelomeric 3.3-kb macrosatellite repeat array (Wijmenga et al. 1992; van Deutekom et al. 1993) composed of one–ten repeats (10–48-kb *Eco*RI fragments) in patients and 11 to more than 100 repeats (fragments larger than 48 kb) in controls (van Deutekom et al. 1993; Wijmenga et al. 1994). There is one *Kpn*I site within each repeat unit (Fig. 9.1). The macrosatellite, named D4Z4, was found to contain an open reading frame within each *Kpn*I repeat that encodes a putative protein with two homeodomains, since named *DUX4* (Hewitt et al. 1994; Lee et al. 1995; Gabriels et al. 1999).

The p13E-11 probe was found also to hybridise to a locus on chromosome 10q26-ter (Wijmenga et al. 1994; Bakker et al. 1995). This 10q26 subtelomeric region shares homology with 4q35, which extends from a truncated, inverted D4Z4 unit proximal to D4Z4 to the distal end of the chromosome (Fig. 9.1) (Hewitt et al. 1994; van Geel et al. 2002). However, even though the 3.3-kb repeat arrays and the flanking regions of these two chromosomes share high sequence and organisational similarity (van Geel et al. 2002; Cacurri et al. 1998), FSHD was found to be associated only with contractions on chromosome 4qter (Lemmers et al. 2001). Because the distance from the D4Z4 array to the telomeric TTAGGG repeats was estimated to be only 25–50 kb (van Geel et al. 2002; Lemmers et al. 2002; Bengtsson et al. 1994), most disease models proposed epigenetic heterochromatic effects related to this sub-telomeric location.

After these initial data were published in the 1990s, it would take a further two decades of research until convincing evidence of a direct link between *DUX4* and FSHD pathogenesis was established.

## 9.2 Historic Context of FSHD Research

Before we move on to discuss the most interesting and recent findings about the 4q35 subtelomeric region that have emerged from this substantial body of work, we will briefly cover the early disease models that were initially proposed to explain the molecular pathogenesis.

The first D4Z4 sequence data came either from truncated patient arrays or from cloned repeats that were susceptible to rearrangements, and the extent of nucleotide conservation between different repeat units along the array was unknown (Hewitt et al. 1994; Lee et al. 1995; Gabriels et al. 1999; Winokur et al. 1994). Despite extensive efforts, several independent groups failed to detect any convincing evidence of *DUX4* transcription (Hewitt et al. 1994; Winokur et al. 2003; Lyle et al. 1995; Yip and Picketts 2003; Osborne et al. 2007). For many years, the *DUX4* open reading frame was thus considered a pseudogene that was not expressed. Instead, early work focused on the region surrounding D4Z4. There are no known genes located between D4Z4 and the telomere, while the region proximal to the array is relatively gene poor (van Geel et al. 2002). The gene sequences closest to D4Z4 are *FRG2*, followed by *DUX4c*, *TUBB4Q*, *FRG1*, *PDLIM3* and *ANT1* (Fig. 9.1) (van Koningsbruggen et al. 2004; van der Maarel et al. 2007). Several of the early disease models hypothesised a D4Z4 contraction-dependent aberrant regulation of these genes in FSHD.

Prompted by the extreme telomeric location and the presence of the *LSau* and *Hhspm3* repeats (known to be associated with heterochromatin), one of these ideas was a *cis*-spreading model based on a loss of heterochromatinisation upon D4Z4 contraction. It was suggested that in its normal state, chromatin at D4Z4 is packaged into heterochromatin and transcriptionally inactive, with this silencing effect spreading to proximal neighbouring genes, a phenomenon that was known as position effect variegation (PEV) in *D. melanogaster.* Array contraction was hypothesised to result in a loss of local heterochromatinisation and therefore aberrant up-regulation of the proximal genes, culminating in FSHD. However, a later study (albeit limited to somatic cell hybrids) showed that chromatin at D4Z4 is of a euchromatic nature and that there is no position-dependent increase in neighbouring gene transcription or a histone 4 acetylation gradient (Jiang et al. 2003), features that would both be expected under a loss-of-PEV model.

Other, similar models which also proposed *cis* or *trans* effects of D4Z4 contraction include DNA looping (Jiang et al. 2003), D4Z4 functioning as a chromatin 'insulator' preventing spread of heterochromatinisation (van Deutekom 1996) or interactions of the D4Z4 region with the nuclear rim (Masny et al. 2004; Tam et al. 2004). The last of these three ideas is linked to the observation that the chromosome

4q and 10q telomeres localise to different parts of the nucleus. Using a 3D-by-2D in situ hybridisation method, Masny et al. showed that 4q35ter always localises to the outermost region of the nucleus. This localisation is dependent on the nuclear rim protein lamin A/C, loss of which randomises the 4q35 nuclear position. 10q26ter does localise to the nuclear rim (Masny et al. 2004), which implies that it is not the D4Z4 array itself, but sequences proximal to D4Z4 that seem to be responsible for this effect and the difference between 4q and 10q. In line with this, local D4Z4 deletions do not seem to alter the 4q35 subtelomere localisation (Tam et al. 2004).

In the context of these models, the expression levels of genes proximal to D4Z4 were investigated in FSHD and control muscle. After an initial study that found no changes in *FRG1* expression using allele-specific RT-PCR (van Deutekom et al. 1996b), it was reported that transcription levels of the D4Z4 proximal genes are progressively elevated in FSHD muscle cells, with the highest gain detected in *FRG2* (closest to D4Z4) and a lower but notable gain in *FRG1* and *ANT1* (further away) (Gabellini et al. 2002). This study also identified a 27-bp recognition sequence (termed D4Z4-binding element, DBE) within D4Z4 that is bound by a D4Z4 repressor complex (DRC), consisting of the proteins YY1 (Ying Yang 1), HMGB2 (a high-mobility group protein family member) and nucleolin. The authors postulated a model of de-repression in which binding of the DRC normally functions to repress genes upstream of D4Z4, with repeat contraction leading to reduced repressor binding and consequently aberrant up-regulation (Gabellini et al. 2002). Controversy arose when these changes in mRNA levels, which were based on non-quantitative endpoint RT-PCR, could not be replicated by later studies that applied more quantitative methods (Winokur et al. 2003; Osborne et al. 2007; Jiang et al. 2003; Klooster et al. 2009).

A further inconsistency with this *cis*-acting model is that in some patients the D4Z4 deletions extend proximally to include p13E-11, *FRG2* and *DUX4c*. As such deletions can still result in FSHD, these loci are considered unlikely to be directly responsible for pathogenesis in a *cis*-acting over-expression/gain-of-function model (Lemmers et al. 1998, 2003). In agreement with this, *FRG2* over-expression in mice does not seem to cause a myopathy phenotype (Gabellini et al. 2006).

*TUBB4Q*, identified by in silico analysis, likely represents an unexpressed pseudogene encoding a putative protein with 86 % identity to β2-tubulin (Fig. 9.1) (van Geel et al. 2000). The actinin-associated LIM protein PDLIM3 is expressed in muscle, but expression levels were found not to vary between healthy individuals and patients, which effectively excluded this gene as a candidate (Xia et al. 1997; Bouju et al. 1999). *ANT1* encodes an adenine nucleotide translocator responsible for ATP transport across the mitochondrial membrane into the cytoplasm and is highly expressed in muscle tissue (Li et al. 1989). Even though it is located several megabases upstream of D4Z4, it has therefore been considered a candidate for the muscular dystrophy phenotype. There is some evidence that ANT1 protein is elevated in FSHD muscle compared with controls and individuals with Duchenne muscular dystrophy (Laoudj-Chenivesse et al. 2005), but transgenic mice over-expressing ANT1 do not suffer muscle damage and the reported elevated levels of *ANT1* (Gabellini et al. 2006) probably reflect correlation rather than causation.

Among the *cis* candidate genes, most effort has been focused on *FRG1*, which at one point was the prime FSHD candidate gene (Gabellini et al. 2002). *FRG1* (FSHD-related gene 1) maps 100 kb proximal to D4Z4 (van Deutekom et al. 1996b). Its sequence is conserved across the animal kingdom (Grewal et al. 1998) and encodes a 30-kDa protein that localises to the nucleolus, splicing speckles and cajal bodies, hinting at a possible function in RNA splicing (van Koningsbruggen et al. 2004). Aberrant RNA processing is a feature of other muscular disorders such as myotonic muscular dystrophy, which initially made *FRG1* a seemingly good candidate (Nicole et al. 2002; Brais et al. 1998). However, data for altered *FRG1* expression in FSHD have been inconsistent, and several later studies did not detect differences between cases and controls (Winokur et al. 2003; Osborne et al. 2007; Jiang et al. 2003; Gabellini et al. 2002; Klooster et al. 2009).

One difficulty in measuring levels of *FRG1* mRNA is that there are many dispersed copies of this gene that could compromise the specificity of RT-PCR for the chromosome 4 locus (Grewal et al. 1999; Ballarati et al. 2002). Indeed, in a recent study that estimated human gene copy numbers by considering the read depth of next-generation sequencing data, *FRG1* ranked number twenty-four, with 23–30 copies (Alkan et al. 2009). As at least some of these additional loci are transcribed (Grewal et al. 1999), RT-PCR studies could be confounded by variation in copy number and quantitative differences in transcription levels of these homologues.

Several independent studies have now largely discredited the D4Z4 proximal *cis*-effect model (Winokur et al. 2003; Jiang et al. 2003; Klooster et al. 2009; Masny et al. 2010). We have seen a shift from the study of these neighbouring loci to the D4Z4 resident *DUX4*, which is now the prime candidate gene. However, the epigenetic status of this subtelomeric region is still relevant to the disease mechanism. Briefly, local D4Z4 chromatin relaxation in FSHD (van Overveld et al. 2003; de Greef et al. 2009; Zeng et al. 2009) is thought to result in aberrant expression of *DUX4* transcripts from the most distal repeat (Lemmers et al. 2010a; Snider et al. 2010). However, for these *DUX4* transcripts to be stable, the array contraction needs to occur on a particular chromosomal haplotype background. This haplotype, known as '4qA161', is defined by sequence variants near p13E-11 that are in linkage disequilibrium with a functional polyadenylation signal distal to the last D4Z4 repeat (Lemmers et al. 2010a; Dixit et al. 2007).

## 9.3 Evolution of 4q35, D4Z4 and DUX4

Before we discuss the FSHD disease mechanism in detail, we will investigate the evolution of chromosome 4q35, D4Z4 and *DUX4*. To a large extent, these data have arisen as interesting secondary findings that have emerged from research primarily motivated by the FSHD connection. However, studying the evolutionary history and population genetics of D4Z4 and its surrounding region directly led to the single biggest breakthrough in our understanding of this disease.

### 9.3.1 D4Z4 is Present on Both 4q35 and 10q26

When researchers used the p13E-11 probe to detect the polymorphic *Eco*RI fragment in FSHD patients, Southern blots showed a second locus that was mapped to chromosome 10q26 by fluorescence in situ hybridisation (Bakker et al. 1995; Wijmenga et al. 1995). When this locus was cloned, restriction mapping hinted at a high sequence similarity between the 10q and 4q loci (Deidda et al. 1996). Sequencing of the p13E-11 region and part of the first D4Z4 repeat unit from chromosome 10q confirmed that it showed more than 98 % sequence identity to the previously published sequence of chromosome 4q35 (Cacurri et al. 1998). Later, further sequence analysis of YACs, BACs and PACs of the region showed that the homology between 4q35 and 10q26 extends over 42 kb from the inverted, truncated *DUX4c* repeat to the very tip of the telomere (Fig. 9.2) (van Geel et al. 2002).

The 4q35 and 10q26 loci do have some sequence differences. Most usefully, most D4Z4 arrays on chromosomes 10q were found to contain a *Bln*I site that was not present on most 4q arrays (Deidda et al. 1996). Later, a *Xap*I site was found to be present on 4q but not on 10q-derived arrays (Lemmers et al. 2001). Thus, *Eco*RI can be used in combination with either of these two enzymes for more informative p13E-11 Southern blots in FSHD diagnosis, since contractions on chromosome 10q26 do not result in disease (Lemmers et al. 2001).

Unfortunately, this *Bln*I/*Xap*I 'diagnostic' difference between 4q- and 10q-derived arrays is not reliable. Some alleles contain apparently 'hybrid' D4Z4 arrays that are not homogenous for the *Bln*I/*Xap*I sites. Such arrays may carry *Bln*I-sensitive D4Z4 repeats on chromosome 4q (frequency of 6 %) or have *Bln*I-insensitive arrays on chromosome 10q (frequency of 9 %) (Lemmers et al. 2001, 1998; Deidda et al. 1996).

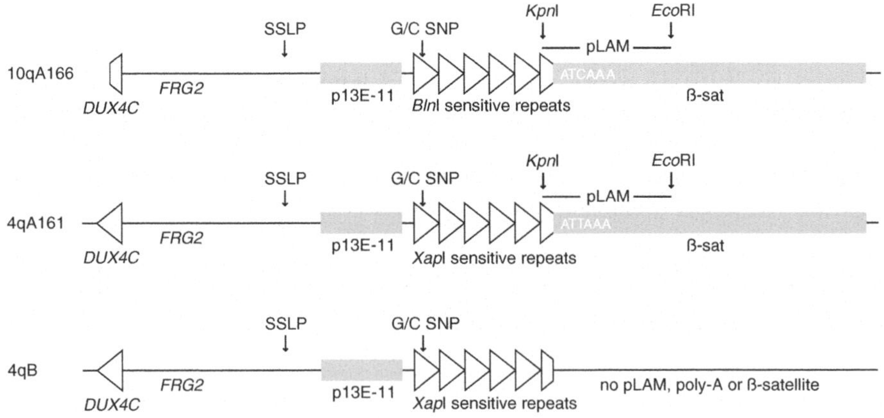


**Fig. 9.2** The homology between chromosomes 4 and 10 extends from *DUX4c* to the telomere. The SSLP and p13E-11 sequence features, the G/C SNP in the first repeat unit and the distal A/B-type telomeres define different haplotypes (see text). The FSHD disease-permissive haplotype 4qA161 has a non-canonical polyadenylation signal (ATTAAA) distal to the last repeat. This sequence is different or absent in alleles that are not disease permissive

Those individuals that have combinations of one *Bln*I-sensitive and three *Bln*I-insensitive arrays have also been classified as being 'monosomic' for a 10q-type array and 'trisomic' for a 4q-type array (van Deutekom et al. 1996a). A study of 106 controls and 70 FSHD probands found that only about two-thirds of people (cases and controls) conform to a 'simple' disomic *Bln*I allele constitution (Rossi et al. 2007). These different allele constitutions do not arise as *de novo* translocations, but exist as separate, ancient haplotypes.

It was further observed that hybrid arrays on 4q tend to be more heterogeneous, carrying a mixture of *Bln*I-sensitive and *Bln*I-insensitive units, while non-canonical 10q arrays tend to be more homogenous and composed entirely of *Bln*I-insensitive repeats. It is also noteworthy that the 4q and 10q arrays have different repeat number distributions, with a tendency for larger arrays on 4q compared to 10q (Rossi et al. 2007).

### *9.3.2 Other D4Z4-Related Loci*

D4Z4-like sequences are not just found on chromosomes 4 and 10. The probe 9B6A hybridises to the homeobox sequences in D4Z4 and thus to each repeat unit (Hewitt et al. 1994). Early data from Southern blots and metaphase FISH hinted at a wide dispersal of D4Z4-related sequences throughout the human genome, which was supported by genomic and cDNA clones (Hewitt et al. 1994). More detailed analysis that included PCR data from somatic cell hybrids confirmed that such sequences could be found on all short arms of the acrocentric chromosomes and at the heterochromatin block at 1q12 (Lyle et al. 1995). Southern blots and linear two-colour fibre FISH experiments showed that these related sequences are arranged as clusters in which the repeats are interspersed with blocks of 68-bp ($\beta$-)satellite DNA, even though they are not organised into discrete, homogenised arrays as at 4q35 and 10q26 (Lyle et al. 1995; Winokur et al. 1996).

D4Z4 sequences were not only found in the human genome, although in the days before the ubiquitous availability of sequence data, inferences had to be made from Southern blot and FISH analyses (Clark et al. 1996). With D4Z4 hybridising probes, *Kpn*I digests yielded strong 3.3-kb bands on Southern blots in most great apes. *Eco*RI digests hybridised with probes to the D4Z4 repeat sequences *LSau* and *hhspm3* also gave signals, which provided evidence that other primate genomes contain related repeat arrays. The high similarity of *Pst*I digest banding patterns in chimpanzee, orang-utan and gorilla further supported a similar organisation of these repeats. Pulsed-field gels proved that just as in humans, D4Z4 arrays in these primates are also arranged in long, polymorphic arrays (Clark et al. 1996).

The wide dispersal of D4Z4 sequences to heterochromatic regions, such as the short arms of the human acrocentric chromosomes, seems to be a relatively recent trend in primate evolution. Unlike humans and great apes, the more distantly related old world monkeys such as macaques show FISH signals on only a few telomeric locations, which is consistent with data from Southerns. The same is true for baboons and the new world monkey marmoset (Clark et al. 1996). Together, this supports the idea that the expansion of D4Z4 repeats probably occurred after the split of old world monkeys and great apes.

In contrast to D4Z4 dispersal, analyses with probes that hybridise proximal of the D4Z4 array indicate that the wider 4q35 subtelomeric region has also undergone duplication and dispersal in primates even before the old world monkey/great ape divergence (Ballarati et al. 2002). In line with this, macaques have a local *FRG1* duplication (Grewal et al. 1999). In gorillas, FISH studies have detected a large tandem duplication of the 4q35 subtelomeric region that includes *FRG1*, *FRG2*, p13E-11 and the D4Z4 repeat (Bodega et al. 2007).

### 9.3.3 Evolution of the DUX Gene Family

*DUX4* is a member of a larger family of *DUX* genes including *DUXA*, *DUXB*, *Duxbl*, *DUXC* and *Dux*. These genes, defined by their two closely spaced homeobox sequences, are only found in mammals (Leidenroth and Hewitt 2010). Most of these genes contain multiple introns within the coding region, but *DUX4* and *Dux* (present in mice and rats) have no such introns and likely represent retrogenes (Leidenroth and Hewitt 2010; Clapp et al. 2007). The *DUX4* open reading frame has been conserved for more than 100 million years, with intact homologues in primates and Afrotheria. The murine *Dux* genes are also arranged in tandem arrays (Clapp et al. 2007).

The genomes of several mammalian species such as *Canis familiaris*, *Bos taurus* and *Equus caballus* lack *DUX4* but contain the closely related intron-containing gene *DUXC*, which shares a conserved C-terminal domain with *DUX4* in addition to the homeobox sequences. *DUX4* and *DUXC* may represent functional homologues, and it is probable that the progenitor *DUX4* retrogene arose from an ancestral *DUXC*. Overall, the distribution of the different *DUX* homologues across the mammalian class is patchy, with reciprocal loss and retention in different lineages that may indicate functional similarities or redundancies. Indeed, we have not found a mammalian genome that retains both *DUX4* and *DUXC*, consistent with the idea of functional redundancy (Clapp et al. 2007; Leidenroth and Hewitt 2010).

Interestingly, *DUX4* may well be one of the human protein-coding genes with the highest overall copy number (Alkan et al. 2009). It is still unclear what selective pressures or mechanisms maintain the intact open reading frame at both chromosomes 4 and 10 at such high copy number.

### 9.3.4 4qA and 4qB Variants and Haplotypes

After the linkage of FSHD to 4q35 had been established, it was found that the region distal of D4Z4 exists in two major variants, dubbed 4qA and 4qB (van Geel et al. 2002). These two alleles both share sequence blocks with chromosome 4pter. Based on their organisation and sequence similarity, the two alleles arose from independent transfers of DNA sequence from 4pter, 4qB being the more recent event. When the chromosome 10q telomere was compared with these two variants, it was found to be of the A type (van Geel et al. 2002), although 10qB alleles were subsequently shown to exist at low frequency (Lemmers et al. 2010b).

The principal difference between the 4qA and 4qB alleles, which are found at roughly equal population frequencies (Lemmers et al. 2010b), is the organisation of the distal end of D4Z4. On 4qA (and 10q), the D4Z4 array terminates in a partial 3.3-kb repeat unit (this can be either 1.25 or 2.9 kb), immediately followed by about 8 kb of 68 bp satellite. This D4Z4/68-bp junction sequence was identified by cloning the distal *Kpn*I/*Eco*RI fragment of patients and is referred to as pLAM (Fig. 9.2) (van Deutekom et al. 1993; van Geel et al. 2002). On 4qB, the most distal D4Z4 repeat unit is truncated after only 570 bp and the sequence immediately distal to D4Z4 contains no $\beta$-satellite (van Geel et al. 2002).

Importantly, it was soon found that FSHD is associated only with D4Z4 contractions on 4qA-type chromosomes (Lemmers et al. 2004, 2002). Contractions on 4qB, just like those on 10q26, were found not to be associated with the disease. Thus, the linkage of FSHD had been narrowed down to a particular chromosome 4 haplotype (Lemmers et al. 2002). The non-pathogenicity of contracted 10qA arrays compared with 4qA, despite their similar genomic organisation, indicated that these two chromosomes must differ in some subtle way. This sparked a new search to identify a genetic change that explains the disease segregation with 4qA haplotype as compared with 4qB and 10qA.

An important clue came in 2007. In a remarkable study in which almost a hundred FSHD patients and more than four hundred controls were analysed, Lemmers et al. 2007) genotyped the sequence of the p13E-11 region, a simple sequence length polymorphism (SSLP) proximal to D4Z4, a SNP in the first D4Z4 repeat unit and the A/B telomeric polymorphism in each individual (Fig. 9.2) (Lemmers et al. 2007). They found that the combination of these markers defined a limited number of haplotypes, only one of which (4qA161, named after its SSLP length of 161 bp) was found to be associated with FSHD. Like previous studies, this paper also described families with short chromosome 4qB alleles and no muscular dystrophy. Interestingly, however, they also described a 4qA haplotype (4qA166) that behaved like a 4qB or 10qA allele: 4qA166 alleles with short D4Z4 alleles were not associated with FSHD. This meant that there was not only something special about 4qA compared with 10qA, but that even within 4qA alleles, there were FSHD disease-permissive and disease non-permissive variants. This was in accordance with an independent report that described the presence of short 4qA alleles in healthy individuals (Rossi et al. 2007).

These haplotype data were significantly expanded 3 years later, which, taken together with another study (Rossi et al. 2007), culminated in a thorough characterisation of the evolutionary history of these subtelomeric regions (Lemmers et al. 2010b). The findings are consistent with the 4qA arrays being ancestral and giving rise to the 10qA array by transfer of material onto the 10q telomere (Rossi et al. 2007; Lemmers et al. 2010b). Haplotype analysis of alleles containing *Bln*I/*Xap*I hybrid arrays indicated that these arose from a small number of ancestral sequence exchanges between 4q and 10q alleles, rather than by *de novo* recombination as implied previously (van Deutekom et al. 1996a; Lemmers et al. 1998).

Population studies showed that the different haplotype configurations gained their own unique single-nucleotide variants before the dispersal of Homo sapiens out of Africa (Lemmers et al. 2010b). Thus, recombination at 4q35 is predominantly intra-allelic, indicating linkage disequilibrium between the sequence variants defining the chromosome 4 haplotype and a putative FSHD-permissive variant in *cis* (Lemmers et al. 2007, 2010b).

This variant was finally identified in 2010, when Lemmers et al. showed that a functional, non-canonical polyadenylation signal (ATTAAA) resides in the 68-bp satellite (within pLAM) on the 4qA161 haplotype and stabilises *DUX4* transcripts in FSHD. In contrast, disease non-permissive haplotypes harbour non-functional variants at this position (Lemmers et al. 2010a; Snider et al. 2010). Although the distal sequence of non-permissive haplotype 4qA166 was not published, it presumably also lacks a functional poly-A signal.

## 9.4 From Repeat Array Contraction to Muscular Dystrophy

Most of the above studies arose primarily out of a desire to improve our understanding of the molecular disease mechanism that causes the FSHD phenotype. Because in the early days of the field no *DUX4* transcripts could be detected, initial disease models were focused on *cis* or *trans* effects. Under these models, D4Z4 contraction was thought to result in local chromatin changes and consequential aberrant regulation of D4Z4 proximal or *trans* target genes, resulting in muscular dystrophy (Jiang et al. 2003; Gabellini et al. 2002). The current disease model does retain a strong epigenetic component, but one that is constrained locally to D4Z4 rather than *cis*-spreading to distant regions. The deregulated target now considered to be the most likely FSHD candidate is *DUX4* (van der Maarel et al. 2011).

Even before this recent paradigm shift, there have been hints at a *DUX4* involvement. For example, complete deletion of all D4Z4 units does not result in disease; it seems that at least one repeat is required for pathology (Rossi et al. 2007; Tupler et al. 1996). Further, D4Z4 deletions extending proximally to p13E-11 have been reported in FSHD patients, which argue against any disease causative variants being located there (Lemmers et al. 2003).

### *9.4.1 DUX4 Transcription and D4Z4 Epigenetics*

It was noted early on that each D4Z4 repeat also housed classic elements of a putative promoter, a GC and a TACAA box (Gabriels et al. 1999), but no canonical polyadenylation signal was found within D4Z4. Despite independent efforts (Hewitt et al. 1994; Winokur et al. 2003; Lyle et al. 1995; Yip and Picketts 2003; Osborne et al. 2007; Alexiadis et al. 2007), endogenous transcription of *DUX4* was not reported until 2007 (Dixit et al. 2007; Kowaljow et al. 2007). Dixit et al. showed that transfected constructs that contained two D4Z4 repeats and the pLAM sequence were transcribed from the distal repeat unit. This study also reported two splice forms for these constructs, with introns in the putative 3′ UTR proximal to the polyadenylation signal within pLAM (Dixit et al. 2007). Introns in the 3′ UTR usually trigger the nonsense-mediated decay pathway and are therefore very rare in most genes, although they have been described in other retrogenes (Fablet et al. 2009).

A more thorough analysis of D4Z4 transcription confirmed these splice forms, the use of the polyadenylation signal and the presence of short discontinuous transcripts originating from D4Z4 (Snider et al. 2009). These studies also contained the first hint that *DUX4* transcription might be increased in FSHD patients compared with controls. Both Dixit et al. and Kowaljow et al. reported full-length *DUX4* transcripts in primary FSHD myoblasts and myotubes, but not in unaffected controls (Kowaljow et al. 2007; Dixit et al. 2007). Initially, Snider et al. could not confirm the presence of those full-length transcripts but found higher transcript levels of fragmented 5′, 3′ and central parts of the *DUX4* ORF in patient myoblasts compared with controls (Snider et al. 2009), although the same group reported full-length transcripts a year later (Snider et al. 2010). If *DUX4* mRNA was present at a higher level in patients, what could account for this increase? Now, several independent studies suggest that this is caused by a local change in D4Z4 chromatin conformation.

DNA methylation and histone modification of the FSHD locus have been extensively investigated. At the histone level, chromatin immunoprecipitation has been employed to show that on chromosome 4q35, the chromatin structure within D4Z4, at p13E-11, and in the promoter regions of *FRG2* and *DUX4c* resembles that of unexpressed euchromatin (Jiang et al. 2003). Using the CpG methylation-sensitive restriction enzymes *Fse*I and *Bsa*AI combined with Southern blotting, van Overveld et al. studied the methylation status of the respective sites of these enzymes within the most proximal *Kpn*I repeat unit (van Overveld et al. 2003). They found significant hypomethylation of this repeat unit in FSHD patients compared with healthy individuals.

Here, we need to introduce an additional genetic subtype of FSHD. A small subset of patients (between 5 and 10 %) is diagnosed with an FSHD phenotype, but does not have a contracted chromosome 4 array (Krasnianski et al. 2003; Yamanaka et al. 2004). These patients are said to have FSHD2 or phenotypic FSHD. While patients with classic FSHD (FSHD1) and non-penetrant carriers show hypomethylation only on the deleted allele, phenotypic FSHD2 patients without a contraction were hypomethylated on both 4q chromosomes (van Overveld et al. 2003). Subsequently, it emerged that loss of methylation in FSHD1 is not uniform, with short alleles below 20 kb (associated with more severe phenotypes) more extensively hypomethylated than alleles between 20 and 31 kb. In the larger disease alleles, the extent of hypomethylation, just as the severity of the phenotype, is more variable (van Overveld et al. 2005).

A thorough follow-up study extended this type of restriction analysis to chromosome 10q. In accordance with previous findings, FSHD1 patients only showed hypomethylation at the contracted 4q allele, while the 10q alleles, like the non-deleted 4q, showed normal methylation. However, in FSHD2 patients, both 10q alleles, just like both 4q alleles, were hypomethylated. This suggested that there are different triggers in FSHD1 and FSHD2 that cause the loss of D4Z4 methylation: contraction of the array in FSHD1, and an unknown mechanism in FSHD2 that is activated further upstream and acts on both 4q and 10q arrays independently of their length (de Greef et al. 2009).

Finally, a ChIP study showed that in FSHD patients, there is a specific loss of the repressive chromatin modification histone 3 lysine 9 tri-methylation (H3K9me3) at D4Z4 (Zeng et al. 2009). This marker, which like DNA methylation is associated with repressive chromatin states, is reduced in both contracted and non-contracted 4q alleles and both 10q alleles of FSHD1 patients. This is in contrast to the loss of DNA methylation that is seen only on the contracted allele in FSHD1. Interestingly, FSHD2 patients without a D4Z4 contraction show the same loss of H3K9me3 on all four D4Z4 alleles (4q and 10q), just as is seen for DNA methylation. Thus, H3K9me3 loss is a further molecular link between FSHD1 and FSHD2, underlining the close relationship of these diseases not just on a phenotypic but also an epigenetic level.

### *9.4.2 A Unifying Model*

In 2010, two important studies finally wove together the various different strands of all these findings. Lemmers et al. identified the pathogenic variant linked to the 4qA161 haplotype (Lemmers et al. 2010a). They analysed the pLAM region on different haplotype backgrounds and found a consistent difference in the status of the polyadenylation signal within the 68-bp satellite. While on the FSHD disease-permissive 4qA161 chromosomes this sequence was ATTAAA (a non-canonical sequence that is commonly used in humans), it was absent or different on chromosomes on which D4Z4 contraction does not result in disease. On chromosome 10q, this sequence is ATCAAA or ATTTAA, neither of which are functional polyadenylation signals in humans. Disease non-permissive chromosomes carrying the 'B'-type telomere lack the 68-bp satellite region and thus the polyadenylation signal entirely. An elegant qPCR assay demonstrated that when the pLAM regions of different haplotypes are cloned into expression vectors, stable polyadenylated *DUX4* transcripts are only produced from alleles containing the ATTAAA sequence (Fig. 9.3) (Lemmers et al. 2010a).

Most importantly, this study also described a special family where, unusually, the disease causative allele is a short chromosome 10 hybrid array. This array begins with '10-type' repeats but has recombined with chromosome 4 and ends in a '4-type' repeat, complete with the ATTAAA polyadenylation signal.

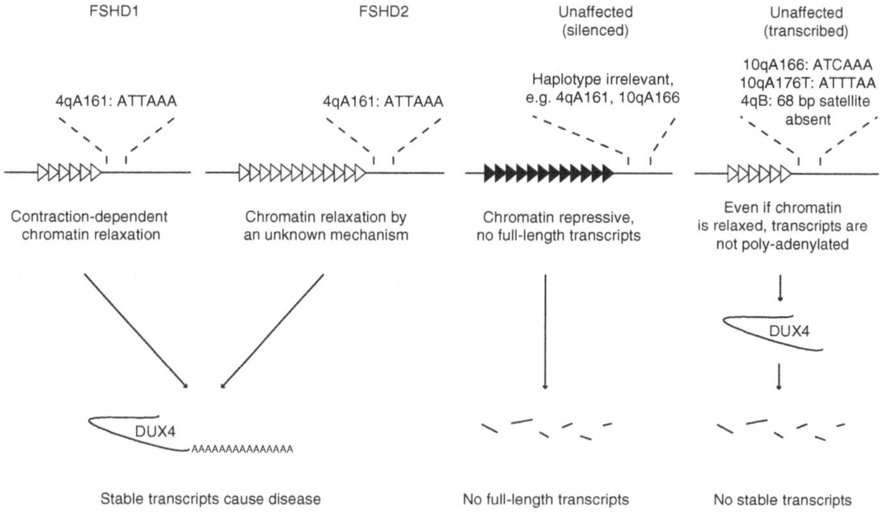


**Fig. 9.3** Summary of the current FSHD disease model. FSHD1 and FSHD2 both lose repressive chromatin marks (DNA and histone methylation, see text). These chromatin changes are thought to arise by independent mechanisms (contraction and unknown), but they converge in aberrant *DUX4* transcription. The current model predicts that contractions on non-4qA161 alleles also result in chromatin de-repression, but because of the lack of the poly-A signal, there is no *DUX4* pathogenic effect

Soon after, the first truly convincing data that full-length *DUX4* transcripts are produced in FSHD patient muscle cells were published. Snider et al. showed that while biopsies of healthy controls only contained transcripts of the short splice form that truncates the *DUX4* open reading frame, most FSHD1 and FSHD2 patient muscle samples contained full-length transcripts (Snider et al. 2010). Similar data were found for myoblasts and myotubes grown in cell culture. Using small-pool PCR experiments, it was shown that the transcripts are only present in approximately one out of every 1,000 FSHD muscle cells. This low frequency is supported by immunocytochemistry data from cultured FSHD muscle cells, which were immunostained with a combination of two new anti-DUX4 antibodies raised against either the N- or C-terminal domain (Snider et al. 2010). Importantly, this protein expression data were collected by scoring only nuclei that showed co-localisation of the two different DUX4 antibodies, greatly increasing confidence compared with previous attempts to show endogenous protein expression (Dixit et al. 2007; Kowaljow et al. 2007). Interestingly, the authors detected full-length *DUX4* transcript and protein in human testis with relative ease, which may provide a clue about the normal wild-type function of this gene.

Together, these two studies present convincing evidence that full-length *DUX4* transcripts are produced in FSHD patients and that the 4qA161 haplotype is disease permissive because it is linked to the ATTAAA polyadenylation sequence. Thus, under the current disease model, D4Z4 chromatin relaxation allows *DUX4* transcripts to escape repression (Fig. 9.3) (Lemmers et al. 2010a; Snider et al. 2010).

If these transcripts are stabilised by the polyadenylation signal and spliced into the full-length form, muscle disease follows. This model is supported by the finding that all FSHD2 patients carry at least one 4qA161 allele (de Greef et al. 2009). If the *DUX4* transcripts are ultimately responsible for the disease phenotype, it also marks the genetic convergence of contraction-dependent and phenotypic FSHD. The identical phenotype would then follow from this shared aberrant *DUX4* expression, which is enabled by chromatin changes that are either triggered by array contraction (FSHD1) or an unidentified mutation (FSHD2) (van der Maarel et al. 2011).

One slightly puzzling aspect of this hypothesis is the lack of any muscular dystrophy phenotype in patients with immunodeficiency, centromeric instability and facial anomalies syndrome (ICF). Half of these patients have mutations in the global DNA methyltransferase *DNMT3b* with loss of DNA methylation across many genomic regions, including D4Z4 (Hansen et al. 1999). As 4qA161 has an allele frequency of almost 40 % in Europeans (Lemmers et al. 2010b), we would expect stabilised full-length *DUX4* transcripts in roughly two-thirds of ICF patients. There are two possible explanations why these patients do not suffer from muscular dystrophy: First, patients rarely live beyond their second decade, allowing little time for FSHD onset (Hansen et al. 1999). Second, while ICF patients share the loss of D4Z4 DNA methylation, they do not lose the H3K9me3 marker (Zeng et al. 2009), loss of which may be required for *DUX4* to escape repression.

It was postulated that H3K9me3 reduction in FSHD1 is triggered by the array contraction and subsequently spreads to the other three alleles. Considering the latest findings on *DUX4* transcription and transcript stabilisation, the change of this histone marker may be correlative with FSHD, but is probably not sufficient. If H3K9me3 loss was sufficient and was solely caused by array contraction, then the shortening of a non-permissive allele (e.g. 4qB) could spread its epigenetic effects to a 4qA161 allele and also trigger FSHD. However, this model is inconsistent with the genetic data that show that only contracted 4qA161 alleles are pathogenic. It is likely that there is a complex interrelationship between chromatin state and transcription at D4Z4.

Another open question is the status of the pLAM poly-A signal on the 4qA166 haplotype. These alleles are reportedly not FSHD disease permissive, but their pLAM sequence has not yet been published. For the current model to be fully consistent, this haplotype should lack a functional poly-A signal.

## 9.5 A Role for the DUX4 Protein?

It remains an open question whether the disease pathogenesis is triggered by a *DUX4* transcript or a protein-mediated mechanism. We still do not know much about the function of *DUX4*, although there are several studies that have contributed to this subject.

Ectopic expression of *DUX4* is certainly toxic in all mammalian cell types tested and induces nuclear foci and apoptosis (Snider et al. 2010; Kowaljow et al. 2007; Bosnakovski et al. 2008). This toxicity of over-expression systems and the low, hard-to-detect endogenous levels make functional studies difficult. Furthermore, the lack of clear orthologues in commonly used model organisms such as *Xenopus laevis* and *Danio rerio* (Leidenroth and Hewitt 2010) means that the relevance of in vitro or in vivo over-expression experiments to FSHD is unclear.

Inspired by the high levels of *DUX4* germ line expression and data from induced pluripotent stem cells, Snider et al. proposed that *DUX4* may have a role in development (Snider et al. 2010). It is important to remember that whatever the normal function of *DUX4* is, it does not depend on the polyadenylation signal associated with FSHD. This is because based on the allele frequency of the 4qB alleles (Lemmers et al. 2010b), around a quarter of Europeans lack this poly-A signal entirely while being healthy. In testis, 3′ RACE analysis shows that *DUX4* transcripts from chromosome 10 can utilise a poly-A signal 6.5 kb distal to that in pLAM (Snider et al. 2010). This alternative poly-A signal is also used for some 4qA transcripts; however, it is absent from 4qB (Lemmers et al. 2010a). Altogether, the transcriptional landscape of D4Z4 is rather complex and includes a number of differently splice and un-spliced transcripts of different lengths, including microRNAs (Snider et al. 2009). Resolving the function of these products—provided they have one—will be important for a complete understanding of D4Z4 function.

The two homeodomains encoded by *DUX4* are classic DNA-binding motifs of the PRD class that can be found in many transcription factors including the developmentally important Hox genes (Gehring et al. 1994). Interestingly, the C-terminal domain of DUX4 has been observed to have transcriptional enhancer activity (Kawamura-Saito et al. 2006). At least 19 cancer cases (Ewing-like sarcomas and round cell tumours) have been shown to be caused by the fusion of this domain to a high-mobility group member DNA-binding protein (CIC) by chromosomal translocation events (Graham et al. 2012; Italiano et al. 2012; Kawamura-Saito et al. 2006; Yoshimoto et al. 2009). In this case, the CIC-DUX4 fusion protein drives increased expression of *PEA3* genes, resulting in tumorigenesis. The first transcriptional target that was proposed for DUX4 is *PITX1*, itself a transcription factor (Dixit et al. 2007). Recently, a ChIP-Seq experiment in transfected control myoblast identified over 1,800 DUX4-binding sites (Geng et al. 2012). This study also confirmed that a number of the associated target genes are differentially expressed between control and FSHD muscle. This will provide a useful framework for understanding the pathogenic effects downstream of *DUX4*.

## 9.6 Conclusions

Why did it take two decades to arrive at the current disease model for FSHD? The slow progress is primarily attributable to the fact that working with this locus is technically very challenging. Because of the high GC content of D4Z4, its repetitive nature, the homologous locus on chromosome 10q and the many dispersed copies, it is difficult to amplify D4Z4 DNA or RNA transcripts and be confident of their genomic origins. The high level of sequence identity between the tandem copies and the dispersed copies also means that D4Z4 will remain inaccessible to short-read sequencing with current Roche and Illumina technologies (Alkan et al. 2011). New sequencing technologies (e.g. Oxford Nanopore) that allow much longer read lengths will open up new opportunities to study this locus in unprecedented detail.

The most significant contribution to the FSHD field in recent years has come from the careful study of the evolution and population genetics of the 4q35 subtelomeric region by Lemmers et al. These data have not only advanced our understanding of the molecular defect of this disease, but have also produced an almost unprecedentedly detailed examination of the recent evolutionary history of a human subtelomeric region. FSHD affects hundreds of thousands of people across the globe. Understanding the genetics of this disease will hopefully enable us to identify suitable treatment strategies that will improve the quality of life of patients. For everyone working in our field, this is the ultimate motivation.

The disease mechanism of FSHD is unique and has been concisely summarised as the 'incomplete suppression of a retrotransposed gene' (Snider et al. 2010) caused by local epigenetic perturbations. It is likely that the extreme subtelomeric location of D4Z4 influences its chromatin packaging, and we speculate that this may have driven the evolution of its epigenetic regulation.

**Acknowledgments** We would like to thank the Muscular Dystrophy Campaign, UK, the Muscular Dystrophy Association, USA, and the FSH Society, USA, for past and present funding. We also thank the Muscular Dystrophy Campaign, UK, for funding A. L. with a Ph.D studentship.

## References

Alexiadis, V., Ballestas, M. E., Sanchez, C., Winokur, S., Vedanarayanan, V., Warren, M., et al. (2007). RNAPol-ChIP analysis of transcription from FSHD-linked tandem repeats and satellite DNA. *Biochimica et Biophysica Acta, 1769*(1), 29–40. doi:10.1016/j.bbaexp.2006.11.006.

Alkan, C., Kidd, J. M., Marques-Bonet, T., Aksay, G., Antonacci, F., Hormozdiari, F., et al. (2009). Personalized copy number and segmental duplication maps using next-generation sequencing. *Nature Genetics, 41*(10), 1061–1067. doi:10.1038/ng.437.

Alkan, C., Sajjadian, S., & Eichler, E. E. (2011). Limitations of next-generation genome sequence assembly. *Nature Methods, 8*(1), 61–65. doi:10.1038/nmeth.1527.

Bakker, E., Wijmenga, C., Vossen, R. H., Padberg, G. W., Hewitt, J., van der Wielen, M., et al. (1995). The FSHD-linked locus D4F104S1 (p13E-11) on 4q35 has a homologue on 10qter. *Muscle Nerve, 2*, S39–S44.

Ballarati, L., Piccini, I., Carbone, L., Archidiacono, N., Rollier, A., Marozzi, A., et al. (2002). Human genome dispersal and evolution of 4q35 duplications and interspersed LSau repeats. *Gene, 296*(1–2), 21–27. doi:S0378-1119(02)00858-2.

Bengtsson, U., Altherr, M. R., Wasmuth, J. J., & Winokur, S. T. (1994). High-resolution fluorescence in situ hybridisation to linearly extended DNA visually maps a tandem repeat associated with facioscapulohumeral muscular-dystrophy immediately to the telomere of 4q. *Human Molecular Genetics, 3*(10), 1801–1805.

Bodega, B., Cardone, M. F., Muller, S., Neusser, M., Orzan, F., Rossi, E., et al. (2007). Evolutionary genomic remodelling of the human 4q subtelomere (4q35.2). *BMC Evolutionary Biology, 7*. doi:10.1186/1471-2148-7-39.

Bosnakovski, D., Xu, Z. H., Gang, E. J., Galindo, C. L., Liu, M. J., Simsek, T., et al. (2008). An isogenetic myoblast expression screen identifies *DUX4*-mediated FSHD-associated molecular pathologies. *EMBO Journal, 27*(20), 2766–2779. doi:10.1038/emboj.2008.201.

Bouju, S., Pietu, G., Le Cunff, M., Cros, N., Malzac, P., Pellissier, J. F., et al. (1999). Exclusion of muscle specific actinin-associated LIM protein (ALP) gene from 4q35 facioscapulohumeral muscular dystrophy (FSHD) candidate genes. *Neuromuscular Disorders, 9*(1), 3–10.

Brais, B., Bouchard, J. P., Xie, Y. G., Rochefort, D. L., Chretien, N., Tome, F. M., et al. (1998). Short GCG expansions in the PABP2 gene cause oculopharyngeal muscular dystrophy. *Nature Genetics, 18*(2), 164–167. doi:10.1038/ng0298-164.

Cacurri, S., Piazzo, N., Deidda, G., Vigneti, E., Galluzzi, G., Colantoni, L., et al. (1998). Sequence homology between 4qter and 10qter loci facilitates the instability of subtelomeric KpnI repeat units implicated in facioscapulohumeral muscular dystrophy. *American Journal of Human Genetics, 63*(1), 181–190.

Clapp, J., Mitchell, L. M., Bolland, D. J., Fantes, J., Corcoran, A. E., Scotting, P. J., et al. (2007). Evolutionary conservation of a coding function for D4Z4, the tandem DNA repeat mutated in facioscapulohumeral muscular dystrophy. *American Journal of Human Genetics, 81*(2), 264–279. doi:10.1086/519311.

Clark, L. N., Koehler, U., Ward, D. C., Wienberg, J., & Hewitt, J. E. (1996). Analysis of the organisation and localisation of the FSHD-associated tandem array in primates: Implications for the origin and evolution of the 3.3 kb repeat family. *Chromosoma, 105*(3), 180–189.

de Greef, J. C., Lemmers, R. J. L., van Engelen, B. G. M., Sacconi, S., Venance, S. L., Frants, R. R., et al. (2009). Common epigenetic changes of D4Z4 in contraction-dependent and contraction-independent FSHD. *Human Mutation, 30*(10), 1449–1459.

Deidda, G., Cacurri, S., Piazzo, N., & Felicetti, L. (1996). Direct detection of 4q35 rearrangements implicated in facioscapulohumeral muscular dystrophy (FSHD). *Journal of Medical Genetics, 33*(5), 361–365.

Dixit, M., Ansseau, E., Tassin, A., Winokur, S., Shi, R., Qian, H., et al. (2007). *DUX4*, a candidate gene of facioscapulohumeral muscular dystrophy, encodes a transcriptional activator of PITX1. *Proceedings of the National Academy of Sciences USA, 104*, 18157–18162. doi:10.1073/pnas.0708659104.

Fablet, M., Bueno, M., Potrzebowski, L., & Kaessmann, H. (2009). Evolutionary origin and functions of retrogene introns. *Molecular Biology and Evolution, 26*(9), 2147–2156. doi:10.1093/molbev/msp125.

Gabellini, D., D'Antona, G., Moggio, M., Prelle, A., Zecca, C., Adami, R., et al. (2006). Facioscapulohumeral muscular dystrophy in mice overexpressing FRG1. *Nature, 439*(7079), 973–977. doi:10.1038/Nature04422.

Gabellini, D., Green, M. R., & Tupler, R. (2002). Inappropriate gene activation in FSHD: A repressor complex binds a chromosomal repeat deleted in dystrophic muscle. *Cell, 110*(3), 339–348.

Gabriels, J., Beckers, M. C., Ding, H., De Vriese, A., Plaisance, S., van der Maarel, S. M., et al. (1999). Nucleotide sequence of the partially deleted D4Z4 locus in a patient with FSHD identifies a putative gene within each 3.3 kb element. *Gene, 236* (1), 25–32.

Gehring, W. J., Affolter, M., & Burglin, T. (1994). Homeodomain proteins. *Annual Review of Biochemistry, 63*, 487–526.

Geng, L. N., Yao, Z., Snider, L., Fong, A. P., Cech, J. N., Young, J. M., et al. (2012). *DUX4* activates germline genes, retroelements, and immune mediators: Implications for facioscapulohumeral dystrophy. *Developmental Cell, 22*(1), 38–51. doi:10.1016/j.devcel.2011.11.013.

Graham, C., Chilton-MacNeill, S., Zielenska, M., & Somers, G. R. (2012). The CIC-*DUX4* fusion transcript is present in a subgroup of pediatric primitive round cell sarcomas. *Human Pathology, 43*(2), 180–189. doi:10.1016/j.humpath.2011.04.023.

Grewal, P. K., Todd, L. C., van der Maarel, S., Frants, R. R., & Hewitt, J. E. (1998). FRG1, a gene in the FSH muscular dystrophy region on human chromosome 4q35, is highly conserved in vertebrates and invertebrates. *Gene, 216*(1), 13–19.

Grewal, P. K., van Geel, M., Frants, R. R., de Jong, P., & Hewitt, J. E. (1999). Recent amplification of the human FRG1 gene during primate evolution. *Gene, 227*(1), 79–88.

Hansen, R. S., Wijmenga, C., Luo, P., Stanek, A. M., Canfield, T. K., Weemaes, C. M., et al. (1999). The DNMT3B DNA methyltransferase gene is mutated in the ICF immunodeficiency syndrome. *Proceedings of the National Academy of Sciences USA, 96*(25), 14412–14417.

Hewitt, J. E., Lyle, R., Clark, L. N., Valleley, E. M., Wright, T. J., Wijmenga, C., et al. (1994). Analysis of the tandem repeat locus D4Z4 associated with facioscapulohumeral muscular-dystrophy. *Human Molecular Genetics, 3*(8), 1287–1295.

Italiano, A., Sung, Y. S., Zhang, L., Singer, S., Maki, R. G., Coindre, J. M., et al. (2012). High prevalence of CIC fusion with double-homeobox (*DUX4*) transcription factors in EWSR1-negative undifferentiated small blue round cell sarcomas. *Genes, Chromosomes and Cancer, 51*(3), 207–218. doi:10.1002/gcc.20945.

Jiang, G. C., Yang, F., van Overveld, P. G. M., Vedanarayanan, V., van der Maarel, S., & Ehrlich, M. (2003). Testing the position-effect variegation hypothesis for facioscapulohumeral muscular dystrophy by analysis of histone modification and gene expression in subtelomeric 4q. *Human Molecular Genetics, 12*(22), 2909–2921. doi:10.1093/hmg/ddg323.

Kawamura-Saito, M., Yamazaki, Y., Kaneko, K., Kawaguchi, N., Kanda, H., Mukai, H., et al. (2006). Fusion between CIC and *DUX4* up-regulates PEA3 family genes in Ewing-like sarcomas with t(4;19)(q35;q13) translocation. *Human Molecular Genetics, 15*(13), 2125–2137. doi:10.1093/hmg/ddl136.

Klooster, R., Straasheijm, K., Shah, B., Sowden, J., Frants, R., Thornton, C., Tawil, R., van der Maarel, S. (2009). Comprehensive expression analysis of FSHD candidate genes at the mRNA and protein level. *European Journal of Human Genetics, 17*, 1615–1624.

Kowaljow, V., Marcowycz, A., Ansseau, E., Conde, C. B., Sauvage, S., Mattotti, C., et al. (2007). The *DUX4* gene at the FSHDIA locus encodes a pro-apoptotic protein. *Neuromuscular Disorders, 17*(8), 611–623. doi:10.1016/j.nmd.2007.04.002.

Krasnianski, M., Neudecker, S., Eger, K., Jakubiczka, S., & Zierz, S. (2003). Typical facioscapulohumeral dystrophy phenotype in patients without FSHD 4q35 deletion. *Journal of Neurology, 250*(9), 1084–1087. doi:10.1007/s00415-003-0158-5.

Laoudj-Chenivesse, D., Carnac, G., Bisbal, C., Hugon, G., Bouillot, S., Desnuelle, C., et al. (2005). Increased levels of adenine nucleotide translocator 1 protein and response to oxidative stress are early events in facioscapulohumeral muscular dystrophy muscle. *Journal of Molecular Medicine, 83*(3), 216–224. doi:10.1007/s00109-004-0583-7.

Lee, J. H., Goto, K., Matsuda, C., & Arahata, K. (1995). Characterization of a tandemly repeated 3.3-kb KpnI unit in the facioscapulohumeral muscular dystrophy (FSHD) gene region on chromosome 4q35. *Muscle and Nerve, 2*, S6–S13.

Leidenroth, A., & Hewitt, J. E. (2010). A family history of *DUX4*: phylogenetic analysis of DUXA, B, C and Duxbl reveals the ancestral DUX gene. *BMC Evolutionary Biology, 10*, 364. doi:10.1186/1471-2148-10-364.

Lemmers, R., de Kievit, P., Sandkuijl, L., Padberg, G. W., van Ommen, G. J. B., Frants, R. R., et al. (2002). Facioscapulohumeral muscular dystrophy is uniquely associated with one of the two variants of the 4q subtelomere. *Nature Genetics, 32*(2), 235–236. doi:10.1038/ng999.

Lemmers, R., de Kievit, P., van Geel, M., van der Wielen, M. J., Bakker, E., Padberg, G. W., et al. (2001). Complete allele information in the diagnosis of facioscapulohumeral muscular dystrophy by triple DNA analysis. *Ann Neurol, 50*(6), 816–819.

Lemmers, R., Osborn, M., Haaf, T., Rogers, M., Frants, R. R., Padberg, G. W., et al. (2003). D4F104S1 deletion in facioscapulohumeral muscular dystrophy—Phenotype, size, and detection. *Neurology, 61*(2), 178–183.

Lemmers, R., van der Maarel, S. M., van Deutekom, J. C. T., van der Wielen, M. J. R., Deidda, G., Dauwerse, H. G., et al. (1998). Inter-and intrachromosomal sub-telomeric rearrangements on 4q35: implications for facioscapulohumeral muscular dystrophy (FSHD) aetiology and diagnosis. *Human Molecular Genetics, 7*(8), 1207–1214.

Lemmers, R., van der Vliet, P. J., Klooster, R., Sacconi, S., Camano, P., Dauwerse, J. G., et al. (2010a). A unifying genetic model for facioscapulohumeral muscular dystrophy. *Science, 329*(5999), 1650–1653. doi:10.1126/science.1189044.

Lemmers, R., van der Vliet, P. J., van der Gaag, K. J., Zuniga, S., Frants, R. R., de Knijff, P., et al. (2010b). Worldwide population analysis of the 4q and 10q Subtelomeres identifies only four discrete interchromosomal sequence transfers in human evolution. *American Journal of Human Genetics, 86*(3), 364–377. doi:10.1016/j.ajhg.2010.01.035.

Lemmers, R., Wohlgemuth, M., Frants, R. R., Padberg, G. W., Morava, E., & van der Maarel, S. M. (2004). Contractions of D4Z4 on 4qB subtelomeres do not cause facioscapulohumeral muscular dystrophy. *American Journal of Human Genetics, 75*(6), 1124–1130. doi:10.1086/426035.

Lemmers, R., Wohlgemuth, M., van der Gaag, K. J., van der Vliet, P. J., van Teijlingen, C. M. M., de Knijff, P., et al. (2007). Specific sequence variations within the 4q35 region are associated with facioscapulohumeral muscular dystrophy. *American Journal of Human Genetics, 81*, 884–894. doi:10.1086/521986.

Li, K., Warner, C. K., Hodge, J. A., Minoshima, S., Kudoh, J., Fukuyama, R., et al. (1989). A human muscle adenine nucleotide translocator gene has four exons, is located on chromosome 4, and is differentially expressed. *Journal of Biological Chemistry, 264*(24), 13998–14004.

Lyle, R., Wright, T. J., Clark, L. N., & Hewitt, J. E. (1995). FSHD-associated repeat, D4Z4, is a member of a dispersed family of homeobox-containing repeats, subsets of which are clustered on the short arms of the acrocentric chromosomes. *Genomics, 28*(3), 389–397.

Masny, P. S., Bengtsson, U., Chung, S. A., Martin, J. H., van Engelen, B., van der Maarel, et al. (2004). Localization of 4q35.2 to the nuclear periphery: Is FSHD a nuclear envelope disease? *Human molecular genetics, 13*(17), 1857–1871. doi:10.1093/hmg/ddh205.

Masny, P. S., Chan, O. Y. A., de Greef, J. C., Bengtsson, U., Ehrlich, M., Tawil, R., et al. (2010). Analysis of allele-specific RNA transcription in FSHD by RNA-DNA FISH in single myonuclei. *European Journal of Human Genetics, 18*(4), 448–456.

Nicole, S., Diaz, C. C., Frugier, T., & Melki, J. (2002). Spinal muscular atrophy: recent advances and future prospects. *Muscle and Nerve, 26*(1), 4–13. doi:10.1002/mus.10110.

Osborne, R. J., Welle, S., Venance, S. L., Thornton, C. A., & Tawil, R. (2007). Expression profile of FSHD supports a link between retinal vasculopathy and muscular dystrophy. *Neurology, 68*(8), 569–577.

Padberg, G. W. (2004). Facioscapulohumeral muscular dystrophy: A clinician's experience. In: M. Upadhyaya DNC (Ed.) *Facioscapulohumeral muscular dystrophy: Clinical medicine and molecular cell biology* (pp. 41–53). Oxon: Garland Science.

Padberg, G. W., Brouwer, O. F., Dekeizer, R. J. W., Dijkman, G., Wijmenga, C., Grote, J. J., Frants, R. R. (1995). On the significance of retinal vascular-disease and hearing-loss in facioscapulohumeral muscular dystrophy. *Muscle Nerve,* S73–S80.

Rossi, M., Ricci, E., Colantoni, L., Galluzzi, G., Frusciante, R., Tonali, P. A., et al. (2007). The facioscapulohumeral muscular dystrophy region on 4qter and the homologous locus on 10qter evolved independently under different evolutionary pressure. *BMC Medical Genetics, 8*. doi:10.1186/1471-2350-8-8.

Sarfarazi, M., Upadhyaya, M., Padberg, G., Pericakvance, M., Siddique, T., Lucotte, G., et al. (1989). An exclusion map for facioscapulohumeral (Landouzy-Dejerine) disease. *Journal of Medical Genetics, 26*(8), 481–484.

Sarfarazi, M., Wijmenga, C., Upadhyaya, M., Weiffenbach, B., Hyser, C., Mathews, K., et al. (1992). Regional mapping of facioscapulohumeral muscular-dystrophy gene on 4q35—combined analysis of an international consortium. *American Journal of Human Genetics, 51*(2), 396–403.

Snider, L., Asawachaicharn, A., Tyler, A. E., Geng, L. N., Petek, L. M., Maves, L., et al. (2009). RNA transcripts, miRNA-sized fragments and proteins produced from D4Z4 units: new candidates for the pathophysiology of facioscapulohumeral dystrophy. *Human Molecular Genetics, 18*(13), 2414–2430. doi:10.1093/Hmg/Ddp180.

Snider, L., Geng, L. N., Lemmers, R. J. L. F., Kyba, M., Ware, C. B., Nelson, A. M., et al. (2010). Facioscapulohumeral dystrophy: Incomplete suppression of a retrotransposed gene. *PLoS Genetics, 6*(10), e1001181.

Tam, R., Smith, K. P., & Lawrence, J. B. (2004). The 4q subtelomere harboring the FSHD locus is specifically anchored with peripheral heterochromatin unlike most human telomeres. *Journal of Cell Biology, 167*(2), 269–279. doi:10.1083/jcb.200403128.

Tupler, R., Berardinelli, A., Barbierato, L., Frants, R., Hewitt, J. E., Lanzi, G., et al. (1996). Monosomy of distal 4q does not cause facioscapulohumeral muscular dystrophy. *Journal of Medical Genetics, 33*(5), 366–370.

Upadhyaya, M., Lunt, P., Sarfarazi, M., Broadhead, W., Farnham, J., & Harper, P. S. (1992). The mapping of chromosome 4q markers in relation to facioscapulohumeral muscular-dystrophy (FSHD). *American Journal of Human Genetics, 51*(2), 404–410.

van der Maarel, S. M., Frants, R. R., & Padberg, G. W. (2007). Facioscapulohumeral muscular dystrophy. *Biochimica et Biophysica Acta (BBA)-Molecular Basis of Disease, 1772*(2), 186–194. doi:10.1016/j.bbadis.2006.05.009.

van der Maarel, S. M., Tawil, R., & Tapscott, S. J. (2011). Facioscapulohumeral muscular dystrophy and *DUX4*: breaking the silence. *Trends in Molecular Medicine, 17*(5), 252–258. doi:10.1016/j.molmed.2011.01.001.

van Deutekom, J. C. (1996). Towards the molecular mechanism of facioscapulohumeral muscular dystrophy. Ph. D. thesis, Leiden University, Leiden.

van Deutekom, J. C., Bakker, E., Lemmers, R., van der Wielen, M. J. R., Bik, E., Hofker, M. H., et al. (1996a). Evidence for subtelomeric exchange of 3.3 kb tandemly repeated units between chromosomes 4q35 and 10q26: Implications for genetic counselling and etiology of FSHD1. *Human molecular genetics, 5*(12), 1997–2003.

van Deutekom, J. C., Lemmers, R. J., Grewal, P. K., van Geel, M., Romberg, S., Dauwerse, H. G., Wright. T. J., Padberg, G. W., Hofker, M. H., Hewitt, J. E., & Frants, R. R. (1996b). Identification of the first gene (FRG1) from the FSHD region on human chromosome 4q35. *Human Molecular Genetics, 5*(5), 581–590.

van Deutekom, J .C., Wijmenga, C., van Tienhoven, E. A .E., Gruter, A .M., Frants, R. R., Hewitt, J .E., et al. (1993). FSHD associated DNA rearrangements are due to deletions of integral copies of a 3.2 kb tandemly repeated unit. *Human molecular genetics, 2*(12), 2037–2042.

van Geel, M., Dickson, M. C., Beck, A. F., Bolland, D. J., Frants, R. R., van der Maarel, S. M., et al. (2002). Genomic analysis of human chromosome 10q and 4q telomeres suggests a common origin. *Genomics, 79*(2), 210–217. doi:10.1006/geno.2002.6690.

van Geel, M., van Deutekom, J. C. T., van Staalduinen, A., Lemmers, R., Dickson, M. C., Hofker, M. H., et al. (2000). Identification of a novel beta-tubulin subfamily with one member (TUBB4Q) located near the telomere of chromosome region 4q35. *Cytogenetics and Cell Genetics, 88*(3–4), 316–321.

van Koningsbruggen, S., Dirks, R. W., Mommaas, A. M., Onderwater, J. J., Deidda, G., Padberg, G. W., et al. (2004a). FRG1P is localised in the nucleolus, cajal bodies, and speckles. *Journal of Medical Genetics, 41*(4), e46.

van Koningsbruggen, S., Frants, R. R., & van der Maarel, S. M. (2004). Identification and characterization of candidate genes in FSHD region. In: M. Upadhyaya DNC (Ed.) *Facioscapulohumeral muscular dystrophy: Clinical medicine and molecular cell biology*. Oxon: Garland Science.

van Overveld, P. G., Enthoven, L., Ricci, E., Rossi, M., Felicetti, L., Jeanpierre, M., et al. (2005). Variable hypomethylation of D4Z4 in facioscapulohumeral muscular dystrophy. *Annals of Neurology, 58*(4), 569–576. doi:10.1002/ana.20625.

van Overveld, P. G., Lemmers, R., Sandkuijl, L. A., Enthoven, L., Winokur, S. T., Bakels, F., et al. (2003). Hypomethylation of D4Z4 in 4q-linked and non-4q-linked facioscapulohumeral muscular dystrophy. *Nature Genetics, 35*(4), 315–317. doi:10.1038/ng1262.

Weiffenbach, B., Bagley, R., Falls, K., Hyser, C., Storvick, D., Jacobsen, S. J., et al. (1992). Linkage analyses of 5 chromosome-4 markers localizes the facioscapulohumeral muscular-dystrophy (FSHD) gene to distal 4q35. *American Journal of Human Genetics, 51*(2), 416–423.

Wijmenga, C., Dauwerse, H. G., Padberg, G. W., Meyer, N., Murray, J. C., Mills, K., et al. (1995). Fish mapping of 250 cosmid and 26 YAC clones to chromosome 4 with special emphasis on the FSHD region at 4q35. *Muscle and Nerve, 2*, S14–S18.

Wijmenga, C., Hewitt, J. E., Sandkuijl, L. A., Clark, L. N., Wright, T. J., Dauwerse, H. G., et al. (1992). Chromosome 4q DNA rearrangements associated with facioscapulohumeral muscular dystrophy. *Nature Genetics, 2*(1), 26–30.

Wijmenga, C., Vandeutekom, J. C. T., Hewitt, J. E., Padberg, G. W., Vanommen, G. J. B., Hofker, M. H., et al. (1994). Pulsed-field gel-electrophoresis of the D4F104S1 locus reveals the size and the parental origin of the facioscapulohumeral muscular-dystrophy (FSHD)-associated deletions. *Genomics, 19*(1), 21–26.

Winokur, S. T., Bengtsson, U., Feddersen, J., Mathews, K. D., Weiffenbach, B., Bailey, H., et al. (1994). The DNA rearrangement associated with facioscapulohumeral muscular dystrophy involves a heterochromatin-associated repetitive element: implications for a role of chromatin structure in the pathogenesis of the disease. *Chromosome Research, 2*(3), 225–234.

Winokur, S. T., Bengtsson, U., Vargas, J. C., Wasmuth, J. J., Altherr, M. R., Weiffenbach, B., et al. (1996). The evolutionary distribution and structural organization of the homeobox-containing repeat D4Z4 indicates a functional role for the ancestral copy in the FSHD region. *Human Molecular Genetics, 5*(10), 1567–1575.

Winokur, S. T., Chen, Y. W., Masny, P. S., Martin, J. H., Ehmsen, J. T., Tapscott, S. J., et al. (2003). Expression profiling of FSHD muscle supports a defect in specific stages of myogenic differentiation. *Human Molecular Genetics, 12*(22), 2895–2907. doi:10.1093/hmg/ddg327.

Xia, H. H., Winokur, S. T., Kuo, W. L., Altherr, M. R., & Bredt, D. S. (1997). Actinin-associated LIM protein: Identification of a domain interaction between PDZ and spectrin-like repeat motifs. *Journal of Cell Biology, 139*(2), 507–515.

Yamanaka, G., Goto, K., Ishihara, T., Oya, Y., Miyajima, T., Hoshika, A., et al. (2004). FSHD-like patients without 4q35 deletion. *Journal of the Neurological Sciences, 219*(1–2), 89–93. doi:10.1016/j.jns.2003.12.010.

Yip, D. J., & Picketts, D. J. (2003). Increasing D4Z4 repeat copy number compromises C2C12 myoblast differentiation. *FEBS Letters, 537*(1–3), 133–138.

Yoshimoto, M., Graham, C., Chilton-MacNeill, S., Lee, E., Shago, M., Squire, J., et al. (2009). Detailed cytogenetic and array analysis of pediatric primitive sarcomas reveals a recurrent CIC-*DUX4* fusion gene event. *Cancer Genetics and Cytogenetics, 195*(1), 1–11. doi:10.1016/j.cancergencyto.2009.06.015.

Zeng, W., de Greef, J. C., Chen, Y.-Y., Chien, R., Kong, X., Gregson, H. C., et al. (2009). Specific loss of histone H3 lysine 9 trimethylation and HP1gamma/cohesin binding at D4Z4 repeats is associated with facioscapulohumeral dystrophy (FSHD). *PLoS Genetics, 5*(7), e1000559.

# Chapter 10
# Characterization of Chromosomal Ends on the Basis of Chromosome-Specific Telomere Variants and Subtelomeric Repeats in Rice (*Oryza sativa* L.)

**Hiroshi Mizuno, Jianzhong Wu and Takashi Matsumoto**

**Abstract** The telomeric sequences in rice are composed of the plant canonical telomeric sequence TTTAGGG and blocks of at least six variants in a chromosome-specific manner. Variants are more common in the proximal region than in the distal region, suggesting that the telomeres in the proximal region have rarely been reconstructed by the action of telomerase on an evolutionary timescale. The chromosome-specific distribution of telomeric variants suggests that they have arisen from the rapid expansion of a single mutation rather than from the gradual accumulation of random mutations. TrsA—a subtelomeric repetitive sequence of rice—is arrayed in tandem on the ends of 5L, 6S, 8L, 9L, and 12L. Rice subtelomeres are composed of discrete clusters of a TrsA-rich region and a gene-rich region with high transcriptional activity. Intra-chromosomal duplications have resulted in a striking degree of variation in the number and distribution of TrsAs, suggesting that the areas near the ends of the chromosomes are dynamic and variable.

**Keywords** Rice genome • Telomere • Subtelomeric repeats

## 10.1 Introduction

Rice has a useful model monocot genome because of its relatively small size and its high synteny with other cereal crops (Devos 2005). In 2004, the International Rice Genome Sequencing Project completed a map-based sequencing of the genome of *Oryza sativa* L. ssp. *japonica* 'Nipponbare' (IRGSP 2005). IRGSP attempted clone-by-clone genomic sequencing to cover the whole genome, but

H. Mizuno · J. Wu · T. Matsumoto (✉)
National Institute of Agrobiological Sciences, 1-2, Kannondai 2-chome, Tsukuba, Ibaraki 305-8602, Japan
e-mail: mat@nias.affrc.go.jp

E. J. Louis and M. M. Becker (eds.), *Subtelomeres*, DOI: 10.1007/978-3-642-41566-1_10, 

the sequencing was unable to reach plant canonical 5′-TTTAGGG-3′ telomeric repeats (Richards and Ausubel 1988; Chen et al. 2002; Wu et al. 2003) by chromosomal walking (IRGSP 2005). Because the restriction enzymes used to construct P1-derived artificial chromosome (PAC) or bacterial artificial chromosome (BAC) libraries could not cut the canonical telomere array $(TTTAGGG)_n$, these libraries did not contain the clones derived from telomeric sequences. To capture the sequences of chromosomal ends, a rice fosmid library constructed by the cloning of random mechanically sheared DNA (Ammiraju et al. 2005) was screened (Mizuno et al. 2006). Use of this library enabled telomeric sequences to be obtained without the constraints imposed by enzyme site preferences. The complete genomic sequencing of those fosmid clones has revealed the detailed structure of telomeric and subtelomeric repetitive sequences and their junctions.

## 10.2 Telomere Variants

### *10.2.1 Accumulation of Telomeric Variants is Higher in the Proximal Region than in the Distal Region*

The rice chromosomal end has tandemly repeated blocks of the sequence 5′-TTTAGGG-3′ (Wu and Tanksley 1993). The 5′-TTTAGGG-3′ sequence is a canonical telomere repetitive sequence in plants (Richards and Ausubel 1988). These telomeric repeats are organized in the order of 5′-TTTAGGG-3′ from the chromosome-specific region (Yang et al. 2005; Mizuno et al. 2006). The seven-nucleotide unit has deletions, insertions, or substitutions of single nucleotides near the junction between the telomere and the chromosome-specific region. The rate of accumulation of telomeric variants is higher in the proximal region than in the distal region (Mizuno et al. 2008b), suggesting that the telomeres in the proximal region has rarely been reconstructed by the action of telomerase on an evolutionary timescale.

### *10.2.2 Chromosome-Specific Substitution of Telomeric Repeats*

The telomeric variants were not derived from random mutations. Copies of ATTAGGG, CTTAGGG, GTTAGGG, TATAGGG, TTCAGGG, or TTGAGGG are arrayed in tandem, or the same subtypes lie close to each other, at the ends of chromosomes 2L, 3L, 7L, and 10S (Mizuno et al. 2008b) (Fig. 10.1a). Inversion of telomeric repeats is observed adjacent to the beginning of the telomere array on the ends of chromosomes 4L, 7S, and 9S (Fig. 10.1b). Therefore, the proximal

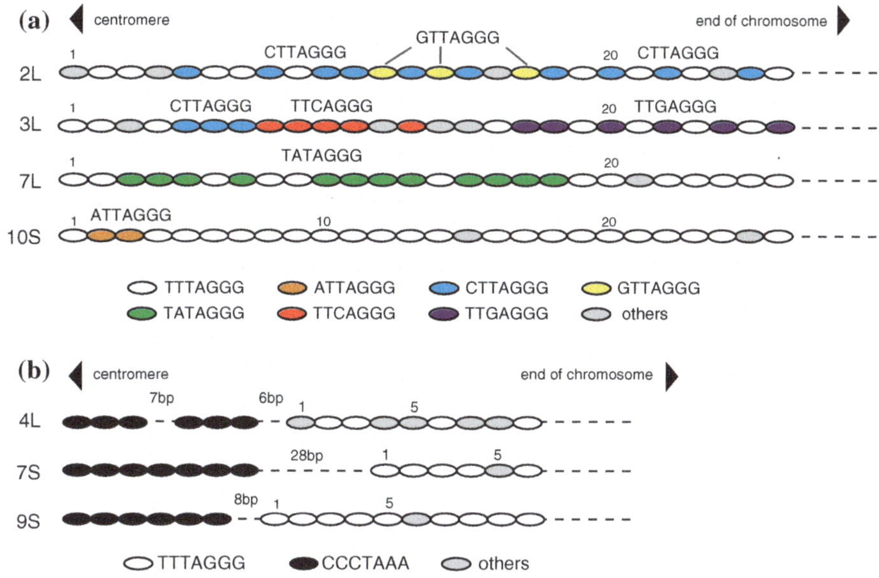


**Fig. 10.1** Nucleotide substitution or inversion in the TTTAGGG repeat. **a** Distribution of TTTAGGG substitution variants. *Each oval* represents the 7-nucleotide unit of the telomeric repeat TTTAGGG (*white*) and the different variants (ATTAGGG, CTTAGGG, GTTAGGG, TATAGGG, TTCAGGG, and TTGAGGG), as shown in the key. *Gray ovals* represent other variants, including deletion (TTAGGG) and insertion (TTTTAGGG) variants. *Numbers* indicate positions of telomere sequences from the junction between the chromosome-specific region and the telomere array. **b** TTTAGGG inversion on chromosomes 4L, 7S, and 9S. *Each box* represents the 7-nucleotide unit of the telomeric repeat TTTAGGG or the inversion (CCCTAAA) variant, as shown in the key

telomeric sequences are composed of blocks of at least six TTTAGGG variants and the canonical sequence in a chromosome-specific manner. The mosaics of blocks of non-canonical telomere sequences could have resulted from polymerase slips during DNA synthesis, a high frequency of DNA recombination, or rapid deletion (Li and Lustig 1996; Watson and Shippen 2007) in the telomere region. The telomeric variants might therefore have arisen from the rapid expansion of a single mutation rather than from the gradual accumulation of random mutations. The functions of these variants remain to be elucidated.

The telomere of rice contains a nucleotide deletion of one T in TTTAGGG: Rice has a 4.9 % content of TTAGGG dispersed throughout the whole of the sequenced region (Mizuno et al. 2008b). TTAGGG is a major sequence in the Asparagales, as it is in vertebrates (Sykorova et al. 2003). The partial or full replacement of the telomeric sequences might have been due to evolutionary changes in the genomic sequence that codes the RNA template or to structural changes in the catalytic subunit of telomerase.

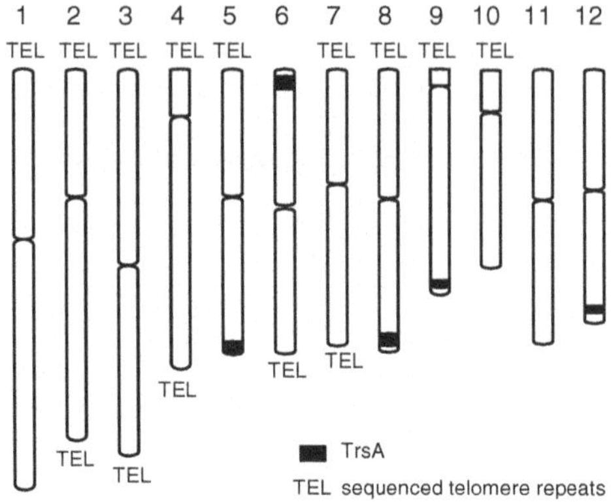


**Fig. 10.2** Distribution of TrsA on the 12 rice chromosomes. *Filled rectangles* on the ends of 5L, 6S, 8L, 9L, and 12L indicate the presence of TrsA clusters. *TEL* indicates the sequenced telomere array; 14 of the 24 chromosome ends (1S, 2S, 2L, 3S, 3L, 4S, 4L, 5S, 6L, 7S, 7L, 8S, 9S, 10S), including their telomeric repeats, have been sequenced

# 10.3 Subtelomeric Repeats

## *10.3.1 Distribution of Subtelomeric Repeats*

Species-specific subtelomeric repeats have been reported in barley (Kilian and Kleinhofs 1992), tobacco (Fajkus et al. 1995), tomato (Ganal et al. 1991), wheat (Mao et al. 1997), and rice (Ohtsubo et al. 1991). Fluorescence in situ hybridization (FISH) analysis has revealed these repeats on almost all ends in barley (13 of 14) (Roder et al. 1993) and tomato (20 of 24) (Zhong et al. 1998). Complete genomic sequencing has revealed that the rice-specific subtelomeric tandem repeat sequence A (TrsA) is found on 5 of 24 ends in rice (Mizuno et al. 2008a). Therefore, the numbers of species-specific subtelomere sequences seem to have expanded, particularly on the ends of the chromosome in each plant.

TrsA is a 355-bp tandemly repeated sequence distributed on the distal ends of chromosomes and is widely distributed in the *Oryza* genus (Ohtsubo et al. 1991). Its distribution has been analyzed by FISH analysis (Ohtsubo and Ohtsubo 1994; Ohmido and Fukui 1997). The diversity of chromosomal loci of TrsA among various rice accessions suggests that there has been dynamic change in the evolutionary history of these loci. Among the 24 chromosome arms of the completely sequenced Nipponbare, TrsA is arrayed in tandem on the ends of only five: 5L, 6S, 8L, 9L, and 12L (Fig. 10.2). TrsA sequences are arranged in discrete clusters of 3–106 copies in a chromosome-specific manner, instead of being distributed uniformly throughout the subtelomeric regions (Fig. 10.3). The canonical telomere array TTTAGGG is repeated at the distal-most part of TrsA-rich ends, with a ~500-bp non-coding sequence junction. Therefore, the distal-most end of the rice chromosome is composed of a telomere array with or without a TrsA junction.

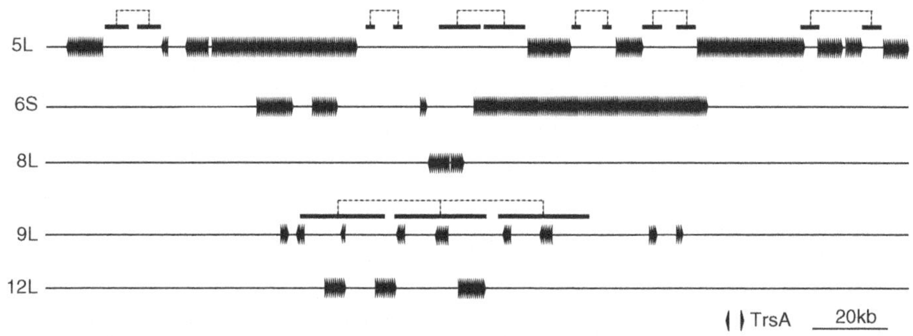


**Fig. 10.3** Structure of TrsA clusters on 5L, 6S, 8L, 9L, and 12L. Clusters are depicted as *arrowheads*. Sequences with the same direction as the canonical TrsA (accession no. D16452) point to the *right*, and those with the opposite direction point *left*. *Dotted lines* indicate regions derived from segmental duplications in 5L and 9L

## *10.3.2 Dynamic Recombination in the Subtelomeric Region*

The structure of the TrsA region suggests the evolutionary history of the amplification of TrsA in the subtelomeric regions. The block of TrsA repeats has been duplicated by intra-chromosomal duplication (Fig. 10.3) (Mizuno et al. 2008a). In addition, TrsA itself has been amplified tandemly after the segmental duplication, as the number of TrsAs differs among duplicated segments on 9L (Mizuno et al. 2008a). The high frequency of DNA recombination at the chromosome ends (Wu et al. 2003; Gaut et al. 2007) might have contributed to the frequent duplications in the subtelomeric region. Moreover, as each duplicated segment is flanked by TrsA, TrsA might have worked as a homologous region required for DNA recombination. Therefore, amplification of TrsA has occurred by segmental duplication and subsequently by tandem duplication of TrsA itself. The amplification has resulted in a striking degree of variation in the number and distribution of TrsAs among the chromosomes, suggesting that the areas near the ends of the chromosomes have a dynamic and variable character.

## *10.3.3 Genes in the Subtelomeric Region*

Genes have been annotated in 500 kb of the distal ends by the Rice Annotation Project (Mizuno et al. 2008a). Expression of most of the annotated genes is supported by the corresponding cDNA data. Expressed genes have not been found between the TrsA arrays; although one gene has been predicted inside the TrsA cluster on 6S, it does not yet have corresponding cDNA data. Thus, rice subtelomeres are composed of discrete clusters of a TrsA-rich region and a gene-rich region with high transcriptional activity. What is the potential role of subtelomeric repeats in rice? Subtelomeric repeats can buffer the spread of gene silencing caused by the telomere position effect (TPE) (Baur et al. 2001). Because the strength of the silencing effect depends on the distance between the gene and the telomere, subtelomeric repeats may help to block the effect of the TPE. TrsA may block the TPE and therefore potentially maintain the expression of subtelomeric genes.

## 10.4 Diversity of Telomere Length

The telomere lengths vary among various accessions of rice. The telomeres of 31 rice accessions (both cultivars and wild species, belonging to the AA, BB, BBCC, CC, CCDD, GG, or HHJJ genome types of *Oryza*) are 5–20 kb long (Mizuno et al. 2006). Marked variation in telomere length is apparent within cultivated rice of the AA genome: The *japonica* cultivar Nipponbare has a relatively low molecular weight pattern, and the *indica* cultivar Kasalath has a relatively high molecular weight pattern. Moreover, variation in telomere length is apparent among chromosomes in Nipponbare. Use of the fiber-FISH method has revealed the diversity of telomere length on each chromosome. Seven telomeres in Nipponbare range from 5.1 to 10.8 kb in length, corresponding to about 730–1,500 copies of the TTTAGGG telomeric repeat (Mizuno et al. 2006). This chromosome-dependent variation might be a consequence of genetic or epigenetic differences among the sequences of subtelomeres; these differences might affect the balance between telomere shortening and telomere elongation. Telomere length has been reported in various plants: 2.5 kb in *Arabidopsis thaliana* (Kotani et al. 1999); 4.5 kb at most in *Melandrium album* (Riha et al. 1998); 60–160 kb (in most cases 90–130 kb) in *Nicotiana tabacum* (Fajkus et al. 1995a); and 1.8–40.0 kb in maize (Burr et al. 1992). Does telomere length differ among different cells? In barley (*Hordeum vulgare*), wide variation in telomere length is apparent during differentiation or aging of cells. The cells that develop in long-term callus cultures have very long telomeres (Kilian et al. 1995). However, the differences in telomere length among different tissues or developmental stages of rice remain to be elucidated.

## 10.5 Conclusions

Rice telomeric sequences are composed of the canonical telomeric sequence TTTAGGG and blocks of at least six TTTAGGG variants in a chromosome-specific manner. This composition suggests that telomeric variants have arisen from the rapid expansion of a single mutation rather than from the gradual accumulation of random mutations.

Subtelomeres are composed of discrete clusters of a TrsA-rich region and a gene-rich region with high transcriptional activity. Intra-chromosomal duplications have resulted in a striking degree of variation in the number and distribution of TrsAs, suggesting that the areas near the ends of the chromosomes are dynamic and variable.

**Acknowledgments** We thank all the members of the Rice Genome Research Program for joining our research and discussions.

## References

Ammiraju, J. S., Yu, Y., Luo, M., Kudrna, D., Kim, H., Goicoechea, J. L., et al. (2005). Random sheared fosmid library as a new genomic tool to accelerate complete finishing of rice (*Oryza sativa* spp. Nipponbare) genome sequence: Sequencing of gap-specific fosmid clones uncovers new euchromatic portions of the genome. *TAG: Theoretical and Applied Genetics, 111*(8), 1596–1607.

Baur, J. A., Zou, Y., Shay, J. W., & Wright, W. E. (2001). Telomere position effect in human cells. *Science, 292*(5524), 2075–2077.

Burr, B., Burr, F. A., Matz, E. C., & Romero-Severson, J. (1992). Pinning down loose ends: Mapping telomeres and factors affecting their length. *Plant Cell, 4*(8), 953–960.

Chen, M., Presting, G., Barbazuk, W. B., Goicoechea, J. L., Blackmon, B., Fang, G., et al. (2002). An integrated physical and genetic map of the rice genome. *Plant Cell, 14*(3), 537–545.

Devos, K. M. (2005). Updating the 'crop circle'. *Current Opinion in Plant Biology, 8*(2), 155–162.

Fajkus, J., Kovarik, A., Kralovics, R., & Bezdek, M. (1995a). Organization of telomeric and subtelomeric chromatin in the higher plant *Nicotiana tabacum*. *Molecular and General Genetics, 247*(5), 633–638.

Fajkus, J., Kralovics, R., Kovarik, A., & Fajkusova, L. (1995b). The telomeric sequence is directly attached to the HRS60 subtelomeric tandem repeat in tobacco chromosomes. *FEBS Letters, 364*(1), 33–35.

Ganal, M. W., Lapitan, N. L., & Tanksley, S. D. (1991). Macrostructure of the tomato telomeres. *Plant Cell, 3*(1), 87–94.

Gaut, B. S., Wright, S. I., Rizzon, C., Dvorak, J., & Anderson, L. K. (2007). Recombination: An underappreciated factor in the evolution of plant genomes. *Nature Reviews Genetics, 8*(1), 77–84.

IRGSP. (2005). The map-based sequence of the rice genome. *Nature, 436*(7052), 793–800.

Kilian, A., & Kleinhofs, A. (1992). Cloning and mapping of telomere-associated sequences from *Hordeum vulgare* L. *Molecular and General Genetics, 235*(1), 153–156.

Kilian, A., Stiff, C., & Kleinhofs, A. (1995). Barley telomeres shorten during differentiation but grow in callus culture. *Proceedings of the National Academy of Sciences, 92*(21), 9555–9559.

Kotani, H., Hosouchi, T., & Tsuruoka, H. (1999). Structural analysis and complete physical map of *Arabidopsis thaliana* chromosome 5 including centromeric and telomeric regions. *DNA Research, 6*(6), 381–386.

Li, B., & Lustig, A. J. (1996). A novel mechanism for telomere size control in *Saccharomyces cerevisiae*. *Genes and Development, 10*(11), 1310–1326.

Mao, L., Devos, K. M., Zhu, L., & Gale, M. D. (1997). Cloning and genetic mapping of wheat telomere-associated sequences. *Molecular and General Genetics, 254*(5), 584–591.

Mizuno, H., Wu, J., Kanamori, H., Fujisawa, M., Namiki, N., Saji, S., et al. (2006). Sequencing and characterization of telomere and subtelomere regions on rice chromosomes 1S, 2S, 2L, 6L, 7S, 7L and 8S. *The Plant Journal, 46*(2), 206–217.

Mizuno, H., Wu, J., Katayose, Y., Kanamori, H., Sasaki, T., & Matsumoto, T. (2008a). Characterization of chromosome ends on the basis of the structure of TrsA subtelomeric repeats in rice (*Oryza sativa* L.). *Molecular Genetics and Genomics, 280*(1), 19–24.

Mizuno, H., Wu, J., Katayose, Y., Kanamori, H., Sasaki, T., & Matsumoto, T. (2008b). Chromosome-specific distribution of nucleotide substitutions in telomeric repeats of rice (*Oryza sativa* L.). *Molecular Biology and Evolution, 25*(1), 62–68.

Ohmido, N., & Fukui, K. (1997). Visual verification of close disposition between a rice A genome-specific DNA sequence (TrsA) and the telomere sequence. *Plant Molecular Biology, 35*(6), 963–968.

Ohtsubo, H., & Ohtsubo, E. (1994). Involvement of transposition in dispersion of tandem repeat sequences (TrsA) in rice genomes. *Molecular and General Genetics, 245*(4), 449–455.

Ohtsubo, H., Umeda, M., & Ohtsubo, E. (1991). Organization of DNA sequences highly repeated in tandem in rice genomes. *Japanese Journal of Genetics, 66*(3), 241–254.

Richards, E. J., & Ausubel, F. M. (1988). Isolation of a higher eukaryotic telomere from *Arabidopsis thaliana. Cell, 53*(1), 127–136.

Riha, K., Fajkus, J., Siroky, J., & Vyskot, B. (1998). Developmental control of telomere lengths and telomerase activity in plants. *Plant Cell, 10*(10), 1691–1698.

Roder, M. S., Lapitan, N. L., Sorrells, M. E., & Tanksley, S. D. (1993). Genetic and physical mapping of barley telomeres. *Molecular and General Genetics, 238*(1–2), 294–303.

Sykorova, E., Lim, K. Y., Kunicka, Z., Chase, M. W., Bennett, M. D., Fajkus, J., et al. (2003). Telomere variability in the monocotyledonous plant order Asparagales. *Proceedings of the Royal Society of London. Series B: Biological Sciences, 270*(1527), 1893–1904.

Watson, J. M., & Shippen, D. E. (2007). Telomere rapid deletion regulates telomere length in *Arabidopsis thaliana. Molecular and Cellular Biology, 27*(5), 1706–1715.

Wu, K. S., & Tanksley, S. D. (1993). Genetic and physical mapping of telomeres and macrosatellites of rice. *Plant Molecular Biology, 22*(5), 861–872.

Wu, J., Mizuno, H., Hayashi-Tsugane, M., Ito, Y., Chiden, Y., Fujisawa, M., et al. (2003). Physical maps and recombination frequency of six rice chromosomes. *The Plant Journal, 36*(5), 720–730.

Yang, T. J., Yu, Y., Chang, S. B., de Jong, H., Oh, C. S., Ahn, S. N., et al. (2005). Toward closing rice telomere gaps: Mapping and sequence characterization of rice subtelomere regions. *TAG: Theoretical and Applied Genetics, 111*(3), 467–478.

Zhong, X. B., Fransz, P. F., Wennekes-Eden, J., Ramanna, M. S., van Kammen, A., Zabel, P., et al. (1998). FISH studies reveal the molecular and chromosomal organization of individual telomere domains in tomato. *The Plant Journal, 13*(4), 507–517.

# Chapter 11
# What is the Specificity of Plant Subtelomeres?

**A. V. Vershinin and E. V. Evtushenko**

**Abstract** According to the current concept, formed through a comprehensive analysis of the molecular structure of human and yeast subtelomeres, these regions are particularly dynamic and variable parts of chromosomes enriched for segmental duplications. This chapter considers to what degree this concept is applicable to the subtelomeres of plant species with different genome sizes paying a special attention on the own results on the rye (*Secale cereale*) subtelomeric heterochromatin. The rye belongs to the species with a large genome size ($8.3 \times 10^9$ bp). The *S. cereale* genome has increased during the evolution mostly through enlargement of the subtelomeric heterochromatic regions. The main components of this heterochromatin are a few multicopy tandemly repeated DNA families. Several arrays of each family localized to separate nonoverlapping domains have been detected in the short arm of the first rye chromosome. They display specific patterns of hierarchical arrangement into multimeric blocks, where the monomers form various higher-order repeat units. In conclusion, the data on a high rate of recombination characteristic of the plant subtelomeres are summarized. The consequence of these recombinations is various types of molecular rearrangements in these chromosomal regions, which contribute to the overall size of the genome.

## 11.1 Introduction

The regions extending from the arrays of specific telomeric DNA sequences inward along the chromosome in the direction of centromere are usually referred to as subtelomeres. The very first notion of molecular structure in these regions

A. V. Vershinin (✉) · E. V. Evtushenko
Institute of Molecular and Cellular Biology, Siberian Branch of the Russian Academy of Sciences, Novosibirsk, Russia 630090
e-mail: avershin@mcb.nsc.ru

E. V. Evtushenko
e-mail: evt@mcb.nsc.ru

E. J. Louis and M. M. Becker (eds.), *Subtelomeres*, DOI: 10.1007/978-3-642-41566-1_11, 

suggested that they contained mostly a rather random mixture of manifold repetitive DNAs, ranging from derivatives of different classes of mobile elements to tandemly organized repeats. Comparative analysis of the first complete sequences of human and yeast subtelomeric DNAs detected a common structure (segmental duplication or duplicon) in several human chromosome ends (Flint et al. 1997). The subsequent intensive research, first and foremost, into the subtelomeric regions of the yeast and human chromosomes, has gradually established the current concept of a large-scale organization and evolutionary dynamics of these regions. According to this concept, the subtelomeres are very plastic and rapidly evolving genomic regions. Multiple translocations between nonhomologous chromosomes, mainly accompanied by a nonhomologous end-joining (NHEJ) mechanism providing for break repair, lead to a remarkably high rate of sequence exchange in the subtelomeres (Louis et al. 1994; Mefford and Trask 2002; Linardopoulou et al. 2005). This brings about the polymorphic patchworks (mosaic) of interchromosomal segmental duplications (Riethman et al. 2004; Linardopoulou et al. 2005). Thus, a large-scale organization of each human subtelomere is largely determined by its specific segmental duplication content and organization, which vary from chromosome to chromosome (Ambrosini et al. 2007).

The postulated high rate of rearrangements taking place in the subtelomeric regions during the evolution has been confirmed by a comprehensive comparison of the X chromosome structure in 87 *Drosophila melanogaster* lines (Anderson et al. 2008). The *drosophila* subtelomeres display a significantly higher level of polymorphism as compared to the adjacent euchromatin regions. However, the question on the generality of the concept on a high plasticity of subtelomeres and their enrichment for segmental duplications, characteristic of the human genome, for the other eukaryotic species is still to be answered. In this chapter, we will try to tackle this problem by the case studies of currently well-characterized subtelomeric regions in plant chromosomes.

## 11.2 Molecular Description of Subtelomeric Regions in Various Plant Species

The most detailed large-scale organization patterns of subtelomeric regions have been obtained for the species with completely sequenced genomes. Among the plants, these are the species with a small genome size, first and foremost, *arabidopsis* and rice.

### *11.2.1 Arabidopsis*

As a rule, the active genes in plants are separated from telomeres by tens of kilobases of repetitive DNA. However, the chromosomes of arabidopsis do not follow this general rule. In contrast, the *Arabidopsis thaliana* subtelomeric regions are remarkably small and simple, in accordance with small genome size and paucity of repetitive sequences of this species (Kuo et al. 2006). So far, a detailed structure has been described for the subtelomeric regions of several chromosomes. Physical mapping and RFLP analysis have shown that the subtelomeres of chromosomes 2 and 4 (left or "northern" arm) house tandemly arranged rDNA genes, NOR2, and NOR4 (Copenhaver and Pikaard 1996). NOR4 is directly associated with the telomeric repeat, and repetitive DNA is absent at the junction between the telomere and rDNA. The presence of short subtelomeric regions (<5 kb) as well as the absence of highly repetitive DNA and transposons was assumed to be a common characteristic of the remaining eight chromosome ends (Arabidopsis Genome Initiative 2000; Heacock et al. 2004).

Unlike the yeast and human subtelomeres, any extended segmental duplications also have not been found in the *arabidopsis* subtelomeres, although some subtelomeric regions do share a few blocks of similarity of low-copy sequences among nonhomologous chromosomes (Kotani et al. 1999; Heacock et al. 2004). For example, analysis of the chromosome 3R subtelomeric region (3RTAS; right or "southern" arm) has demonstrated that while the centromere-proximal portion of 3RTAS contains two potential genes, the telomere-proximal portion contains duplicated fragments, which are also present in the chromosomes 1–3 of Columbia ecotype and chromosome 5 of Wassilewskaja ecotype (Wang et al. 2010). The size of these fragments varies from several hundred to over one thousand base pairs. The structure of these duplicated fragments was similar to the so-called filler DNA, captured by NHEJ during double-strand break repair.

Despite the absence of extended subtelomeres enriched for highly repetitive DNA in arabidopsis, characteristic of the terminal regions of its chromosomes is a dynamic nature. This has been found in a comprehensive study of the structural variations in chromosome 1 subtelomeric region (arm N or 1L) with a length of 3.5 kb involving 35 wild accessions (Kuo et al. 2006). An increased level of large-scale rearrangements relative to the proximal part of this region was observed in its distal part, adjacent to telomeric repeat. These rearrangements were frequently accompanied by deletions exceeding 30 bp, associated with the NHEJ repair mechanism. The proximal part contained a short (104 bp) insertion of mitochondrial DNA as well as the traces of inversions and insertions of LTR retrotransposons. These results allowed Kuo et al. (2006) to suggest a large diversity of genomic events that had taken place in this short subtelomeric region during the evolution.

### *11.2.2 Rice*

The rice genome is approximately 3.5-fold larger than that of *arabidopsis* ($3.9 \times 10^8$ bp vs. $1.1 \times 10^8$ bp; http://data.kew.org/cvalues/); however, it is packed into 12 chromosome pairs. Thus, the average size of a rice chromosome only 1.5-fold exceeds that of arabidopsis. However, the subtelomeric regions of rice chromosomes, unlike the arabidopsis subtelomeres, house several families of tandemly arranged repetitive DNA sequences. Among the 24 chromosome arms, one rice genome-specific 355-bp repeat TrsA is arrayed in tandem on the ends of eight chromosome pairs of *indica* rice (*Oryza sativa* ssp. *Indica;* Ohmido and Fukui 1997) and of five chromosome arms—5L, 6S, 8L, 9L, and 12L, of *japonica* rice (*O. sativa* ssp. *Japonica;* Mizuno et al. 2008). The TrsA sequences in *japonica* subtelomeres are arranged in discrete clusters of 3–106 copies in a chromosome-specific manner. Speculating about the possible mechanisms underlying the origin of such discrete segments of a TrsA-rich region, the authors based on a high similarity of the genomic sequences flanking the TrsA clusters assume a recent duplication around the TrsA-rich region (Mizuno et al. 2008). They also consider that segmental duplications could lead to TrsA amplification along with a tandem duplication of TrsA itself. The amplification has resulted in a striking degree of variation in the number and distribution of TrsAs among the chromosomes, suggesting that the areas near the ends of the chromosomes have a dynamic and variable nature.

Along with the TrsA family, the chromosome-specific telomere-associated tandem repeats have been found in *japonica* rice at both ends of chromosome 7 (TATR7) and on the short arm of chromosome 10 (TATR10) (Yang et al. 2005). The contiguous TATR7 and TATR10 arrays are interrupted by other repetitive elements, for example, parts of LTR retrotransposons, RIRE3, and RIRE9. Thus, this pattern of a large-scale organization, characteristic of the large plant genomes and large chromosomes with the tandem repeat arrays interrupted (flanked) by derivatives of mobile elements of various types, is present already in the small rice chromosomes. Note here that the retrotransposons observed in the subtelomeres are dispersed throughout the entire rice genome, including the centromeric and pericentromeric regions (Wu et al. 2004).

In addition to the repeats of various classes, the rice subtelomeres contain unique DNA sequences and expressed genes (Yang et al. 2005; Mizuno et al. 2008; Fan et al. 2008). Further comprehensive analysis has made it possible to predict that 500-kb regions inward from seven chromosome ends contain putative 598 genes of total 3,500 kb (Mizuno et al. 2006). Thus, an average gene density is one gene per 5.9 kb. This is a much higher rate as compared to the average gene density of the whole genome, amounting to one gene per 9.9 kb (International Rice Genome Sequencing Project 2005). Even the annotation program FGENESH, used in analyzing the complete rice genome, gave an average gene density of one gene per 7.4 kb in the 500-kb subtelomeric regions. A total of 303 genes among 598 predicted genes matched rice full-length cDNAs, suggesting that the rice chromosome ends are gene-rich and display a high transcriptional activity (Mizuno et al. 2006). This assumption is supported by discovery of 12 new genes in the subtelomeric region of *O. sativa* chromosome 3 that have recently originated through independent recombination and transposition events (Fan et al. 2008). Nine of these genes are functional and five have a chimeric structure caused by multiple recombination events in numerous parental precursor genes.

### *11.2.3 Other Species*

Apart from *arabidopsis* and rice, the data about the structural organization of subtelomeric regions in other plant species are considerably sparser, even for such species as maize, the genome of which is intensively studied because of its economic importance.

The size of the maize genome (2.4–2.7 $\times$ $10^9$ bp) exceeds the rice genome approximately 6.5-fold and is comparable to the human genome. Ten pairs of maize chromosomes display a specific pattern of heterochromatin packaging. In addition to the pericentromeric heterochromatin, large heterochromatin blocks form the so-called knobs dispersed over the chromosomes and able to change both their number and localizations. Two families of tandemly repeated DNA sequences have been identified within the maize knobs. Peacock et al. (1981) has shown that a 180-bp tandem repeat is the main component of these knobs. Ananiev et al. (1998) has found another tandem repeat, 350-bp TR-1, is also associated with knobs. Short stretches of DNA sequences within the two repeats show some level of homology, suggesting a common evolutionary origin for the two repeat families (Ananiev et al. 1998). The probes of these families give strong signals at the knobs in FISH (fluorescence in situ hybridization) assay. However, an improved sensitive FISH technique with long exposures detects smaller sites of hybridization at the ends of almost every chromosome arm, adjacent to the telomere tract (Lamb et al. 2007). The frequency and intensity of TR-1 hybridization at the ends of chromosomes were less than in the 180-bp knob satellite. It is known that approximately 80 % of the maize genome is composed of repetitive DNA (Hake and Walbot 1980), mainly LTR retrotransposons (SanMiguel et al. 1996). Thus, it is rather likely that some families of LTR retrotransposons that are distributed along the chromosomes to their very ends (Lamb et al. 2007) are also present in subtelomeres. Note also that analysis of the recombination frequencies along maize chromosomes has demonstrated that they are much higher than is expected from their EST frequencies (Anderson et al. 2006).

The families of tandemly arranged repeats have been most frequently identified as structural elements in the subtelomeric regions of plant chromosomes. They have been described in a wide diversity of species with various genome sizes, namely tobacco (Koukalova et al. 1989), barley (Belostotsky and Ananiev 1990; Killian and Kleinhofs 1992), tomato (Ganal et al. 1991), wheat (Mao et al. 1997), white campion (Buzek et al. 1997), and potato (Torres et al. 2011). We apologize for not citing other papers in this list. The monomer arrays of these families either directly join the telomeric repeat array (Killian and Kleinhofs 1992; Fajkus et al. 1995; Torres et al. 2011) or are separated from it by a short spacer DNA (Zhong et al. 1998). Both variants can present in different chromosomes of the same karyotype (Sykorova et al. 2003). Other classes of repetitive sequences dispersed over other chromosomal regions in addition to the subtelomeres have been also described (Mao et al. 1997; Horakova and Fajkus 1999). However, the description of individual families of different repeats have not yet allowed the structural organization of extended subtelomeric regions of all or the majority of chromosomes constituting plant karyotypes to be characterized in a manner similar to that of subtelomeric regions in the human chromosomes (Riethman et al. 2004; Linardopoulou et al. 2005; Ambrosini et al. 2007).

## 11.3 Heterogeneity of the Internal Organization of Tandem Repeat Arrays is Characteristic of the Rye Subtelomeric Heterochromatin

The rye *Secale cereale* genome is among the largest plant genomes. The average genome size of the flowering plant species is $5.6 \times 10^9$ bp (Rabinowicz and Bennetzen 2006), whereas the rye genome amounts to $8.3 \times 10^9$ bp. Note for comparison that the genome sizes of the closest rye relatives, barley and diploid wheat, are approximately $5 \times 10^9$ and $6 \times 10^9$ bp, respectively. The rye karyotype consists of seven pairs of large chromosomes, insignificantly differing in their size. A specific feature of the rye chromosomes is large heterochromatin blocks at their ends. The presence of heterochromatin blocks is a rather uncertain cytological characteristic of subtelomeres. The wheat (Gill and Kimber 1974) and barley (Linde-Laursen 1978) chromosomes display an intricate pattern of heterochromatin localization; however, their subtelomeres lack large heterochromatin blocks. Even within the genus *Secale,* there is a variation in the size of subtelomeric heterochromatin blocks, accounting for 20 % interspecies variation in the total heterochromatin content in chromosomes (Bennett et al. 1977).

Molecular description of the rye subtelomeric heterochromatin regions has demonstrated that they are enriched for a few multicopy tandemly organized DNA families (Bedbrook et al. 1980; Appels et al. 1981; McIntyre et al. 1990). The molecular organization of the two families most abundant in rye, pSc200 and pSc250, has been studied in most details (Vershinin et al. 1995). Their monomer lengths are 379 and 571 bp, respectively, and they account for ~2.5 and ~1 % of the *S. cereale* genome. As a rule, the signals of these families overlap in a FISH assay on the rye metaphase chromosomes. Note that pSc200 family is localized to the ends of all chromosome arms, while pSc250 signals are absent in some arms. The presence/absence of the signals and their intensities vary in individual rye cultivars (Alkhimova et al. 1999). No pSc200 and pSc250 copies are detectable in the wheat genome by hybridization assays, thereby allowing wheat–rye substitution and addition lines to be used for studying molecular organization of these repeat families in individual rye chromosomes.

An insufficient FISH resolution on metaphase chromosomes prevents from estimating the mutual arrangement of pSc200 and pSc250 families in subtelomeric regions. This can be achieved when applying FISH to more stretched meiotic chromosomes. Figure 11.1 shows the hybridization of pSc200 (green signals) and pSc250 (red signals) to the zygotene chromosomes of wheat–rye monosomic

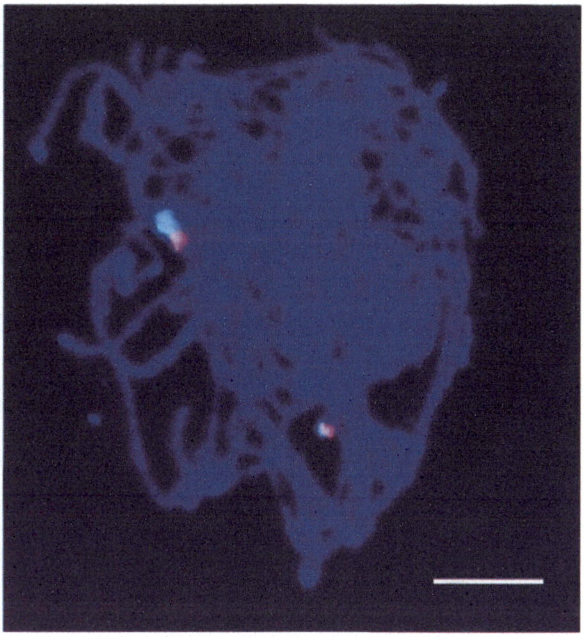

**Fig. 11.1** FISH on meiotic early prophase chromosomes of wheat–rye monosomic 1R line with rye-specific tandemly organized DNA families, pSc200 (*green*) and pSc250 (*red*). *Bar* represents 10 μm

addition line, which contains a single rye chromosome 1R in its genome. It is evident that the extended blocks of pSc200 and pSc250 are located on 1R separately from each other, overlapping only in a limited region (yellow signal) due to an insufficient degree of chromosome stretching. Here, a considerable length of the blocks suggests that each of these families is represented by several arrays of monomers. This assumption was confirmed by pulse-field electrophoresis of high molecular weight DNA isolated from the protoplasts obtained from the wheat–rye line carrying a translocation of the short arm, 1RS, and hydrolyzed with infrequently cutting restriction enzymes, namely *Apa*I, *Bst*XI, *Dra*I, *Not*I, *Pme*I, *Sex*AI, *Sfi*I, and *Swa*I. After the separation, blotting onto nitrocellulose filter, and hybridization to labeled pSc250 probe, six hybridization fragments with a size of 40–300 kb were identified in the 1RS (data not shown).

Along with multicopy tandemly organized families, the rye subtelomeres contain tandemly organized families with considerably smaller copy numbers, which can be regarded as chromosome-specific. Figure 11.2 shows an example of such family, demonstrating the localization of an *Tai*I repeat with a monomer length of 576 bp. This family is localized to the ends of only one arm of two chromosome pairs, one of which is the chromosome 1 short arm (denoted with arrows), which is indicated by the presence of a constriction in this arm. To understand a more detailed molecular structure of subtelomeres, including the mutual arrangement of pSc200 and pSc250 arrays and their molecular environment, we have analyzed the BAC library of chromosome 1 short arm (1RS), constructed at the Institute of Experimental Botany (Olomouc, Czech Republic) under the guidance of

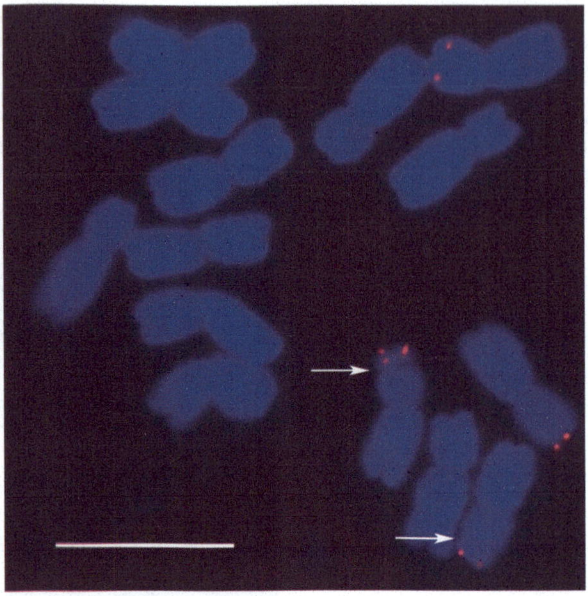

**Fig. 11.2** FISH of *Tai*I tandem repeat family (monomer length, 576 bp) on metaphase rye chromosomes (cultivar Imperial). *Arrows* denote the localization of *Tai*I repeat on the short arm of rye chromosome 1. *Bar* represents 10 μm

Dr. J. Dolezel and kindly provided by Dr. J. Safar. Hundreds of clones containing copies of monomers from the pSc200 and pSc250 families have been identified in the library with seven–eightfold 1RS DNA coverage. However, none of these BAC clones simultaneously carried copies of the monomers belonging to both families. This confirms the above-mentioned FISH data for meiotic chromosomes that each of these tandemly arranged families is localized to separate nonoverlapping domains and that their arrays do not directly adjoin each other.

We have conducted restriction mapping of some BAC clones carrying arrays of pSc250 monomers. After hydrolysis with *Pst*I restriction enzyme and pulse-field gel electrophoresis (PFGE), DNA of four clones (17/C17, 19H13, 1/G22, and 1/K6) displayed a ladder pattern of pSc250 hybridization fragments (Fig. 11.3a), which is typical of the tandemly arranged repeats and comprises monomers, dimers, and elements of higher order, to pentamers and even hexamers (in BAC 17/C17 and BAC 19H/13). Another restriction enzyme, *Hin*dIII (Fig. 11.3b), generates dimers and tetramers within the tandem arrays of 17/C17, 1/G22, and 1/K6 and only tetramers, for 19H/13. A considerable similarity in the restriction and hybridization patterns of 17/C17, 19H/13, 1/G22, and 1/K6 clones suggest that they are likely to originate from the same 1RS region.

The patterns of two other BAC clones, 122/F3 and 12/I5, considerably differ from both each other and the above-described group of four clones. Moreover, while the pSc250 DNA within BAC 12/I5 after *Pst*I hydrolysis does not give a ladder pattern at all (forms no oligomers), DNA of BAC 122/F3 contains both pSc250 dimers and trimers. On the contrary, the pSc250 monomer arrays within BAC 122/F3 lack any *Hin*dIII sites, while the corresponding array in BAC 12/I5 is arranged

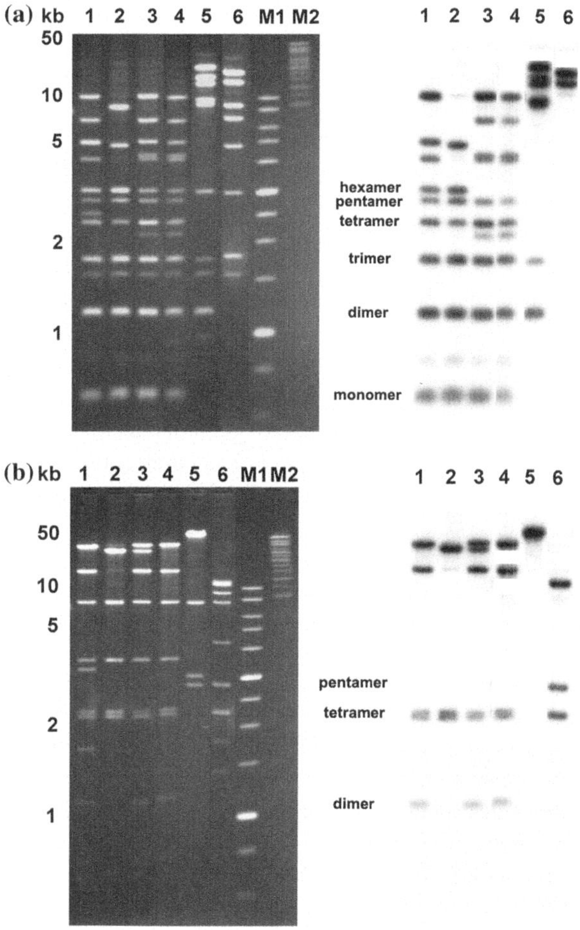


**Fig. 11.3** PFGE of the BAC clones containing pSc250 arrays; DNA is hydrolyzed with (**a**) *Pst*I and (**b**) *Hin*dIII; *left*, staining with ethidium bromide and *right*, blot hybridization with pSc250: (*1*) 17/C17; (*2*) 19/H13; (*3*) 1/K6; (*4*) 1/G22; (*5*) 122/F3; and (*6*) 12/I5. Markers: *M1*, 1 kb and *M2*, 8–48 kb; fragment lengths are shown (kb)

into *Hin*dIII tetramers and pentamers. Further sequencing of all the six BAC clones has demonstrated that 122/F3 and 12/I5 contain extended arrays of pSc250 monomers with lengths of 57 and 38 kb, respectively, flanked from both sides by the nonarray DNA mainly composed of derivatives of various classes of mobile elements. Thus, the analyzed set of BAC clones contains the representatives of at least three distinct arrays of pSc250 monomers with different higher-order internal arrangement of their monomers. A similar high heterogeneity is also characteristic of the internal organization of pSc200 arrays (data not shown).

According to the model proposed by Kimura and Ohta (1979), mutations, random drift, sister-chromatid unequal crossing-over, and meiotic interchromosomal crossing-over contribute to the variation of repetitive DNA sequences during the evolution. The nonreciprocal exchanges between nonhomologous chromosomes through breakage–fusion–bridge (BFB) cycles of a chromosome and the chromatid types, first described by McClintock (1941), represent another mechanism for generation of significant differences in the structural organization of subtelomeric regions. As we see it, a considerable heterogeneity of the higher-order internal structure of the monomer arrays of pSc200 and pSc250 tandem repeat families is another evidence for the high intensity of the described processes during the evolution and formation of the rye subtelomeric regions. The monomers of these families might have worked as the homologous regions required for DNA recombination, which is the key event in both intra- and interchromosomal exchanges. A high level of an intrafamily homology between the DNA sequences of monomers, which exceeds 90 %, and their high copy numbers are also the factors enhancing intensity of the recombination processes. Presumably, these processes included both homologous recombinations and multiple translocations between nonhomologous chromosomes accompanied by NHEJ. Cytological evidences for interchromosomal connections between rye chromosomes have been obtained as early as the 1980s using C-banding (Viinikka and Nokkala 1981); these connections were observed at the diplotene and disappeared before metaphase I. Later, a direct involvement of pSc200 DNA in the association of subtelomeric regions between two or more bivalents was demonstrated by FISH (Gonzalez-Garcia et al. 2006). Such association could result in ectopic recombination between subtelomeres of different bivalents, if crossing-over took place between homologous sequences of nonhomologous chromosomes.

## 11.4 What is the Specificity of Subtelomeric Regions?

The centromeres and telomeres are attributed to specialized chromosomal regions, first and foremost, because they are destined to fulfill certain universal functions of paramount importance in the cells of all eukaryotic species. Another characteristic feature of these regions is in that they contain specialized molecular components. The telomeres of the overwhelming majority of species comprise extended arrays of a short DNA repeat and the proteins bound to them, while characteristic of the centromeres is the presence of a special histone H3 modification in the chromatin. This makes it possible to determine the boundaries, although relative, of the specialized regions according to abundance of the mentioned molecular components. None of the listed characteristics can be currently applied to the subtelomeres. So far, no conserved elements indicative of an important function have been discovered in the subtelomeres. Numerous copies of the telomeric repeat $(TTAGGG)_n$ and its derivatives detected in the subtelomeres of the chromosomes of human, arabidopsis, wheat (Uchida et al. 2002; Ambrosini et al. 2007), and other species imply that subtelomeres may contain internal binding/interaction sites for some telomere-binding proteins (Ambrosini et al. 2007). The DNA composition of subtelomeres displays a considerable diversity. As has been noted above, they can contain DNA sequences of any known molecular nature, including tandem repeats; manifold mobile elements; and various coding sequences, including NOR2 and NOR4 in arabidopsis (Copenhaver and Pikaard 1996) or numerous genes in rice (Fan et al. 2008). Since the subtelomeres lack any specific conserved molecular components, we can speak only about the boundary of subtelomeres in the region adjacent to the telomere, regarding it as the outermost position of the telomeric repeat or equivalent DNA sequences, such as HeT-A, TART, or TAHRE retrotransposons in drosophila.

A significant structural diversity of subtelomeres has brought forth various hypotheses on the functional role of these regions in stability, formation, and behavior of chromosomes. According to one of the hypotheses, which is based on the plasticity of subtelomeres and their gene richness (as in rice chromosome 3), subtelomeres may facilitate gene recombination and transposon insertions and serve as hot spots for new gene origination in rice genomes (Fan et al. 2008). On the other hand, the subtelomeres enriched for noncoding repetitive DNA sequences have been regarded as a sort of "airbags" for protecting the distally located genes in the case of a telomere shortening, for example, caused by telomerase activity loss (Lundblad and Blackburn 1993). Evidently, this assumption is inapplicable to the above-described arabidopsis and rice telomeres.

Studies into the structural arrangement of subtelomeric chromosomal regions in many species suggest that the most general characteristic of these regions is their high plasticity determined by intensive recombination events. Cytological observations have demonstrated that the homologous chromosome synapsis is initiated from the chromosome ends (Schwarzacher 2003; Harper et al. 2004). This implies frequent homologous recombinations in subtelomeric regions, being most likely enhanced by the presence of a large copy number of identical monomers in tandem repeat arrays. However, it has been repeatedly noted that another major mechanism—illegitimate (nonhomologous) recombination through NHEJ (Gorbunova and Levy 1999; Puchta 2005)—is a more important contributor to the plant evolutionary plasticity. In large genomes containing large heterochromatin block at the chromosome ends, the presence of homologous tandem repeats in high copy numbers, similar to the rye chromosomes, should undoubtedly enhance, as in the case of homologous recombinations, a frequent sequence exchange between nonhomologous chromosome ends. This is also suggested by the data on comparative studies of the subtelomeres in tomato chromosomes, where a heterogeneous structure of arrays composed of a species-specific subtelomeric tandem repeat, TGR1, was observed in 20 of the 24 chromosomes (Zhong et al. 1998).

Despite the absence of heterochromatin regions and, correspondingly, enrichment for homologous copies of tandem repeats at the chromosome ends, the human subtelomeres, nonetheless, retain a high level of translocations between the chromosome ends, generating subtelomeric duplications via NHEJ (Linardopoulou et al. 2005). This is suggested by a quantitative estimation made at Riethman's lab (Riethman et al. 2004; Ambrosini et al. 2007), which demonstrates that the human subtelomeric segmental duplication regions comprise about 25 % of the most distal 500-kb segments and 80 % of the most distal 100-kb segments in human DNA. In addition, a high-resolution analysis of the subtelomeric duplication sequence content and organization shows significant differences in the levels of sequence similarity between distinct subtelomere duplicon families as well as large variations in the types and sequence organization of duplicons present in particular subtelomeres (Ambrosini et al. 2007).

In plants, the experimental systems inducing double-strand breaks (DSBs) have allowed various processes accompanying NHEJ to be revealed. In particular, many DSB repair events in tobacco involve deletions and addition of filler DNA (Salomon and Puchta 1998). A comparison of the NHEJ processing at DSB in arabidopsis and tobacco has shown that the average length of deletions recovered in tobacco is relatively smaller and is often accompanied by sequence insertions, whereas deletions in *A. thaliana* are frequently larger and lack insertions (Kirik et al. 2000). Since the tobacco genome is approximately 20-fold larger in size as compared to arabidopsis, it has been assumed that the NHEJ mechanism of DSB repair has a species-specific nature and, in particular, can be a cause of the interspecific differences in the genome size (Kirik et al. 2000). The above-briefed data on the structural variations in the subtelomeric region of chromosome 1 from 35 wild accessions of arabidopsis (Kuo et al. 2006) match well this concept. The distal part of this region, adjacent to the telomeric repeat, displays an increased level of large-scale rearrangements, frequently accompanied by deletions exceeding 30 bp in length.

A comprehensive analysis of the structure and evolution of subtelomeric regions similar to that conducted for the human subtelomeres (Riethman et al. 2004; Linardopoulou et al. 2005; Ambrosini et al. 2007; Rudd et al. 2009) yet has not been performed for plant species. Therefore, it is currently reasonable to forgo any broad generalizations. According to our opinion, only one characteristic for subtelomeric regions in plant chromosomes that may be regarded as general for all eukaryotic species has taken shape. As we mentioned above, this is a high-plasticity stemming from a high rate of recombinations. This can lead to various types of large-scale rearrangements, appearing as segmental duplications (as in human), amplification and formation of extended tandem repeat arrays (as in rye), extended deletions and the corresponding reduction in the junction between a tandem repeat array and coding genome regions (as in arabidopsis), and, presumably, still undiscovered processes. Here, as in the telomeres, the conserved functions may be provided by various sequences involved in different molecular processes (Louis and Vershinin 2005).

**Acknowledgments** We are grateful to Drs. J. Dolezel and J. Safar, Institute of Experimental Botany, Olomouc, Czech Republic, who kindly provided the 1RS BAC library. The research of authors is supported by the Siberian Branch of the Russian Academy of Sciences (integration project 51) and Russian Foundation for Basic Research (Grants 08-04-00784 and 12-04-00512).

## References

Alkhimova, E. G., Heslop-Harrison, J. S., Shchapova, A. I., & Vershinin, A. V. (1999). Rye chromosome variability in wheat–rye addition and substitution lines. *Chromosome Research, 7*, 205–212.

Ambrosini, A., Paul, S., Hu, S., & Riethman, H. (2007). Human subtelomeric duplicon structure and organization. *Genome Biology, 8*, R151.

Ananiev, E. V., Phillips, R. L., & Rines, H. W. (1998). A knob-associated tandem repeat in maize capable of forming fold-back DNA segments: Are chromosome knobs megatransposons? *Proceedings of the National Academy of Sciences of the United States of America, 95*, 10785–10790.

Anderson, J. A., Song, Y. S., & Langley, C. H. (2008). Molecular population genetics of *Drosophila* subtelomeric DNA. *Genetics, 178*, 477–487.

Anderson, L. K., Lai, A., Stack, S. M., Rizzon, C., & Gaut, B. S. (2006). Uneven distribution of expressed sequence tag loci on maize pachytene chromosomes. *Genome Research, 16*, 115–122.

Appels, R., Dennis, E. S., Smith, D. R., & Peacock, W. J. (1981). Two repeated DNA sequences from the heterochromatic regions of rye *Secale cereale* chromosomes. *Chromosoma, 84*, 265–277.

Arabidopsis Genome Initiative. (2000). Analysis of the genome sequence of the flowering plant *Arabidopsis thaliana*. *Nature, 408*, 796–815.

Bedbrook, J. R., Jones, J., O'Dell, M., Tompson, R., & Flavell, R. (1980). A molecular description of telomeric heterochromatin in *Secale* species. *Cell, 19*, 545–560.

Belostotsky, D. A., & Ananiev, E. V. (1990). Characterization of relic DNA from barley genome. *Theoretical and Applied Genetics, 80*, 374–380.

Bennett, M. D., Gustafson, J. P., & Smith, J. B. (1977). Variation in nuclear DNA in the genus *Secale*. *Chromosoma, 62*, 149–176.

Buzek, J., Koutnikova, H., Houben, A., et al. (1997). Isolation and characterization of X chromosome-derived DNA sequences from a dioecious plant *Melandrium album*. *Chromosome Research, 5*, 57–65.

Copenhaver, G. P., & Pikaard, C. S. (1996). RFLP and physical mapping with an rDNA-specific endonuclease reveals that nucleolus organizer regions of *Arabidopsis thaliana* adjoin the telomeres on chromosomes 2 and 4. *The Plant Journal, 9*, 259–272.

Fajkus, J., Kralovics, R., Kovarik, A., & Fajkusova, L. (1995). The telomeric sequence is directly attached to the HRS60 subtelomeric tandem repeat in tobacco chromosomes. *FEBS Letters, 364*, 33–35.

Fan, C., Zhang, Y., Yu, Y., Rounsley, S., Long, M., & Wing, R. A. (2008). The subtelomere of *Oryza sativa* chromosome 3 short arm as a hot bed of new gene origination in rice. *Molecular Plant, 1*, 839–850.

Flint, J., Bates, G. P., Clark, K., Dorman, A., Willington, D., Roe, B. A., et al. (1997). Sequence comparison of human and yeast telomeres identifies structurally distinct subtelomeric domains. *Human Molecular Genetics, 6*, 1305–1314.

Ganal, M. W., Lapitan, N. L. V., & Tanksley, S. D. (1991). Macrostructure of the tomato telomeres. *The Plant Cell, 3*, 87–94.

Gill, B. S., & Kimber, G. (1974). Giemsa C-banding and the evolution of wheat. *Proceedings of the National Academy of Sciences of the United States of America, 71*, 4086–4090.

Gonzalez-Garcia, M., Gonzalez-Sanchez, M., & Puertas, M. J. (2006). The high variability of subtelomeric heterochromatin and connections between nonhomologous chromosomes, suggest frequent ectopic recombination in rye meiocytes. *Cytogenetic and Genome Research, 115*, 179–185.

Gorbunova, V. V., & Levy, A. A. (1999). How plants make ends meet: DNA double-strand break repair. *Trends in Plant Science, 4*, 263–269.

Hake, S., & Walbot, V. (1980). The genome of *Zea mays*, its organization and homology to related grasses. *Chromosoma, 79*, 251–270.

Harper, L., Golubovskaya, I., & Cande, W. Z. (2004). A bouquet of chromosomes. *Journal of Cell Science, 117*, 4025–4032.

Heacock, M., Spangler, E., Riha, K., Puizina, J., & Shippen, D. E. (2004). Molecular analysis of telomeric fusions in Arabidopsis: Multiple pathways for chromosome end-joining. *The EMBO Journal, 23*, 2304–2313.

Horakova, M., & Fajkus, J. (1999). *TAS49*—a dispersed repetitive sequence isolated from subtelomeric regions of *Nicotiana tomentosiformis* chromosomes. *Genome, 43*, 273–284.

International Rice Genome Sequencing Project. (2005). The map-based sequence of the rice genome. *Nature, 436*, 793–800.

Killian, A., & Kleinhofs, A. (1992). Cloning and mapping of telomere-associated sequences from *Hordeum vulgare* L. *Molecular and General Genetics, 235*, 153–156.

Kimura, M., & Ohta, T. (1979). Population genetics of multigene family with special reference to decrease of genetic correlation with distance between gene members on a chromosome. *Proceedings of the National Academy of Sciences of the United States of America, 76*, 4001–4005.

Kirik, A., Salomon, S., & Puchta, H. (2000). Species-specific double-strand break repair and genome evolution in plants. *The EMBO Journal, 19*, 5562–5566.

Kotani, H., Hosouchi, T., & Tsuruoka, H. (1999). Structural analysis and complete physical map of *Arabidopsis thaliana* chromosome 5 including centromeric and telomeric regions. *DNA Research, 6*, 381–386.

Koukalova, B., Reich, J., Matyasek, R., Kuhrova, V., & Bezdek, M. (1989). A *Bam*HI family of highly repeated DNA sequences of *Nicotiana tabacum. Theoretical and Applied Genetics, 78*, 77–80.

Kuo, H.-F., Olsen, K. M., & Richards, E. J. (2006). Natural variation in a subtelomeric region of *Arabidopsis*: Implications for the genomic dynamics of a chromosome end. *Genetics, 173*, 401–417.

Lamb, J. C., Meyer, J. M., Corcoran, B., Kato, A., Han, F., & Birchler, J. A. (2007). Distinct chromosomal distribution of highly repetitive sequences in maize. *Chromosome Research, 15*, 33–49.

Linardopoulou, E. V., Williams, E. M., Fan, Y., Friedman, C., Young, J. M., & Trask, B. J. (2005). Human subtelomeres are hot spots of interchromosomal recombination and segmental duplication. *Nature, 437*, 94–100.

Linde-Laursen, I. (1978). Giemsa C-banding of barley chromosomes. I. Banding pattern polymorphism. *Hereditas, 88*, 55–64.

Louis, E. J., Naumova, E. S., Lee, A., Naumov, G., & Haber, J. E. (1994). The chromosome end in yeasts: Its mosaic nature and influence on recombinational dynamics. *Genetics, 136*, 789–802.

Louis, E. J., & Vershinin, A. V. (2005). Chromosome ends: Different sequences may provide conserved functions. *BioEssays, 27*, 685–697.

Lundblad, V., & Blackburn, E. H. (1993). An alternative pathway for yeast telomere maintenance rescues est1-senescence. *Cell, 73*, 347–360.

Mao, L., Devos, K. M., Zhu, L., & Gale, M. D. (1997). Cloning and genetic mapping of wheat telomere-associated sequences. *Molecular and General Genetics, 254*, 584–591.

McClintock, B. (1941). The stability of broken ends of chromosomes in *Zea mays*. *Genetics, 26*, 234–282.

McIntyre, C. L., Pereira, S., Moran, L. B., & Appels, R. (1990). New *Secale cereale* (rye) DNA derivatives for the detection of rye chromosome segments in wheat. *Genome, 33*, 317–323.

Mefford, H. C., & Trask, B. J. (2002). The complex structure and dynamic evolution of human subtelomeres. *Nature Reviews Genetics, 3*, 91–102.

Mizuno, H., Wu, J., Kanamori, H., Fujisawa, M., Namiki, N., Saji, S., et al. (2006). Sequencing and characterization of telomere and subtelomere regions on rice chromosomes 1S, 2S, 2L, 6L, 7S, 7L and 8S. *The Plant Journal, 46*, 206–217.

Mizuno, H., Wu, J., Katayose, Y., Kanamori, H., Sasaki, T., & Matsumoto, T. (2008). Characterization of chromosome ends on the basis of the structure of TrsA subtelomeric repeats in rice (*Oryza sativa* L.). *Molecular Genetics and Genomics, 280*, 19–24.

Ohmido, N., & Fukui, K. (1997). Visual verification of close disposition between a rice A-genome specific DNA sequence (TrsA) and the telomere sequences. *Plant Molecular Biology, 35*, 963–968.

Peacock, W. J., Dennis, E. S., Roades, M. M., & Pryor, A. (1981). Highly repeated DNA sequence limited to knob heterochromatin in maize. *Proceedings of the National Academy of Sciences of the United States of America, 78*, 4490–4494.

Puchta, H. (2005). The repair of double-strand breaks in plants: Mechanisms and consequences for genome evolution. *Journal of Experimental Botany, 56*, 1–14.

Rabinowicz, P. D., & Bennetzen, J. L. (2006). The maize genome as a model for efficient sequence analysis of large plant genomes. *Current Opinion in Plant Biology, 9*, 149–156.

Riethman, H., Ambrosini, A., Castaneda, C., Finklestein, J., Hu, X.-L., Mudunuri, U., et al. (2004). Mapping and initial analysis of human subtelomeric sequence assemblies. *Genome Research, 14*, 18–28.

Rudd, M. K., Endicott, R. M., Friedman, C., Walker, M., et al. (2009). Comparative sequence analysis of primate subtelomeres originating from a chromosome fission event. *Genome Research, 19*, 33–41.

Salomon, S., & Puchta, H. (1998). Capture of genomic and T-DNA sequences during double-strand break repair in somatic plant cells. *The EMBO Journal, 17*, 6086–6095.

SanMiguel, P., Tikhonov, A., Jin, Y. K., et al. (1996). Nested retrotransposons in the intergenic regions of the maize genome. *Science, 274*, 765–768.

Schwarzacher, T. (2003). Meiosis, recombination and chromosomes: A review of gene isolation and fluorescent in situ hybridization data in plants. *Journal of Experimental Botany, 54*, 11–23.

Sykorova, E., Cartagena, J., Horakova, M., Fukui, K., & Fajkus, J. (2003). Characterization of telomere-subtelomere junctions in *Silene latifolia. Molecular Genetics and Genomics, 269*, 13–20.

Torres, G. A., Gong, Z., Iovene, M., et al. (2011). Organization and evolution of subtelomeric satellite repeats in the potato genome. *Genes Genomes Genet, 1*, 85–92.

Uchida, W., Matsunaga, S., Sugiyama, R., & Kawano, S. (2002). Interstitial telomere-like repeats in the *Arabidopsis thaliana* genome. *Genes and Genetic Systems, 77*, 63–67.

Vershinin, A. V., Schwarzacher, T., & Heslop-Harrison, J. S. (1995). The large-scale organization of repetitive DNA families at the telomeres of rye chromosomes. *The Plant Cell, 7*, 1823–1833.

Viinikka, Y., & Nokkala, S. (1981). Interchromosomal connections in meiosis of *Secale cereale. Hereditas, 95*, 219–224.

Wang, C.-T., Ho, C.-H., Hseu, M.-J., & Chen, C.-M. (2010). The subtelomeric region of the *Arabidopsis thaliana* chromosome IIIR contains potential genes and duplicated fragments from other chromosomes. *Plant Molecular Biology, 74*, 155–166.

Wu, J., Yamagata, H., Hayashi-Tsugane, M., et al. (2004). Composition and structure of the centromeric region of rice chromosome 8. *The Plant Cell, 16*, 967–976.

Yang, T.-J., Yu, Y., Chang, S.-B., de Jong, H., Oh, C.-S., Ahn, S.-N., et al. (2005). Toward closing rice telomere gaps: Mapping and sequence characterization of rice subtelomere regions. *Theoretical and Applied Genetics, 111*, 467–478.

Zhong, X.-B., Fransz, P. F., Wennekes-Eden, J., et al. (1998). FISH studies reveal the molecular and chromosomal organization of individual telomere domains in tomato. *The Plant Journal, 13*, 507–517.

# Chapter 12
# Subtelomeres in *Drosophila* and Other Diptera

**James M. Mason and Alfredo Villasante**

**Abstract** While *Drosophila* telomeres are considered unusual, because they lack short, telomerase-generated telomeric repeat, in other ways they are similar to telomeres found in most eukaryotes. Subtelomeric repeated DNA sequences, for example, exist between the terminal DNA array and the unique sequences found in the euchromatic chromosome arms. Subtelomeric sequences in *Drosophila* consist of complex repeat motifs that are shared among chromosome ends, although the arrangement of the motifs varies considerably. While these motifs diverge rapidly, similarities can still be found in sibling species. Surprisingly, the subtelomeres seem to be able to communicate with the rest of the genome; deletions of the *2*L subtelomere suppress telomere position effect at other chromosome ends, and insertions of transposable elements into the *X*L subtelomere can silence homologous sequences in an RNAi-dependent manner. While *Drosophila* telomeres are maintained by targeted transposition of a small group of retrotransposons, telomeres in nematoceran flies seem to be maintained by gene conversion among telomeric sequences that in many ways resemble complex subtelomeric repeats.

## 12.1 Introduction

Most eukaryotic chromosomes end in arrays of simple telomerase-generated G-rich repeats. However, as a dipteran ancestor lost the telomerase gene, telomeres of nematoceran flies are elongated by recombination, while *Drosophila* telomeres

J. M. Mason (✉)
Laboratory of Molecular Genetics, National Institute of Environmental Health Sciences, Research Triangle Park, NC, USA
e-mail: masonj@niehs.nih.gov

A. Villasante
Centro de Biología Molecular "Severo Ochoa" (CSIC-UAM), Universidad Autónoma de Madrid, Madrid, Spain

E. J. Louis and M. M. Becker (eds.), *Subtelomeres*, DOI: 10.1007/978-3-642-41566-1_12, 

are primarily maintained by retrotransposition of telomeric elements to chromosome ends.

Comparative analysis of the telomeres from 12 *Drosophila* species has identified many phylogenetically distinct, telomere-specific, non-long terminal repeat (LTR) retrotransposons belonging to the *jockey* clade. Since the phylogenetic relationships among the telomeric element agrees with the species phylogeny, these elements seem to derive from an ancestral element that was recruited to perform telomere maintenance (Villasante et al. 2007). In *D. melanogaster*, head-to-tail arrays of telomeric retrotransposons, *HeT-A*, *TART* and *TAHRE* (collectively abbreviated as HTT), form the chromosome termini (Mason et al. 2008). Attachment of these elements to chromosome termini by their 3′ oligo(A) ends, coupled with end erosion due to the end replication problem, results in variably 5′ truncated elements in the terminal array. To overcome this truncation, transcription of *HeT-A* elements starts at the 3′ end and reads into the downstream element and possibly into the subtelomeric sequences.

The overall structure of the telomeres in the species of the melanogaster subgroup resembles the structure of other eukaryotic telomeres: The most distal regions are made of HTT arrays, and between these arrays and the unique chromosomal sequences, there are subtelomeric regions, often called telomere-associated sequences (TAS), composed of various tandem repeats. A remarkable property of the subtelomeric regions of yeast, primates, and plants is their evolutionary plasticity (Broun et al. 1992; Brown et al. 1990; Louis and Haber 1992; Mefford and Trask 2002). *Drosophila* subtelomeres are also dynamic and vary both within and between species (Anderson et al. 2008; Kern and Begun 2008).

## 12.2 Structure and Evolution of *Drosophila* Subtelomeric Regions

Most information about *Drosophila* DNA sequences, including the subtelomeric regions, comes from an isogenic strain used as the reference for the *D. melanogaster* genome, and in which each TAS array covers about 20 kb (Abad et al. 2004). However, as TAS from the left arm of the *X* chromosome (*X*L) is absent in this and some other strains, the sequence of *X*L TAS was obtained from the *Dp1187* chromosome (Karpen and Spradling 1992). The *4*R TAS is also absent from the reference strain (Abad et al. 2004). Similarly, many laboratory stocks lack either the *2*L (Mason et al. 2004) or the *3*L TAS repeats (JMM, unpublished data), and individuals that lack *2*L or *3*L subtelomeric repeats have been collected from natural populations (Kern and Begun 2008; Mechler et al. 1985). Thus, it appears that at least some of the subtelomeric arrays may be deleted without a strongly deleterious effect.

The *X*L subtelomeric region consists of 1.8-kb distal repeats and two copies of a proximal 0.9-kb repeat. The 1.8-kb repeat contains three tandem copies of defective *invader4* LTRs (Fig. 12.1, red bars). These LTRs also appear in tandem within the 1,000- and 984-bp distal repeats of the *2*R and *3*R TAS arrays, respectively.

Thus, *X*L TAS shares regions of homology with *2*R and *3*R TAS (up to 88 % identity), and *2*R TAS shares homology with *3*R TAS (up to 98 % identity) principally through these *invader4* LTRs. The *invader4* LTR clusters are hot spots for transposable *P* element insertions (Karpen and Spradling 1992; Phalke et al. 2009).

2L TAS is comprised of tandem arrays of a 458-bp repeat unit (Walter et al. 1995), which shares high homology (99.5 % identity) with the 458-bp repeat that forms the *3*L subtelomeric region. Moreover, the 1.8-kb and 0.9-kb repeats of *X*L TAS contain a 160-bp region (green in Fig. 12.1) that shares homology with the

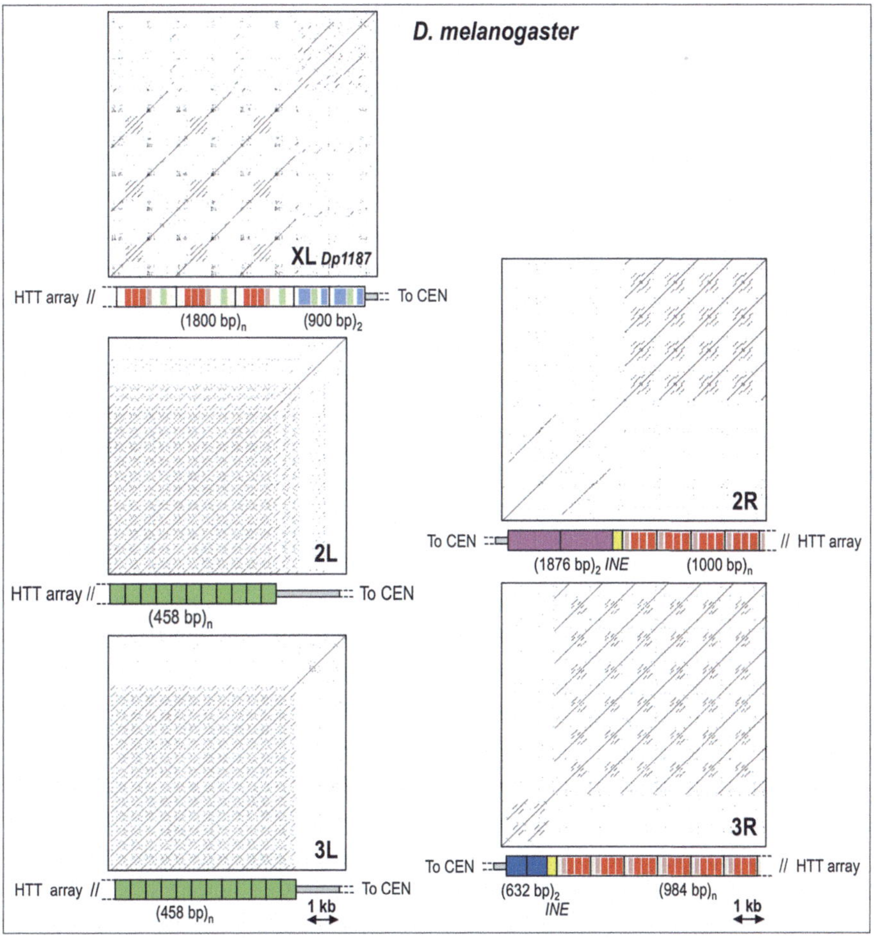


**Fig. 12.1** *D. melanogaster* subtelomeric regions. The diagram of *X*L TAS is based on sequences derived from the *Dp1187* chromosome. The other subtelomeric regions are derived from the strain used to generate the *Drosophila* reference genomic sequence. Dot-matrix comparisons of a subtelomeric region with itself are shown at the top of each diagram. *Each dot* in the diagram represents a short region of homology. The average size of the repeats is indicated. INE element sequences are *yellow*

458-bp repeats in 2L and *3*L TAS (up to 77 % identity), and the 0.9-kb repeat has two regions (blue in Fig. 12.1) that share homology with the 632-bp proximal repeat of *3*R TAS (up to 78 and 81 % identity). Thus, although each subtelomeric array is different, they share similar sequence motifs.

To investigate the variability of *Drosophila* subtelomeres, it is essential to analyze telomere sequences from multiple species with various degrees of divergence. All the subtelomeric regions in *D. melanogaster* are well assembled, but this is not the case for the other 11 *Drosophila* genomes that have been sequenced (Clark et al. 2007). Yet, comparison of *D. melanogaster* TAS arrays with the homologous regions available from the sibling species has revealed a high rate of sequence evolution. In *D. simulans,* the *X*L, 2L, and *3*R subtelomeric regions consist of 519-, 1,215-, and 1,837-bp repeat units, respectively. The principal difference between these repeats is the presence of sequences from the transposon *hopper*. The 519-bp repeat does not have *hopper* sequences, and the 1,215-bp repeat has less *hopper* sequence than the 1,837-bp repeat. These repeats share homology (up to 96 % identity) with the 854-bp repeats found at 2L and *3*R TAS of *D. sechellia.* Since *D. simulans* and *D. sechellia* have only diverged 0.6–0.9 million years ago (Mya), the homology found between their subtelomeric repeats was expected.

Much more significant is the clear homology (up to 85 % identity) between the 2L and *3*L subtelomeres of *D. melanogaster* and the subtelomeres at homologous chromosome ends of *D. simulans* and *D. sechellia* (diverged 5.4 Mya). This homology means that in a common ancestor of these species subtelomeres consisted of the same type of repeats. The *invader4* LTR cluster in *D. melanogaster* appeared at subtelomere after the divergence of *D. melanogaster* and its sibling species and then spread by recombination to non-homologous telomeres. Similarly, an insertion of the *hopper* element into a TAS repeat of *D. simulans* occurred after the divergence of *D. simulans* and *D. sechellia.*

Comparison of the subtelomeres of *D. yakuba* and *D. erecta*, separated from *D. melanogaster* 13 Mya, does not show any homology, among themselves nor with the subtelomeres of *D. melanogaster.* This lack of homology between TAS of distant species is likely due to progressive substitution of subtelomeric repeats via unequal recombination. *D. yakuba* has an uneven distribution of TAS sequences, from absent at the *3*L telomere to unusually complex at the *X*L and 2L telomeres. *X*L TAS is made of three repeats of 241-, 31-, and 660-bp, and 2L TAS contains these three repeats plus two additional repeats of 785- and 169-bp (Fig. 12.2). In contrast, *D. erecta* shows a simple pattern of TAS sequences, with 1.2-kb repeats in distal positions and proximal 171–190-bp repeats that belong to a family of complex repeats interspersed in the genome. Outside the melanogaster group, it is also possible to find repeats in the subtelomeres that belong to families of interspersed repeats, for example the 180-bp TAS repeats of *D. ananassae* and the 370-bp TAS repeats of *D. virilis* (Biessmann et al. 2000).

During the study of *Drosophila* subtelomeres, it has become apparent that TAS and the long 3′ UTRs of many telomeric retrotransposable elements are structurally similar, with internal subrepeats (Fig. 12.3). This fact, together with the occasional juxtaposition of both sequences at the telomeres, suggests that the

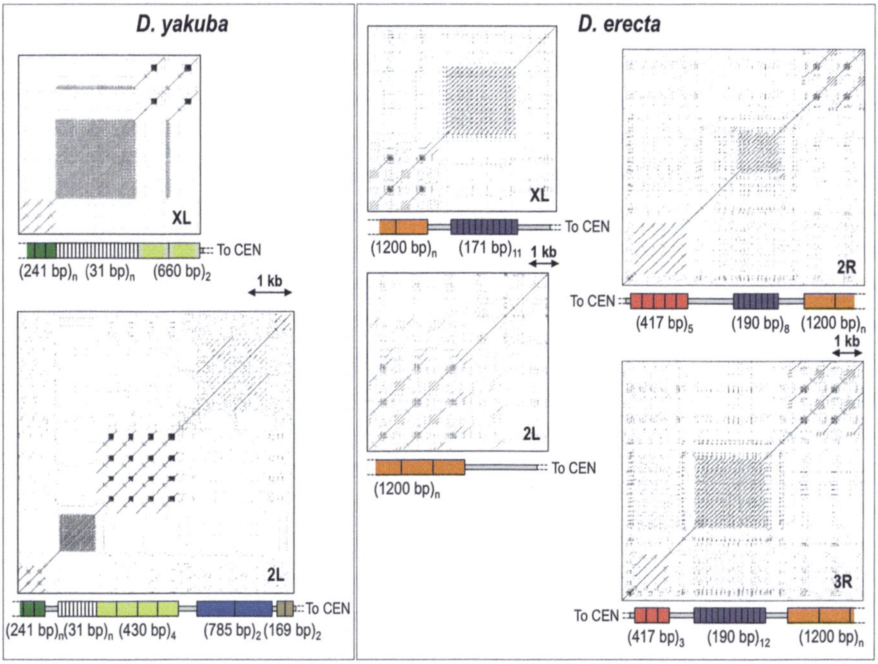


**Fig. 12.2** *D. yakuba* and *D. erecta* subtelomeric regions. The dot-matrix comparison of each region with itself is shown at the *top* of each diagram. The average size of the repeats is indicated

acquisitions of long 3′ UTRs might have occurred by 3′ transduction of TAS. This evolutionary innovation may have conferred heterochromatic properties throughout the telomeric transposon arrays.

## 12.3 Telomeric Proteins

Telomeres in *D. melanogaster* can be divided into three domains based on their proteins and DNA sequences (Biessmann et al. 2005). A complex of proteins at the extreme terminus of a chromosome arm protects the terminal DNA sequences from degradation and distinguishes the natural chromosome end from a DNA double-strand break. In *Drosophila*, the terminal complex is termed 'terminin' (Raffa et al. 2009) by analogy with the shelterin complex in mammals. The terminin complex proteins are not related to shelterin proteins, in part because the shelterin complex binds telomeres in a sequence-specific manner, while the terminin complex binds telomeres independently of DNA sequence, and in part because the terminin proteins are among the fastest-evolving proteins in *Drosophila* (Raffa et al. 2010; Schmid and Tautz 1997). The terminin complex interferes with the activity of transcriptional promoters and enhancers within 4 kb of the terminus (Melnikova and Georgiev 2005), although some of the terminin-associated proteins bind sequences 10 kb or more from the chromosome end (Gao et al. 2010).

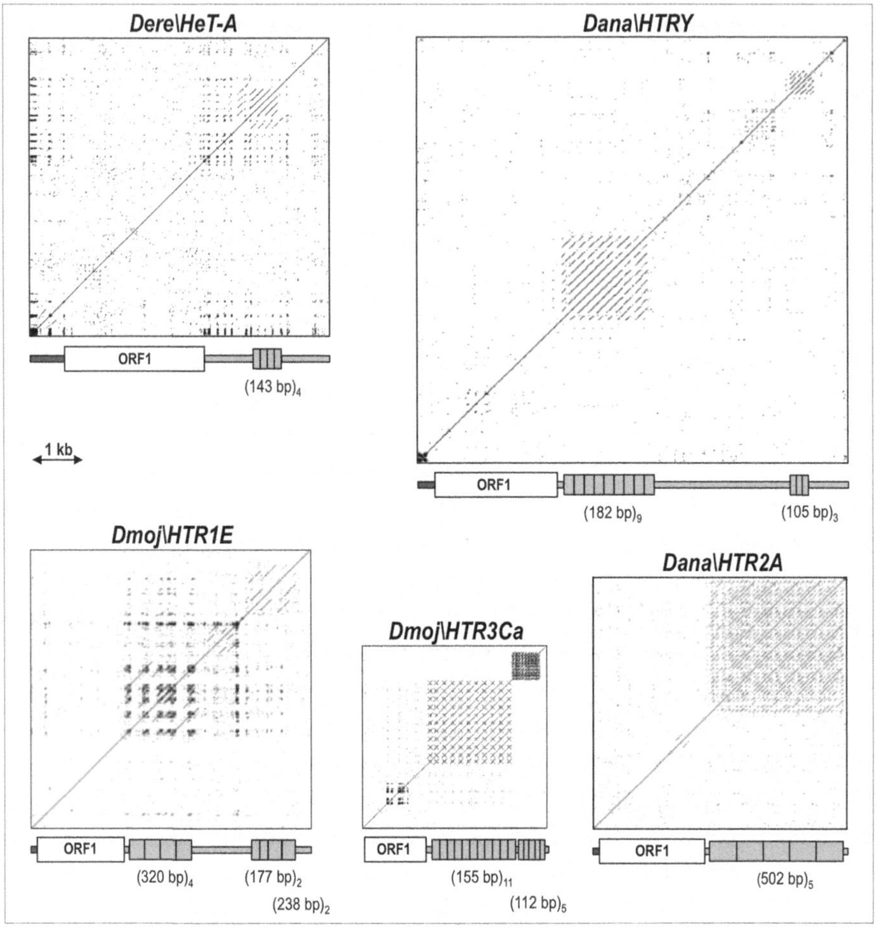


**Fig. 12.3** Structure of several telomeric retrotransposons from *D. erecta*, *D. ananassae*, and *D. mojavensis*. The dot-matrix comparison of each element with itself appears on *top* of the diagram of each element. The average size of the 3′ UTR repeats is indicated. Dere indicates *D. erecta*; Dana, *D. ananassae*; Dmoj, *D. mojavensis*

Two cytological techniques have been used to distinguish the protein-binding partners of individual DNA components of the *D. melanogaster* telomere. In one, a portion of TAS is inserted into the middle of a chromosome arm (Boivin et al. 2003). Additional binding to this site, above that seen without the insertion, can identify proteins with affinity to that sequence. In the second procedure, comparison of protein binding on long HTT arrays generated by the *Tel*$^1$ mutation (Siriaco et al. 2002) to binding on short HTT arrays normally found in wild-type strains allows a distinction to be made between proteins that bind the HTT array and those that bind other telomeric components (Andreyeva et al. 2005). It can thus be seen that the HTT array outside of the terminin complex exhibits both open and closed chromatin marks similar to the 'active' and 'silent' marks at active heterochromatic genes (Riddle et al. 2011). Subtelomeric DNA sequences bind Polycomb group (PcG) proteins, which are involved in the developmentally regulated silencing of genes, and modified histones associated with closed chromatin (Andreyeva et al. 2005; Boivin et al. 2003). Even though subtelomeric sequences vary from one chromosome end to another, the proteins that bind to these sequences are consistent between telomeres.

Telomeres are often considered to be heterochromatic, but in *Drosophila* this may be an oversimplification. While closed chromatin marks were found at both the HTT and TAS arrays, the levels of these marks differ between the two telomeric domains and between these domains and either euchromatin or pericentric heterochromatin (Capkova Frydrychova et al. 2008). This is consistent with the proposition that chromatin can be categorized into multiple chromatin types (Filion et al. 2010). In general, the chromatin marks indicate that the HTT array is more closely related to transcriptionally active genes in pericentric heterochromatin, while TAS is more closely related to PcG heterochromatin.

## 12.4 Genetic Interactions

Two subtelomeres in *Drosophila* have been observed to control the activity of genes in other locations, although the phenomenology of these two effects is different. In the first, deficiencies of 2L TAS suppress telomeric position effect (TPE) at homologous as well as non-homologous telomeres. The mechanism of this interaction is unknown. In the second interaction, insertions into *X*L TAS repress transgenes with similar sequences through an RNAi-based mechanism termed *trans*-silencing. Surprisingly, both of these effects seem to be restricted to single chromosome ends.

### *12.4.1 Suppression of Telomeric Silencing*

*P* transposable elements carrying a *w* reporter gene, which is responsible for pigment deposition in the eye, have been used widely for genetic studies in *Drosophila*. The reporter when in euchromatin is relatively highly expressed and produces a uniform orange-to-red eye color, while the same element in heterochromatin produces a light eye color that is variegated. The initial transposon insertions into telomeres were identified by eye color variegation (reviewed by Biessmann et al. 2005). Surprisingly, genetic mutations that suppressed *w* variegation in pericentric heterochromatin had little if any effect on a similar variegating *w* reporter in a telomere, suggesting that the structures of these two forms of heterochromatin are different. All of these variegating telomeric insertions were found to be located in, or adjacent to, TAS rather than in HTT. Insertions into the HTT array occur, but *w* is not repressed and gives the same phenotype it would if it were in euchromatin (Biessmann et al. 2005). Thus, based on the expression pattern of the *w* reporter, the HTT array appears to be euchromatic, and TAS appears to be heterochromatic independent of which chromosome tip carries the transgene.

Two TAS sequences were tested for their effects on expression of an adjacent *w* gene when at non-telomeric positions. In both cases, a vector carrying both the TAS sequence and the reporter was inserted randomly into the genome. In the first case, 6 kb of the 458-bp *2*L TAS repeat unit was placed upstream of the *w* gene (Kurenova et al. 1998). This caused a uniform decrease in *w* expression that was dependent on the orientation of TAS, and silencing was proportional to the length of TAS. In the second case, 1.2 kb of the *X*L TAS was placed upstream of the *w* reporter (Boivin et al. 2003). This appeared to cause an increase in *w* expression when hemizygous, but decreased or variegated expression in homozygous animals. As above, the effects were orientation dependent. Thus, both TAS elements seem to repress gene activity in non-telomeric positions, either in one copy or in a pairing-dependent manner.

An analogous situation arose with the identification of a variegating *w* transgene inserted between the HTT array and the subtelomeric region at the *2*L telomere (Golubovsky et al. 2001). The *w* transgene was transcribed from distal to proximal, and expression depended on its surroundings. If the homolog carried a full-length TAS array the transgene was strongly repressed (Fig. 12.4), while if the homolog was deficient for *2*L TAS the degree of expression depended on the length of the HTT array in *cis*, with longer arrays giving stronger *w* expression. Further, *w* transgenes inserted into the *2*R and *3*R subtelomeres responded to deficiencies for *2*L TAS, although the reciprocal interaction was not found (Capkova Frydrychova et al. 2007). Surprisingly, deficiencies for *3*L TAS did not have the same effect as deficiencies for *2*L TAS, even though their repeat sequences are very similar. Thus, *2*L TAS deficiencies are suppressors of TPE, while deficiencies for other TAS arrays are not.

It has been proposed that transgene expression was silenced by TAS and activated by *HeT-A* in *cis* and that this competition was mediated by interactions between TAS arrays. In the absence of the intervening transgene, TAS would reduce transcription and possibly transposition of the terminal retrotransposons and thus regulate the length of the HTT array (Mason et al. 2003). While initially tantalizing, this model fails on two counts. First, the silencing effect of TAS on transgenes inserted into HTT only extends a short distance into the terminal array (Biessmann et al. 2005). Second, the level of HeT-A transcript does not respond to deficiencies for *2*L TAS (Capkova Frydrychova et al. 2007). Thus, TPE does not seem to be a mechanism for regulating the transcription of the terminal retrotransposons.

Two reports suggesting that mutations in PcG genes are dominant suppressors of TPE (Boivin et al. 2003; Cryderman et al. 1999) lead to a large-scale screen using overlapping deficiencies for more suppressors of TPE (Mason et al. 2004). About two-thirds of the autosomes were tested with no solid evidence for dominant genic suppressors of TPE, with the possible exception of mutations in *gpp*, which encodes a histone H3 lysine 27 methyl transferase homologous to DOT1, and which had been identified independently as TPE suppressors (Shanower et al. 2005). Instead,

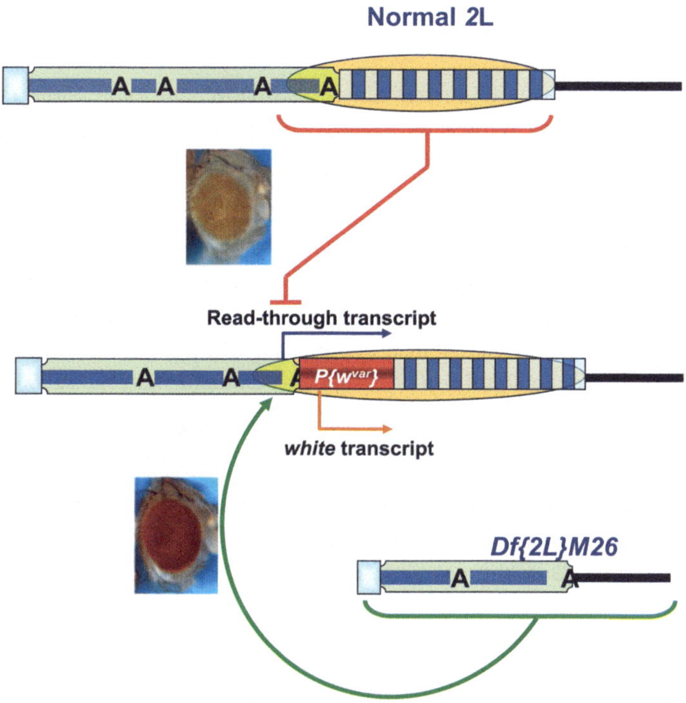


**Fig. 12.4** *Trans*-effect of TAS deletions on the expression of telomeric transgene insertions. *w* transgene expression of the 2L telomere occurs primarily from the *w* promoter (*orange bent arrow*) with a small contribution from the adjacent *HeT-A* promoter (*blue bent arrow*). This expression is repressed by the adjacent TAS in *cis* (*orange oval*) in the presence of a full-length TAS array (*red line*). Deletion of 2L TAS in *trans* (*Df(2L)M26*) causes higher expression (*green round arrow*) of the *w* transgene. Alternating *light* and *dark stripes* represent the subtelomere; *blue bars* alternating with 'A' represent terminal retrotransposons attached to the chromosome end by their oligo(A) tails; *blue rectangle at the left end* represents the protective terminin protein complex; *red rectangle* labeled $P\{w^{var}\}$ represents the *w* transgene insertion (taken from Mason et al. 2008)

the great majority of suppressors identified turned out to be false positives, i.e., the suppressor could be separated from the deficiency. In most cases, the suppressor turned out to be a deficiency for 2L TAS. Even the previously identified suppressors of TPE carried 2L TAS deficiencies on the same chromosome, and the suppressor mapped to the 2L telomere rather than the PcG gene.

As telomeric silencing does not spread more than a few kilobases from TAS and thus only affects the closest telomeric transposon, it seems likely that its purpose is to control transcription of the TAS sequences themselves. While the piRNA pathway has been implicated in *HeT-A* and *TART* transcription (Savitsky et al. 2006), it is surprising that none of the RNAi genes were identified in the deficiency screen for modifiers of TPE (Mason et al. 2004). Specific PcG proteins, trithorax group proteins, and components of pericentric heterochromatin were also excluded as important components of telomeric silencing.

### *12.4.2 Trans-silencing*

In a second example of subtelomeres playing a role in gene expression, *P* elements in the TAS region of *X*L appear to regulate the expression of insertions in other regions of the genome. *P* elements are DNA transposons that move by a cut-and-paste mechanism. In general, when a *P* element is introduced to a naïve genome, the copy number increases for a time, until the transposase gene is repressed and transposition frequency decreases. The degree of repression depends on the number and position of the *P* elements present. Regulation of *P* element transposition is complex, depending on both the presence of a full-length element encoding a transposase and a repressor, and a cytoplasmic element, termed cytotype, that can be inherited independently of a competent element (Engels 1979). A survey of strains from natural populations with full length repressed *P* elements found an accumulation of *P* elements at the *X*L tip, but not other chromosome ends (Ronsseray et al. 1997). The elements were within the *Invader4* LTRs of the *X*L subtelomere (Karpen and Spradling 1992). Further, one or two *P* elements at the *X*L telomere were stronger suppressors of *P* element activity than 10–20 complete *P* elements elsewhere in the genome (Ronsseray et al. 2003). Incomplete *P* elements in *X*L TAS that do not make the repressor protein are strong suppressors, but only when they are maternally inherited (Marin et al. 2000; Stuart et al. 2002). A limited repressor phenotype may be inherited in the absence of the telomeric *P* element, but this depends on transmission through the mother. Repression from *X*L TAS inserts was sensitive to dosage of the heterochromatin protein, HP1, and the RNAi protein, AUB, while *P* element repression due to full-length *P* elements elsewhere was not (Haley et al. 2005; Reiss et al. 2004). Taken together, these data suggest that repression of *P* elements by other elements in the *X*L TAS may depend on transmission of a small RNA to the offspring through the maternal cytoplasm.

In a similar phenomenon known as *trans*-silencing, a *P* element vector in *X*L TAS, e.g., *P*{*lacZ*}, which carries *lacZ* sequences, but lacks most *P* element sequences outside of the terminal repeats, can repress transgenes with similar sequences located elsewhere in the genome (Roche and Rio 1998), although it does not repress *P* element activity. The effect depends on the location of the silencer; only *P*{*lacZ*} insertions into *X*L TAS are effective (Roche and Rio 1998; Ronsseray et al. 2003). Repression also depended on homology between silencer and target, on transmission through the female germ line, and on the dosage of HP1 and the rasiRNA proteins, AUB, SPN-E, ARMI, and PIWI, but not on the dosage of proteins in the siRNA or miRNA pathways (Josse et al. 2007). Cytotype thus appears to be a form of (sub)telomeric *trans*-silencing, and *trans*-silencing a form of cosuppression. The role, if any, of PcG proteins, which bind subtelomeres on several chromosome ends and may interact with the RNAi machinery to regulate pairing-sensitive silencing (Grimaud et al. 2006), is poorly understood.

## 12.5 Complex Terminal Arrays in Nematocera

Telomerase and the short telomeric repeats generated by telomerase have not been found in any dipteran species. Two sister orders, Siphonaptera and Mecoptera also lack canonical telomeric repeats (Frydrychova et al. 2004). It is thus possible that telomerase may have been lost as much as 260–280 Mya (Wiegmann et al. 2009). Nevertheless, dipterans as a group are very successful, accounting for some 11 % of known animal species. Thus, loss of telomerase does not seem to have been a major impediment to survival.

Replacement of the short telomerase-generated repeats with long repeat sequences is reported in lower dipteran species (Reviewed by Mason et al. 2011). Chromosome tips in *Chironomus* consist of large, 50–200 kb, blocks of complex, tandemly repeated sequences that can be classified into subfamilies based on sequence similarities. Different telomeres display different sets of subfamilies, and the distribution of subfamilies differs between different individuals in a stock. The variation of the satellite sequences supports the proposal that telomeres in *Chironomus* are elongated by a gene conversion mechanism involving these long blocks of complex repeat units. The origin of these new non-canonical telomeric repeats is not understood. It is possible that they were originally subtelomeric repeats that became terminal with the loss of the telomerase-generated sequence; however, one 350-bp repeat unit in *Chironomus tentans* contains 15 copies of a rearranged and degenerate canonical telomeric repeat.

Long telomeric repeats have also been observed in the mosquito, *Anopheles gambiae*, with the aid of a plasmid insertion into the complex telomeric sequences at the tip of chromosome 2L. The telomeric repeat unit was 820-bp in length and restricted to the 2L tip. The plasmid sequence was used as a marker to follow the specific telomere, which was found to engage in frequent recombination events to extend the array length, although non-homologous sequences could also be added to this chromosome end (Mason et al. 2011).

A similar situation has been reported in *Rhynchosciara americana*, and only in this species a true subtelomeric satellite, with a repeat unit of 414-bp, has been found at the five non-telocentric chromosome ends (Madalena and Gorab 2005). Tandem arrays of relatively short repeats, 16- and 22-bp in length, were found distal to the subtelomeric repeat (Rossato et al. 2007), and these short repeats extend to the chromosome ends (Madalena et al. 2010). Although telomere elongation has not been assayed in this case, it seems likely that the mechanism is similar to that seen in other dipterans.

Given that telomeric gene conversion has been identified as an alternative elongation mechanism in species as diverse as yeast and humans (Kass-Eisler and Greider 2000) and has been found in nematocerans, it seems likely that this mechanism took over telomere maintenance when telomerase was lost from the lineage that lead to Diptera. Telomere maintenance by retrotransposition, on the other hand, arose sometime after the separation of *Rhynchosciara* from *Drosophila* 230 Mya (Wiegmann et al. 2011) and before the divergence of *Drosophila* species 60 Mya.

## 12.6 Conclusion

As in other eukaryotes, complex, subtelomeric repeated DNA sequences are found between the terminal DNA array and the unique sequences in the euchromatic regions of the chromosome arms. Subtelomeric sequences in *Drosophila* consist of complex repeat motifs that are shared among chromosome ends, although the arrangement of the motifs varies considerably from one telomere to another, even to the point that in situ hybridization studies fail to find homology among some telomeres. The question arises, what is the selective pressure that maintains subtelomeres, despite the fact that individual TAS arrays can be lost in laboratory strains and in wild populations? One possibility, given that the protective cap complex does not bind chromosome ends based on DNA sequence, is that these sequences act as a buffer protecting the euchromatic genes from degradation and loss. Although appealing, there is little direct evidence for or against this proposition in *Drosophila* species, although sequences resembling TAS arrays in nematoceran flies are found at the chromosome ends and appear to be actively involved in chromosome length maintenance through a gene conversion mechanism. A second possibility, that the subtelomeres control the transcription and transposition of the telomeric retrotransposons, is not supported by the available data. Surprisingly, a third possible function stems from the observation that the subtelomeres seem to be able to communicate with the rest of the genome; deletions of 2L TAS suppress telomere silencing at other chromosome ends, and insertions into *X*L TAS can inactivate expression of homologous sequences in an piRNA-dependent manner. It remains to be seen whether other subtelomeres in *Drosophila* have a similar ability to interact with the rest of the genome. As the subtelomeres bind PcG proteins, which are known to affect long range genetic interactions in combination with RNAi pathways, they may play a wider role in genetic regulation.

**Acknowledgments** JMM was supported by the Intramural Research Program, US National Institutes of Health. AV was supported by the Ministerio de Economia y Competitividad (BFU2011-30295-C02-01) and an institutional grant from Fundación Ramón Areces to the CBMSO.

## References

Abad, J. P., de Pablos, B., Osoegawa, K., de Jong, P. J., Martin-Gallardo, A., et al. (2004). Genomic analysis of *Drosophila melanogaster* telomeres: Full-length copies of *HeT-A* and *TART* elements at telomeres. *Molecular Biology and Evolution, 21*, 1613–1619.

Anderson, J. A., Song, Y. S., & Langley, C. H. (2008). Molecular population genetics of *Drosophila* subtelomeric DNA. *Genetics, 178*, 477–487.

Andreyeva, E. N., Belyaeva, E. S., Semeshin, V. F., Polkholkova, G. V., & Zhimulev, I. F. (2005). Three distinct chromatin domains in telomere ends of polytene chromosomes in *Drosophila melanogaster Tel* mutants. *Journal of Cell Science, 118*, 5465–5477.

Biessmann, H., Zurovcova, M., Yao, J. G., Lozovskaya, E., & Walter, M. F. (2000). A telomeric satellite in *Drosophila virilis* and its sibling species. *Chromosoma, 109*, 372–380.

Biessmann, H., Prasad, S., Walter, M. F., & Mason, J. M. (2005). Euchromatic and heterochromatic domains at *Drosophila* telomeres. *Biochemistry and Cell Biology, 83*, 477–485.
Boivin, A., Gally, C., Netter, S., Anxolabehere, D., & Ronsseray, S. (2003). Telomere associated sequences of *Drosophila* recruit Polycomb-group proteins in vivo and can induce pairing-sensitive repression. *Genetics, 164*, 195–208.
Broun, P., Ganal, M. W., & Tanksley, S. D. (1992). Telomeric arrays display high levels of heritable polymorphism among closely related plant varieties. *Proceedings of the National Academy of Sciences of the United States of America, 89*, 1354–1357.
Brown, W. R., MacKinnon, P. J., Villasante, A., Spurr, N., Buckle, V. J., et al. (1990). Structure and polymorphism of human telomere-associated DNA. *Cell, 63*, 119–132.
Capkova Frydrychova, R., Biessmann, H., Konev, A. Y., Golubovsky, M. D., Johnson, J., et al. (2007). Transcriptional activity of the telomeric retrotransposon *HeT-A* in *Drosophila melanogaster* is stimulated as a consequence of subterminal deficiencies at homologous and nonhomologous telomeres. *Molecular and Cellular Biology, 27*, 4991–5001.
Capkova Frydrychova, R., Mason, J. M., & Archer, T. K. (2008). HP1 is distributed within distinct chromatin domains at telomeres. *Genetics, 180*, 121–131.
Clark, A. G., Eisen, M. B., Smith, D. R., Bergman, C. M., Oliver, B., et al. (2007). Evolution of genes and genomes on the *Drosophila* phylogeny. *Nature, 450*, 203–218.
Cryderman, D. E., Morris, E. J., Biessmann, H., Elgin, S. C. R., & Wallrath, L. L. (1999). Silencing at *Drosophila* telomeres: Nuclear organization and chromatin structure play critical roles. *EMBO Journal, 18*, 3724–3735.
Engels, W. R. (1979). Extrachromosomal control of mutability in *Drosophila melanogaster*. *Proceedings of the National Academy of Sciences of the United States of America, 76*, 4011–4015.
Filion, G. J., van Bemmel, J. G., Braunschweig, U., Talhout, W., Kind, J., et al. (2010). Systematic protein location mapping reveals five principal chromatin types in *Drosophila* cells. *Cell, 143*, 212–224.
Frydrychova, R., Grossmann, P., Trubac, P., Vitkova, M., & Marec, F. E. (2004). Phylogenetic distribution of TTAGG telomeric repeats in insects. *Genome, 47*, 163–178.
Gao, G. J., Walser, J. C., Beaucher, M. L., Morciano, P., Wesolowska, N., et al. (2010). HipHop interacts with HOAP and HP1 to protect *Drosophila* telomeres in a sequence-independent manner. *EMBO Journal, 29*, 819–829.
Golubovsky, M. D., Konev, A. Y., Walter, M. F., Biessmann, H., & Mason, J. M. (2001). Terminal retrotransposons activate a subtelomeric *white* transgene at the 2L telomere in *Drosophila*. *Genetics, 158*, 1111–1123.
Grimaud, C., Bantignies, F., Pal-Bhadra, M., Ghana, P., Bhadra, U., et al. (2006). RNAi components are required for nuclear clustering of Polycomb group response elements. *Cell, 124*, 957–971.
Haley, K. J., Stuart, J. R., Raymond, J. D., Niemi, J. B., & Simmons, M. J. (2005). Impairment of cytotype regulation of P-element activity in *Drosophila melanogaster* by mutations in the *Su(var)205* gene. *Genetics, 171*, 583–595.
Josse, T., Teysset, L., Todeschini, A. L., Sidor, C. M., Anxolabehere, D., et al. (2007). Telomeric *trans*-silencing: An epigenetic repression combining RNA silencing and heterochromatin formation. *PLoS Genetics, 3*, 1633–1643.
Karpen, G. H., & Spradling, A. C. (1992). Analysis of subtelomeric heterochromatin in the *Drosophila* minichromosome *Dp1187* by single *P* element insertional mutagenesis. *Genetics, 132*, 737–753.
Kass-Eisler, A., & Greider, C. W. (2000). Recombination in telomere-length maintenance. *Trends in Biochemical Sciences, 25*, 200–206.
Kern, A. D., & Begun, D. J. (2008). Recurrent deletion and gene presence/absence polymorphism: Telomere dynamics dominate evolution at the tip of 3L in *Drosophila melanogaster* and *D. simulans*. *Genetics, 179*, 1021–1027.
Kurenova, E., Champion, L., Biessmann, H., & Mason, J. M. (1998). Directional gene silencing induced by a complex subtelomeric satellite from *Drosophila*. *Chromosoma, 107*, 311–320.

Louis, E. J., & Haber, J. E. (1992). The structure and evolution of subtelomeric-Y′ repeats in *Saccharomyces cerevisiae*. *Genetics, 131*, 559–574.
Madalena, C. R. G., & Gorab, E. (2005). A chromosome end satellite of *Rhynchosciara americana* (Diptera: Sciaridae) resembling nematoceran telomeric repeats. *Insect Molecular Biology, 14*, 255–262.
Madalena, C. R. G., Amabis, J. M., & Gorab, E. (2010). Unusually short tandem repeats appear to reach chromosome ends of *Rhynchosciara americana* (Diptera: Sciaridae). *Chromosoma, 119*, 613–623.
Marin, L., Lehmann, M., Nouaud, D., Hassan, I., Anxolabéhère, D., et al. (2000). *P*-element repression in *Drosophila melanogaster* by a naturally occurring defective telomeric *P* copy. *Genetics, 155*, 1841–1854.
Mason, J. M., Konev, A. Y., & Biessmann, H. (2003). Telomeric position effect in *Drosophila melanogaster* reflects a telomere length control mechanism. *Genetica, 117*, 319–325.
Mason, J. M., Ransom, J., & Konev, A. Y. (2004). A deficiency screen for dominant suppressors of telomeric silencing in *Drosophila*. *Genetics, 168*, 1353–1370.
Mason, J. M., Frydrychova, R. C., & Biessmann, H. (2008). *Drosophila* telomeres: An exception providing new insights. *BioEssays, 30*, 25–37.
Mason, J. M., Reddy, H. K., & Capkova Frydrychova, R. (2011). Telomere maintenance in organisms without telomerase. In H. Seligmann (Ed.), *DNA replication: Current advances*. Rijeka: InTech.
Mechler, B. M., McGinnis, W., & Gehring, W. J. (1985). Molecular cloning of *lethal(2)giant larvae*, a recessive oncogene of *Drosophila melanogaster*. *EMBO Journal, 4*, 1551–1557.
Mefford, H. C., & Trask, B. J. (2002). The complex structure and dynamic evolution of human subtelomeres. *Nature Reviews Genetics, 3*, 91–102.
Melnikova, L., & Georgiev, P. (2005). *Drosophila* telomeres: The non-telomerase alternative. *Chromosome Research, 13*, 431–441.
Phalke, S., Nickel, O., Walluscheck, D., Hortig, F., Onorati, M. C., et al. (2009). Retrotransposon silencing and telomere integrity in somatic cells of *Drosophila* depends on the cytosine-5 methyltransferase DNMT2. *Nature Genetics, 41*, 696–702.
Raffa, G. D., Siriaco, G., Cugusi, S., Ciapponi, L., Cenci, G., et al. (2009). The *Drosophila modigliani* (*moi*) gene encodes a HOAP-interacting protein required for telomere protection. *Proceedings of the National Academy of Sciences of the United States of America, 106*, 2271–2276.
Raffa, G. D., Raimondo, D., Sorino, C., Cugusi, S., Cenci, G., et al. (2010). Verrocchio, a *Drosophila* OB fold-containing protein, is a component of the terminin telomere-capping complex. *Genes & Development, 24*, 1596–1601.
Reiss, D., Josse, T., Anxolabehere, D., & Ronsseray, S. (2004). *aubergine* mutations in *Drosophila melanogaster* impair *P* cytotype determination by telomeric *P* elements inserted in heterochromatin. *Molecular Genetics and Genomics, 272*, 336–343.
Riddle, N. C., Minoda, A., Kharchenko, P. V., Alekseyenko, A. A., Schwartz, Y. B., et al. (2011). Plasticity in patterns of histone modifications and chromosomal proteins in *Drosophila* heterochromatin. *Genome Research, 21*, 147–163.
Roche, S. E., & Rio, D. C. (1998). *Trans*-silencing by *P* elements inserted in subtelomeric heterochromatin involves the *Drosophila* Polycomb group gene, *Enhancer of zeste*. *Genetics, 149*, 1839–1855.
Ronsseray, S., Lehmann, M., Nouaud, D., & Anxolabehere, D. (1997). *P* element regulation and *X*-chromosome subtelomeric heterochromatin in *Drosophila melanogaster*. *Genetica, 100*, 95–107.
Ronsseray, S., Josse, T., Boivin, A., & Anxolabehere, D. (2003). Telomeric transgenes and *trans*-silencing in *Drosophila*. *Genetica, 117*, 327–335.
Rossato, R. M., Madalena, C. R. G., & Gorab, E. (2007). Unusually short tandem repeats in the chromosome end structure of *Rhynchosciara* (Diptera: Sciaridae). *Genetica, 131*, 109–116.

Savitsky, M., Kwon, D., Georgiev, P., Kalmykova, A., & Gvozdev, V. (2006). Telomere elongation is under the control of the RNAi-based mechanism in the *Drosophila* germline. *Genes & Development, 20*, 345–354.
Schmid, K. J., & Tautz, D. (1997). A screen for fast evolving genes from *Drosophila*. *Proceedings of the National Academy of Sciences of the United States of America, 94*, 9746–9750.
Shanower, G. A., Muller, M., Blanton, J. L., Honti, V., Gyurkovics, H., et al. (2005). Characterization of the *grappa* gene, the *Drosophila* histone H3 lysine 79 methyltransferase. *Genetics, 169*, 173–184.
Siriaco, G. M., Cenci, G., Haoudi, A., Champion, L. E., Zhou, C., et al. (2002). *Telomere elongation* (*Tel*), a new mutation in *Drosophila melanogaster* that produces long telomeres. *Genetics, 160*, 235–245.
Stuart, J. R., Haley, K. J., Swedzinski, D., Lockner, S., Kocian, P. E., et al. (2002). Telomeric *P* elements associated with cytotype regulation of the *P* transposon family in *Drosophila melanogaster*. *Genetics, 162*, 1641–1654.
Villasante, A., Abad, J. P., Planello, R., Mendez-Lago, M., Celniker, S. E., et al. (2007). *Drosophila* telomeric retrotransposons derived from an ancestral element that was recruited to replace telomerase. *Genome Research, 17*, 1909–1918.
Walter, M. F., Jang, C., Kasravi, B., Donath, J., Mechler, B. M., et al. (1995). DNA organization and polymorphism of a wild-type *Drosophila* telomere region. *Chromosoma, 104*, 229–241.
Wiegmann, B. M., Trautwein, M. D., Kim, J. W., Cassel, B. K., Bertone, M. A., et al. (2009). Single-copy nuclear genes resolve the phylogeny of the holometabolous insects. *BMC Biology, 7*, 34.
Wiegmann, B. M., Trautwein, M. D., Winkler, I. S., Barr, N. B., Kim, J. W., et al. (2011). Episodic radiations in the fly tree of life. *Proceedings of the National Academy of Sciences of the United States of America, 108*, 5690–5695.

# Chapter 13
# Accumulation of Telomeric-Repeat-Specific Retrotransposons in Subtelomeres of *Bombyx mori* and *Tribolium castaneum*

**Haruhiko Fujiwara**

**Abstract** Arthropod telomeres are generally constituted by TTAGG pentanucleotide repeats, which are synthesized by telomerase. However, all species in Diptera examined to date have lost TTAGG repeats and are suggested to recruit telomerase-independent telomere maintenance. In contrast, the silkworm *Bombyx mori* retains TTAGG telomeric repeats, but the telomerase activity is repressed in quite a low level in all investigated tissues. In addition, the flour beetle *Tribolium castaneum*, which contains unconventional TCAGG telomeric repeats, also shows a weak telomerase activity. Telomerase reverse transcriptase (TERT) genes for *B. mori (BmoTERT)* and *T. castaneum* (*TcTERT*) have several unusual features; both *TERT* genes without introns have upstream ATG codons and no N-terminal GQ motifs, which possibly explain their repressed telomerase activity. In subtelomeres of *Bombyx* and *Tribolium*, telomeric-repeat-specific non-long terminal repeat (LTR) retrotransposons (or long interspersed nuclear elements (LINEs)), SARTBm and SARTTc, are accumulated. Respective retrotransposons prefer telomeric repeats of their hosts. This chapter focuses on subtelomere, TERT, and telomeric-repeat-specific LINEs in *Bombyx* and *Tribolium* and discusses mechanisms and evolution for telomere maintenance in higher insects.

## 13.1 Background

Telomeres are defined as regions at the end of chromosome that are essential for the complete replication, meiotic pairing, and stability of chromosomes. The telomeres of most eukaryotes are composed of simple repeated sequences called telomeric repeats. One strand of the repeats is G rich with its 3′ end toward the end

H. Fujiwara (✉)
Department of Integrated Biosciences Graduate School of Frontier Sciences,
The University of Tokyo, Kashiwa, Chiba 277–8562, Japan
e-mail: haruh@k.u-tokyo.ac.jp

E. J. Louis and M. M. Becker (eds.), *Subtelomeres*, DOI: 10.1007/978-3-642-41566-1_13, 

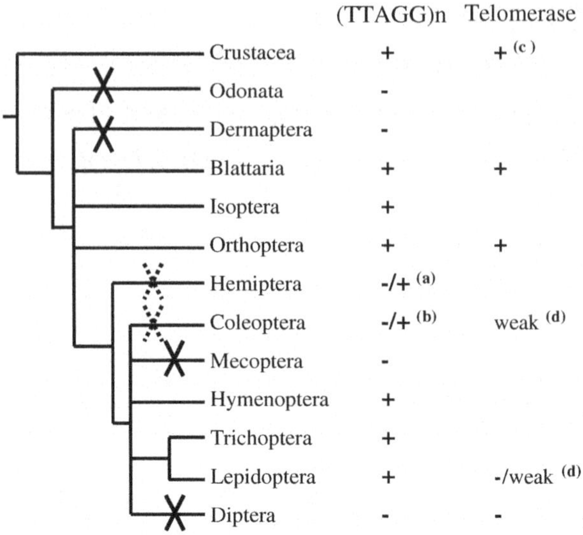


**Fig. 13.1** Distribution patterns of telomeric repeats $(TTAGG)_n$ and telomerase in insects. −/+: Some species in the order show signals. −/*weak*: Some species show *weak* telomerase activities. (*a*) Mohan et al. 2011, Monti et al. 2011; (*b*) Frydrychova et al. 2004, Mravinac et al. 2011; (*c*) Klapper et al. 1998; (*d*) Sasaki and Fujiwara 2000

of the chromosome. The G-rich strand is synthesized by a specialized enzyme telomerase, which is composed of two subunits, the telomerase reverse transcriptase (TERT) and the telomere RNA component (TERC; as a template) (Blackuburn 1991). It is hypothesized that the terminal sequences of chromosomal ends shorten with each cell division, and a reduction in telomere length causes cellular senescence or the cessation of cell division. In many organisms, therefore, the addition of telomeric repeats is essential to compensate for the critical telomere shortening.

## 13.2 Structure and Distribution of Telomeric Repeats in Insects

It has been shown that vertebrates and some species of fungi and protozoa have TTAGGG hexanucleotide telomeric repeats and that many other species retain a telomeric sequence resembling TTAGGG (Zakian 1995). In the early 1990s, our group used $(TTNGGG)_5$ as a cross-hybridization probe to identify the telomeric repeats in the genome of the silkworm *Bombyx mori* and found many TTAGG pentanucleotide repeats in this genome (Okazaki et al. 1993). Fluorescent in situ hybridization (FISH) and Bal31 exonuclease experiments showed that the telomere of *B. mori* consists of a 6–8-kb stretch of $(TTAGG)_n$. This pentanucleotide repeat sequence also exists in the genome of a wide variety of insects, including Lepidoptera, Hymenoptera, Coleoptera, termites, and in prawns (Okazaki et al. 1993; Fujiwara et al. 2005) (Fig. 13.1). Marec's group analyzed exhaustively the distribution of TTAGG telomeric motifs in insects (Frydrychova et al. 2004) and in arthropods and their close relatives (Vitkova et al. 2005).

They observed that TTAGG repeats are conserved among a wide variety of species of Arthropoda, but not in Tardigrada and Onychophora, and suggested that the vertebrate TTAGGG motif is an ancestral motif of telomeres in bilaterian animals. It is of great interest that some orders of insects lack typical $(TTAGG)_n$ repeats, while most order of insects have the repeats (Fig. 13.1, (TTAGG) +). All the species in Odonata, Dermaptera, Mecoptera, and Diptera (Fig. 13.1, −) and some species in Hemiptera and Coleoptera (Fig. 13.1, −/+) tested to date have no TTAGG repeats in their genome (Okazaki et al. 1993; Frydrychova et al. 2004). Hemiptera is divided into two groups, Heteroptera and Homoptera. Two species of Heteroptera have no TTAGG repeats (Frydrychova et al. 2004), whereas the mealybug (Mohan et al. 2011) and aphids (Monti et al. 2011) in Homoptera harbor these repeats, suggesting that a loss or change in the repeats has occurred in Heteroptera. Twelve species of Coleoptera carry the TTAGG repeats, but another nine species have lost the repeats (Frydrychova et al. 2004). Our group has shown that the flour beetle *Tribolium castaneum* has TCAGG telomeric repeats instead of TTAGG repeats (Osanai et al. 2006). Recently, Mravinac et al. (2011) analyzed thirty coleopteran beetles and found that all nineteen species in the Tenebrionoidea superfamily have TCAGG repeats, and not TTAGG repeats, whereas eight species in the Chrysomeloidea superfamily have TTAGG, and not TCAGG repeats, and some are TTAGG-/TCAGG-negative species. These data suggest that the TTAGG telomeric repeats have been lost independently several times during the evolution of insects and in different branches of the phylogenetic tree.

## 13.3 Accumulations of Non-LTR Retrotransposons in Telomere/Subtelomere Regions of *Drosophila* Species and the Silkworm *Bombyx mori*

Although $(TTAGG)_n$ is often undetectable in many insect species, it has not been proven in many cases whether their telomeres are composed of telomeric repeats other than $(TTAGG)_n$ or are preserved by other telomere maintenance systems. However, in the case of Dipteran insects examined thus far, all of which have lost the $(TTAGG)_n$ repeats, they seem to have lost telomerase activity. We did not find TERT-like genes in the whole-genome sequence information for the fruit fly *Drosophila melanogaster* and the malaria mosquito *Anopheles gambiae* (Adams et al. 2000; Holt et al. 2002), nor telomerase activity in *D. melanogaster* and *Sarcophaga* cells using the telomeric repeat amplification protocol (TRAP) (Sasaki and Fujiwara 2000). These facts support the existence in dipteran insects of the unusual telomere maintenance system mentioned above.

*Drosophila* telomeres are composed of three specialized non-long terminal repeat (LTR) retrotransposons (otherwise known as long interspersed nuclear elements (LINEs)) TART, HeT-A, and TAHRE and maintained by retrotransposition of these retrotransposons, in addition to recombination and gene conversion (Bieessmann et al. 1992; Levis et al. 1993; Pardue and De Baryshe 2003, 2011a, b; George et al. 2006; Melnikova et al. 2005; Mason et al. 2008). In other dipteran classes (the mosquito *A. gambiae* and the chironomids *Chironomus tentans*), telomere regions are composed of other types of longer repeats, such as 350-bp repeats in *C. tentans* (Nielsen et al. 1990; Nielsen and Edstrom 1993), rather than pentanucleotide short telomeric repeats. These telomeres in *C. tentans* and *A. gambiae* are hypothesized to be maintained by gene conversion and recombination (Cohn and Edstrom 1992; Roth et al. 1997). The accumulated evidences for dipteran insects, especially for *D. melanogaster*, show clearly that the necessity of telomerase is weakened in these insects and the telomerase system could be replaced by other unusual mechanisms of telomere maintenance.

In contrast with dipteran insects, the silkworm *B. mori* retains the (TTAGG) n repeats in its chromosomal ends. In the process of analyzing subtelomeric structure, we found that the silkworm telomeric repeats are inserted with over 1,000 copies of two telomeric-repeat-specific non-LTR retrotransposon families, TRAS and SART (Okazaki et al. 1995; Takahashi et al. 1997; Kubo et al. 2001; Fujiwara et al. 2005; Osanai-Futahashi et al. 2008). TRAS1, a major element in the TRAS family, is 7.8 kb in length and is inserted specifically into the exact site between the C and T of the CCTAA strand of telomeric repeats (Anzai et al. 2001; Maita et al. 2004). SART1 of *B. mori* (SART1Bm), which is 6.7 kb in length, is a major element in the SART family of the silkworm genome and is inserted specifically into another site of telomeric repeats, between the G and T of the TTAGG strand (Osanai-Futahashi and Fujiwara 2011) (Fig. 13.4). The TRAS and SART families occupy 3 % of the silkworm genome and more than 300 kb of each chromosomal end (Okazaki et al. 1995; Takahashi et al. 1997; Kubo et al. 2001). These elements are transcribed actively both in somatic and in germ line cells, whereas usual retrotransposons are rarely expressed (Takahashi and Fujiwara 1999). Based on an in vivo assay system for SART1 and TRAS1 using a baculovirus expression system, we have shown the exact retrotransposition of these elements into the telomeric repeats in cultured cells and larva of *B. mori* (Takahashi and Fujiwara 2002; Matsumoto et al. 2004; Osanai et al. 2004; Kawashima et al. 2007; Osanai-Futahashi and Fujiwara 2011). As described later in more detail, we could not detect telomerase activity in any tissues of *B. mori* nor in three *Bombyx* cell lines (Sasaki and Fujiwara 2000); thus, these data suggest that retrotransposition of the TRAS and SART non-LTR elements into the telomeric repeats possibly backs up telomere elongation by weak activity of the silkworm telomerase. A chromosomal fragment caused by X-ray irradiation, of the genetic mosaic mutant *pSm788*, is unstable and is often lost from somatic and germ line cells (Fujiwara et al. 1991, 1994, 2000). The broken ends of this chromosomal fragment produced by X-ray irradiation do not contain SART and TRAS elements in the (TTAGG) short repeats added *de novo*, suggesting a possible functional role for these elements in chromosome stability (Fujiwara et al. 2000).

## 13.4 Telomerase Activity in Insects

To determine whether TTAGG repeats are truly lost from the genome of TTAGG-negative insects or whether they are only changed into another similar repeat, it is essential to analyze telomerase activity in these species. Using a modified TRAP method, we measured the telomerase activity in various insects (Sasaki and Fujiwara 2000). Using PCR with an appropriate set of forward and reverse primers, we detected strong telomerase activity in crickets (*Teleogryllus taiwanemma*) and cockroaches (*Periplaneta americana*), both of which carry TTAGG repeats (Fig. 13.1). Telomerase activity in the insects requires dATP, dGTP, and dTTP, but not dCTP, as substrates. In addition, the sequence analysis of TRAP products showed that the TTAGG repeats are generated by telomerase activity in these insects. Although complicated subtelomeric structures are observed in *T. taiwanemma* (Kojima et al. 2002), these data demonstrate that the extreme end of chromosomes of this insect is generated by telomerase. Using similar methods, Klapper et al. (1998) reported that telomerase activity adds the TTAGG repeats in a lobster (*Homarus americana*), suggesting that the TTAGG repeats that are observed widely in the Arthropod telomeres are synthesized by the activity of telomerase. The telomerase activity in *P. americana* and *H. americana* is observed not only in germ line cells but also in several somatic cells, which are in contrast with the restricted expression of telomerase in germ line cells in humans and other mammals.

In two cell lines (s-2 and mbn) of *D. melanogaster* and a cell line (sape 4) of the fresh fly *Sarcophaga peregrina,* in contrast, no telomerase activity was detected using the modified TRAP approach mentioned above (Sasaki and Fujiwara 2000), which is consistent with the absence of the short telomeric repeats in Diptera (Fig. 13.1). In addition, we found no detectable levels of telomerase activity in three cell lines (BmN4, L30, and DK10) and three tissues (testis, silk gland, and fat body) of *Bombyx mori*. Interestingly, we found a short tract of TTAGG repeats on a broken end of a chromosomal fragment that had been induced by X-ray irradiation more than 50 years ago (Fujiwara et al. 2000). This observation indicates that the silkworm telomerase has a potential to add TTAGG repeats over a long period, but its activity is extraordinarily low and under the level that is detectable by conventional TRAP. We expressed the *TERT* gene from *Bombyx mori* and from another Lepidoptera (*Spodoptera frugiperda*, fall armyworm) in BmN and Sf9 cells, respectively, using a baculovirus expression system and prepared cell extracts (Yaguchi et al. unpublished data). However, telomerase activity was not detected even in these extracts, suggesting that its weak activity is caused not only by lower expression of the TERT gene in the body, but rather depends on its structural deficiency.

The activity of insect telomerase of adding telomeric repeats seems to be originally weak (Sasaki and Fujiwara, 2000). TRAP showed that 10 μg of protein from a cricket (*Teleogryllus*) extract yielded a band that had approximately one-third of the intensity observed using 2 μg of protein from a HeLa cell extract. Moreover,

0.01 μg of protein from the cricket extract did not generate a detectable TRAP result. Recently, we also analyzed (using TRAP) the capacity of the telomerase from *Tribolium* to add TCAGG repeats and found that it exhibited only a few ladder bands, suggesting its weak activity for the elongation steps. These results indicate that telomerase activity is weaker in insects compared with vertebrates, even when the activity is demonstrated by TRAP, such as crickets and *Tribolium* (Osanai-Futahashi unpublished data, Mitchell et al. 2010).

## 13.5 Structural Features of TERT Genes in the Silkworm and Flour Beetle

To determine why telomerase activity was not detected, or was very weak, using TRAP in some insects, we analyzed the gene structure of TERT in insects. The amino acid sequences of TERT exhibit low conservation among distantly related species, with the exception of a few catalytic sites; thus, it is hard to search for homologous sequence of TERT using regular methods, such as PCR with degenerate primer sets designed based on the conserved sequence. However, a recent advance of whole-genome sequencing in several insects provided an opportunity to screen for TERT genes. By screening the TERT homologous sequences from the silkworm genome database, which was available after 2004 (Mita et al. 2004; Xia et al. 2004), we succeeded in finding a fragmental sequence of TERT and identified the complete sequence of putative TERT in the silkworm (*BmoTERT*) using 5′ and 3′ RACE (Osanai et al. 2006). We also found the TERT sequences from *Bombyx mandarina* (*BmaTERT*) (Osanai et al. 2006), *T. castaneum* (*TcTERT*) (Osanai et al. 2006), and *Spodoptera frugiperda* (*SfTERT*) (Yaguchi et al. unpublished data). The *TERT* genes from the honey bee *Apis mellifera* (*AmTERT*) (Robertson and Gordon 2006) and from the aphid *Acyrthosiphon pisum* (Monti et al. 2011) have also been reported recently.

The silkworm *TERT* gene, which is 2,532 bp with 28 bp poly-A tail, encodes open reading frame (ORF) of 703 amino acids. Compared with the general *TERT* genes, *BmoTERT* has several unusual structural features, as follows.

### *13.5.1 Intronless Gene*

Interestingly, the comparison of the cDNA sequence of *BmoTERT* with the genomic sequence showed that *BmoTERT* has no introns (Osanai et al. 2006). Most *TERT* genes reported to date include many introns: 15 introns in human *TERT*; 17 introns in *Tetrahymena TERT*; 15 introns in *Saccharomyces TERT* (Fig. 13.2a). Intronless genes are sometimes processed pseudogenes. If the *BmoTERT* was processed pseudogenes, there would be another *TERT* gene in the

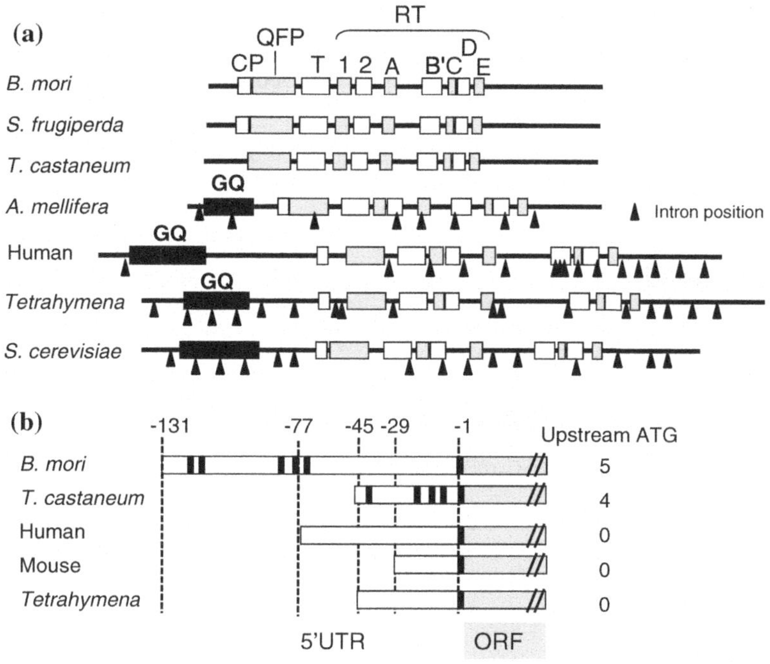


**Fig. 13.2** Structures of TERT genes in various organisms. **a** Diagram of TERT structure. Motifs are shown as *boxes*. Intron positions are shown as filled triangles on the domain structures. **b** The position of upstream ATG in several TERT genes. Numbers of upstream ATG in 5′UTR are shown on the *right*

genome of *Bombyx mori*. Southern hybridization experiments, however, showed that the *BmoTERT* is not a pseudogene and a single-copy gene without introns. We also found that *TcTERT* and *SfTERT* are intronless (Yaguchi et al. unpublished data), whereas *AmTERT* of the honeybee has eight introns similar to other usual *TERT* genes.

## 13.5.2 3′-Terminal Poly-A Tails of the Genomic Copy

The comparison of cDNA and the genomic sequences of *BmoTERT* also showed that the genomic sequence of *BmoTERT* (and not only its cDNA) contains a poly-A tail at the 3′-terminal region (Fig. 13.3a) (Osanai et al. 2006). The poly-A tail structure was also found at the 3′ end both in cDNA and in genomic sequences of *SfTERT* (Yaguchi et al. unpublished data). This type of poly-A-tail in the genomic sequence is often observed in the retrotransposed copy of non-LTR retrotransposons, which recognize poly-A tail of their mRNA in the initial step of

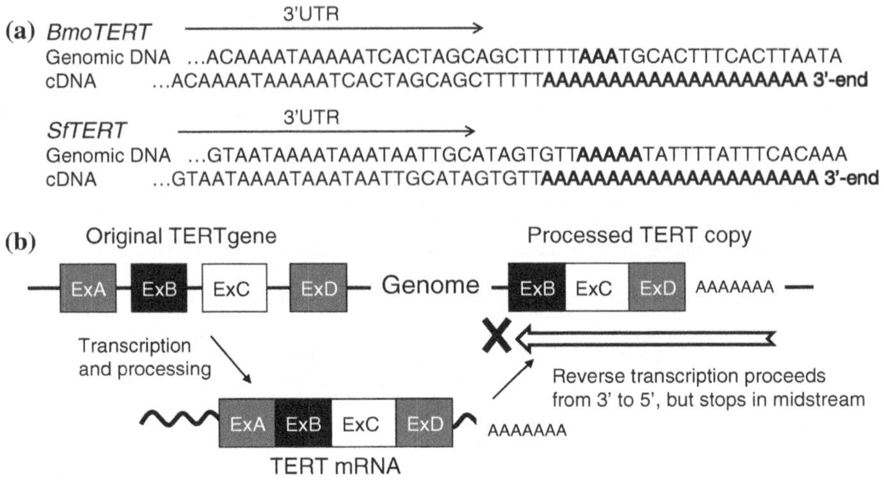


**Fig. 13.3** TERT genes of *Bombyx* and *Spodoptera* are processed gene. **a** Polyadenylation site of *BmoTERT* and *SfTERT*. Polyadenylation sites in genomic DNA and cDNA are in *bold*. **b** The schematic model for generation of processed TERT gene by reverse transcription (see text). Four exons are shown as ExA to ExD

target-primed reverse transcription. Thus, it is possible that the intronless TERT genes are originally generated by reverse transcription of the processed mRNA with reverse transcriptase of some internal retrotransposable elements (Fig. 13.3b). Because the *BmoTERT* and *SfTERT* are single-copy genes, the original TERT-containing introns should have been lost from the genome.

### *13.5.3 Upstream ATG Codons*

Another unusual feature of *BmoTERT* is that it has five ATGs in the 5′UTR, although this feature is not found in usual TERT genes reported to date (Fig. 13.2b). It is presumed that the ribosome initiates translation at the first ATG; thus, these codons possibly reduce the translation efficiency of *BmoTERT* proteins. It is reported that this type of upstream ATG (or upper AUG) actually reduces the translation of some genes (Jin et al. 2003). We also found four upstream ATG codons in the 5′UTR of *TcTERT* (Fig. 13.2b).

### *13.5.4 Loss of N-Terminal GQ Motif*

The putative protein encoded in ORF of *BmoTERT* is estimated to have a molecular weight of only 84 kDa, which is quite smaller than other TERT proteins reported to date. The TERT proteins of some vertebrates, *Arabidopsis,* and *Tetrahymena* are about 130 kDa, and the TERT of *S. pombe* and *S. cerevisiae* are 116 and 103 kDa, respectively. We characterized the conserved motifs of *BmoTERT* and found that it does not contain the N-terminal TERT-specific GQ motif, whereas all other motifs, RT domains (1, 2, A, B′, C, D, and E) and TERT-specific motifs (CP, QFP, and T) (Bosoy et al. 2003) are retained (Fig. 13.2a) (Osanai et al. 2006). The comparison of the schematic domain structure of TERT shows clearly that the smaller size of the *BmoTERT* protein is caused mainly by loss of the N-terminal region, especially of the GQ motif. The GQ motif is shown to be involved in the telomeric repeat processivity in yeast and human TERT (Moriarty et al. 2004; Lue 2005), and loss of the GQ motif may cause inefficient elongation of telomeric repeats. We found that the *TcTERT* protein, which has a molecular weight of only 70 kDa, also lost the GQ and CP motifs (Osanai et al. 2006) and that *SfTERT* protein exhibits a structure that is similar to that of *BmoTERT*, lacking the GQ motif (Yaguchi et al. unpublished data). The loss of GQ motifs in TERT proteins is consistent with marginal elongation of $(TCAGG)_n$ repeats by *TcTERT* (Osanai-Futahashi, unpublished) and the very low telomerase activity of *BmoTERT* and *SfTERT* (Yaguchi et al., unpublished data).

### *13.5.5 Repressed Transcription*

Northern hybridization showed that *BmoTERT* is transcribed at very low level in all tissues tested (testis, ovary, trachea, silk glands, and nerves) (Osanai et al. 2006).

These unusual characters of *BmoTERT* gene may explain the weak telomerase activity of *BmoTERT*, *SfTERT*, and *TcTERT*, in every step of transcription, translation, and enzymatic action.

## 13.6 Coevolution of Telomeric Repeats and Telomeric-Repeat-Specific Retrotransposons

Because there are many similarities between the structure and activity of TERT in *Bombyx mori* and *T. castaneum*, we searched the telomeric-repeat-specific retrotransposons, such as TRAS and SART families (which accumulate in the subtelomeres of *B. mori* chromosomes), in the genome of *T. castaneum*. The telomeres of *T. castaneum* consist of TCAGG repeats, instead of TTAGG repeats found in many other insects and arthropod species. The screening of insertion elements in the TCAGG repeats of the *T. castaneum* genome showed that seven types of non-LTR retrotransposons are inserted between the T and G of the CCTGA strand of TCAGG telomeric repeats, which is similar to the insertion

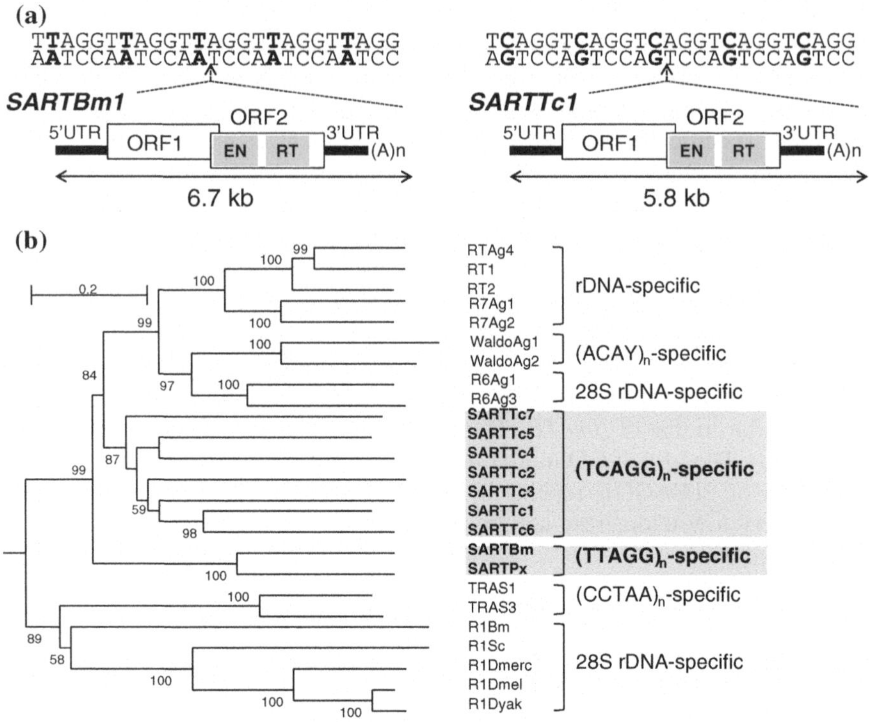


**Fig. 13.4** Structural features of SART families of *B. mori* and *T. castaneum*. **a** Target sites of SARTBm and SARTTc in telomeric repeats of each species and **b** phylogenetic relationships of SART families and other target-specific LINEs of R1 clade

site of SARTBm (between the T and A of the CCTAA strand of TTAGG telomeric repeats) (Fig. 13.4) (Osanai-Futahashi and Fujiwara 2011). The orientation of all these elements in the telomeric repeats was the same as that observed for SARTBm. In addition, phylogenetic analyses showed that they are closely related to the SART families of *B. mori*; they were named as SARTTc1 to SARTTc7 (Fig. 13.4b). In contrast to a *Drosophila* telomere-specific retrotransposons TART which belongs to the Jockey clade, the SART families are categorized into the R1 clade. Although most non-LTR retrotransposons integrate randomly into the genome, many R1 clade elements are inserted into specific sequences (Kojima and Fujiwara 2003). Target specificity in the R1 clade was changed several times, and SART families may have derived from some R1 clade elements or vice versa.

A modified ex vivo retrotransposition assay using baculovirus expression vector (Takahashi and Fujiwara 2002; Osanai et al. 2004; Matsumoto et al. 2006) showed that SARTBm1 had a target sequence preference for TTAGG repeats and that SARTTc1 had a target sequence preference to TCAGG repeats (Osanai-Futahashi and Fujiwara 2011). Although both SARTBm and SARTTc prefer their host telomeric repeat sequences, they also retrotransposed into TTAGG and TCAGG repeats. If the sequence specificity of these elements was very strict, they would not be able to accommodate to alterations of telomeric repeats. In addition, swapping experiments indicated that the endonuclease domain of SARTBm and SARTTc is involved in recognizing the target telomeric sequences (Osanai-Futahashi and Fujiwara 2011). Moreover, the SARTBm1 protein was able to retrotranspose 3′UTR sequence of SARTTc1, in addition to its own 3′UTR, whereas the SARTTc1 protein could only retrotranspose its own 3′UTR. These findings suggest that SARTTc changed the target specificity from TTAGG to unconventional TCAGG repeats mainly via the functional change of its endonuclease domain during evolution. The telomeric repeat is complementary to the partial sequence of telomeric RNA components (TERC); thus, the TERC sequence that is responsible for generating telomeric repeats should have changed in *T. castaneum*, whereas no group succeeded in identifying TERC in insects.

The highly specific integration of SARTTc elements into the TCAGG telomeric repeats should contribute to the generation of the subtelomere of *T. castaneum* and possibly back up the weak *TcTERT* activity of telomerase; this situation is very similar to the silkworm system of maintenance of telomere and subtelomere structures.

## 13.7 Evolution of Telomeric and Subtelomeric Structures in Higher Insects

Why did the TERT of the silkworm and the flour beetle become so fragile and why did unusual telomeric/subtelomeric structures appear in these insects? Hymenoptera such as *Apis mellifera,* which is thought to be an ancestral group among higher insects, have the usual *TERT* structure with many introns and long telomeric repeats undisturbed by retrotransposons (Fig. 13.5). However, both Coleoptera *T. castaneum* and Lepidoptera *B. mori,* which are groups that branched more recently phylogenetically compared with Hymenoptera, carry intronless TERT, suggesting that a TERT mRNA was reverse transcribed in a common ancestor of Coleoptera and Lepidoptera after branching from Hymenoptera and integrated into the genome as a processed gene. Reverse transcription occurs from the 3′ to the 5′ end of mRNA and sometimes stops before completion of the process, which yields a 5′-truncated processed gene (Fig. 13.3b). Thus, the N-terminal deletion including GQ motif observed in *BmoTERT*, *SfTERT*, and *TcTERT* is hypothetically explained by this 5′ truncation in the process of reverse transcription. In the more recently evolved insect group Diptera, it is hypothesized that the telomerase activity mediated by such a fragile TERT is attenuated and the TERT gene itself disappeared from the genomes of these insects and that alternative systems other than telomerase, such as telomere-specific retrotransposons in *Drosophila melanogaster*, save the insects from crisis.

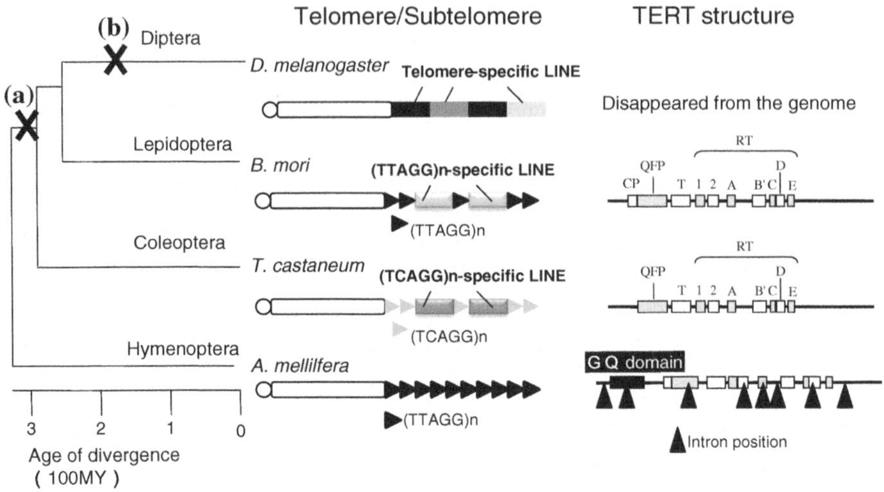


**Fig. 13.5** Evolution of telomere/subtelomere and TERT structures in higher insects. **a** Loss of introns and GQ domains from TERT genes. Appearance of telomeric-repeat-specific LINEs. **b** Loss of TERT gene from the genome

It is surprising that higher insects, which are the largest and most thriving species on earth, survive telomerase-negative conditions. To confirm the above hypothetical evolutionary scenario, we need to analyze telomeric and subtelomeric structures and the TERT structure in additional species in Lepidoptera, Coleoptera, and Diptera. We are also interested in the loss of telomeric repeats in lower insects: What kind of alternative telomere maintenance mechanisms are used in these insects? What happens to the TERT gene or telomerase activity in lower insects? Further analyses aimed at answering these questions may clarify novel aspects of telomere maintenance and structure.

## References


Adams, M. D., Celniker, S. E., Holt, R. A., et al. (2000). The genome sequence of *Drosophila melanogaster*. *Science, 287*, 2185–2195. doi:10.1126/science.287. 5461.2185.

Anzai, T., Takahashi, H., & Fujiwara, H. (2001). Sequence-specific recognition and cleavage of telomeric repeat $(TTAGG)_n$ by endonuclease of non-long terminal repeat retrotransposon TRAS1. *Molecular and Cellular Biology, 21*, 100–108. doi:10.1128/MCB.21.1.100-108.2001.

Biessmann, H., Valgeirsdottir, K., Lofsky, A., Chin, C., Ginther, B., Levis, R. W., et al. (1992). HeT-A, a transposable element specifically involved in "healing" broken chromosome ends in *Drosophila melanogaster*. *Molecular and Cellular Biology, 12*, 3910–3918.

Blackburn, E. H. (1991). Structure and function of telomeres. *Nature, 350*, 569–573. doi:10.1038/350569a0.

Bosoy, D., Peng, Y., Mian, I. S., & Lue, N. F. (2003). Conserved N-terminal motifs of telomerase reverse transcriptase required for ribonucleoprotein assembly in vivo. *Journal of Biological Chemistry, 278*, 3882–3890. doi:10.1074/jbc.M210645200.

Cohn, M., & Edstrom, J. E. (1992). Telomere-associated repeats in *Chironomus* form discrete subfamilies generated by gene conversion. *Journal of Molecular Evolution, 35*, 114–122. doi:10.1007/BF00183222.

Frydrychova, R., Grossmann, P., Trubac, P., Vitkova, M., & Marec, F. (2004). Phylogenetic distribution of TTAGG telomeric repeats in insects. *Genome, 47*, 163–178. doi:10.1139/g03-100.

Fujiwara, H., Ninaki, O., Kobayashi, M., Kusuda, J., & Maekawa, H. (1991). Chromosomal fragment responsible for genetic mosaicism in larval body marking of the silkworm, *Bombyx mori*. *Genetical Research, 57*, 11–16. doi:10.1017/S0016672300028974.

Fujiwara, H., Yanagawa, M., & Ishikawa, H. (1994). Mosaic formation by developmental loss of a chromosomal fragment in a "mottled striped" mosaic strain of the silkworm, *Bombyx mori*. *Roux's Arch Dev Biol, 203*, 389–396. doi:10.1007/BF00188687.

Fujiwara, H., Nakazato, Y., Okazaki, S., & Ninaki, O. (2000). Stability and telomere structure of chromosomal fragments in two different mosaic strains of the silkworm, *Bombyx mori*. *Zool Sci, 17*, 743–750. doi:10.2108/zsj.17.743.

Fujiwara, H., Osanai, M., Matsumoto, T., & Kojima, K. K. (2005). Telomere-specific non-LTR retrotransposons and telomere maintenance in the silkworm, *Bombyx mori*. *Chromosome Research, 13*, 455–467. doi:10.1007/s10577-005-0990-9.

George, J. A., DeBaryshe, P. G., Traverse, K. L., Celniker, S. E., & Pardue, M. L. (2006). Genomic organization of the *Drosophila* telomere retrotransposable elements. *Genome Research, 16*, 1231–1240. doi:10.1101/gr.5348806.

Holt, R. A., Subramanian, G. M., Halpern, A., et al. (2002). The genome sequence of the malaria mosquito *Anopheles gambiae*. *Science, 298*, 129–149. doi:10.1126/science.1076181.

Jin, X., Turcott, E., Englehardt, S., Mize, G. J., & Morris, D. R. (2003). The two upstream open reading frames of oncogene mdm2 have different translational regulatory properties. *Journal of Biological Chemistry, 278*, 25716–25721. doi:10.1074/jbc.M300316200.

Kawashima, T., Osanai, M., Futahashi, R., Kojima, T., & Fujiwara, H. (2007). A novel target-specific gene delivery system combining baculovirus and sequence-specific LINEs. *Virus Research, 127*, 49–60. doi:10.1016/j.virusres.2007.03.014.

Klapper, W., Kuhne, K., Singh, K. K., Heidorn, K., Parwaresch, R., & Krupp, G. (1998). Longevity of lobsters is linked to ubiquitous telomerase expression. *FEBS Letters, 439*, 143–146. doi:10.1016/S0014-5793(98)01357-X.

Kojima, T. T., & Fujiwara, H. (2003). Evolution of target specificity in R1 clade non-LTR retrotransposons. *Molecular Biology and Evolution, 20*, 351–361. doi:10.1093/molbev/msg031.

Kojima, K. K., Kubo, Y., & Fujiwara, H. (2002). Complex and tandem repeat structure of subtelomeric regions in the Taiwan cricket, Teleogryllus taiwanemma. *Journal of Molecular Evolution, 54*, 474–485. doi:10.1007/s00239-001-0038-5.

Kubo, Y., Okazaki, S., Anzai, T., & Fujiwara, H. (2001). Structural and phylogenetic analysis of TRAS, telomeric repeat-specific non-LTR retrotransposon families in Lepidopteran insects. *Molecular Biology and Evolution, 18*, 848–857.

Levis, R. W., Ganesan, R., Houtchens, K., Tolar, L. A., & Sheen, F. M. (1993). Transposons in place of telomeric repeats at a *Drosophila* telomere. *Cell, 75*, 1083–1093. doi:10.1016/0092-8674(93)90318-K.

Lue, N. F. (2005). A physical and functional constituent of telomerase anchor site. *Journal of Biological Chemistry, 280*, 26586–26591. doi:10.1074/jbc.M503028200.

Maita, N., Anzai, T., Aoyagi, H., Mizuno, H., & Fujiwara, H. (2004). Crystal structure of the endonuclease domain encoded by the telomere-specific long interspersed nuclear element, TRAS1. *Journal of Biological Chemistry, 24*, 41067–41076. doi:10.1074/jbc.M406556200.

Mason, J. M., Frydrychova, R. C., & Biessmann, H. (2008). Drosophila telomeres: an exception providing new insights. *Bioessays, 30*, 25–37. doi:10.1002/bies.20688.

Matsumoto, T., Takahashi, H., & Fujiwara, H. (2004). Targeted nuclear import of open reading frame 1 is required for in vivo retrotransposition of a telomere-specific non-long terminal repeat retrotransposon, SART1. *Molecular and Cellular Biology, 24*, 105–122. doi:10.1128/MCB.24.1.105-122.2004.

Matsumoto, T., Hamada, M., Osanai, M., & Fujiwara, H. (2006). Essential domains for ribonucleoprotein complex formation required for retrotransposition of telomere-specific non-long terminal repeat retrotransposon SART1. *Molecular and Cellular Biology, 26*, 5168–5179. doi:10.1128/MCB.00096-06.

Melnikova, L., Biessmann, H., & Georgiev, P. (2005). The Ku protein complex is involved in length regulation of *Drosophila* telomeres. *Genetics, 170*, 221–235. doi:10.1534/genetics.104.034538.

Mitchell, M., Gillis, A., Futahashi, M., Fujiwara, H., & Skordalakes, E. (2010). Structural basis for telomerase catalytic subunit TERT binding to RNA template and telomeric DNA. *Nature Structural and Molecular Biology, 17*, 513–518. doi:10.1038/nsmb.1777.

Mita, K., Kasahara, M., Sasaki, S., et al. (2004). The genome sequence of silkworm, *Bombyx mori*. *DNA Research, 11*, 27–35.

Mohan, K. N., Rani, B. S., Kulashreshta, P. S., & Kadandale, J. S. (2011). Characterization of TTAGG telomeric repeats, their interstitial occurrence and constitutively active telomerase in the mealybug *Planococcus* lilacinus (Homoptera; Coccoidea). *Chromosoma, 120*, 165–175. doi:10.1007/s00412-010-0299-0.

Monti, V., Giusti, M., Bizzaro, D., Manicardi, G. C., & Mandrioli, M. (2011). Presence of a functional (TTAGG)(n) telomere-telomerase system in aphids. *Chromosome Research, 19*, 625–633. doi:10.1007/s10577-011-9222-7.

Moriarty, T. J., Marie-Egyptienne, D. T., & Autexier, C. (2004). Functional organization of repeat addition processivity and DNA synthesis determinants in the human telomerase multimer. *Molecular and Cellular Biology, 24*, 3720–3733. doi:10.1128/MCB.24.9.3720-. 3733.2004.

Mravinac, B., Meštrović, N., Cavrak, V. V., & Plohl, M. (2011). TCAGG, an alternative telomeric sequence in insects. *Chromosoma, 120*, 367–376. doi:10.1007/s00412-011-0317-x.

Nielsen, L., & Edstrom, J. E. (1993). Complex telomere-associated repeat units in members of the genus *Chironomus* evolve from sequences similar to simple telomeric repeats. *Molecular and Cellular Biology, 13*, 1583–1589. doi:10.1128/MCB.13.3.1583.

Nielsen, L., Schmidt, E. R., & Edstrom, J. E. (1990). Subrepeats result from regional DNA sequence conservation in tandem repeats in *Chironomus* telomeres. *Journal of Molecular Biology, 216*, 577–584. doi:10.1016/0022-2836(90)90385-Y.

Okazaki, S., Tsuchida, K., Maekawa, H., Ishikawa, H., & Fujiwara, H. (1993). Identification of a pentanucleotide telomeric sequence, (TTAGG)n, in the silkworm *Bombyx mori* and in other insects. *Molecular and Cellular Biology, 13*, 1424–1432.

Okazaki, S., Ishikawa, H., & Fujiwara, H. (1995). Structural analysis of TRAS1, a novel family of telomeric repeat-associated retrotransposons in the silkworm, *Bombyx mori*. *Molecular and Cellular Biology, 15*, 4545–4552.

Osanai, M., Takahashi, H., Kojima, K. K., Hamada, M., & Fujiwara, H. (2004). Essential motifs in the 3′ untranslated region required for retrotransposition and the precise start of reverse transcription in non-long-terminal-repeat retrotransposon SART1. *Molecular and Cellular Biology, 24*, 7902–7913. doi:10.1128/MCB.24.18.7902-. 7913.2004.

Osanai, M., Kojima, K. K., Futahashi, R., Yaguchi, S., & Fujiwara, H. (2006). Identification and characterization of the telomerase reverse transcriptase of *Bombyx mori* (silkworm) and *Tribolium castaneum* (flour beetle). *Gene, 376*, 281–289. doi:10.1016/j.gene.2006.04.022.

Osanai-Futahashi, M., & Fujiwara, H. (2011). Coevolution of telomeric repeats and telomeric-repeat-specific non-LTR retrotransposons in insects. *Molecular Biology Evolution,*. doi:10.1093/molbev/msr135.

Osanai-Futahashi, M., Suetsugu, Y., Mita, K., & Fujiwara, H. (2008). Genome-wide screening and characterization of transposable elements and their distribution analysis in the silkworm, *Bombyx mori*. *Insect Biochemistry and Molecular Biology, 38*, 1046–1057. doi:10.1016/j.ibmb.2008.05.012.

Pardue, M. L., & DeBaryshe, P. G. (2003). Retrotransposons provide an evolutionarily robust non-telomerase mechanism to maintain telomeres. *Ann Rev Genet, 37*, 485–511. doi:10.1146/annurev.genet.38.072902.093115.

Pardue, M. L., & De Baryshe, P. G. (2011a). Adapting to life at the end of the line: how *Drosophila* telomeric retrotransposons cope with their job. *Mob Genet Elements, 1*, 128–134. doi:10.4161/mge/1.2.16914.

Pardue, M. L., & De Baryshe, P. G. (2011b). Retrotransposons that maintain chromosome ends. *Proceedings of the National academy of Sciences of the United States of America,*. doi:10.1073/pnas.1100278108.

Robertson, H. M., & Gordon, K. H. (2006). Canonical TTAGG-repeat telomeres and telomerase in the honey bee, *Apis mellifera*. *Genome Research, 16*, 1345–1351. doi:10.1101/gr.5085606.

Roth, C. W., Kobeski, F., Walter, M. F., & Biessmann, H. (1997). Chromosome end elongation by recombination in the mosquito *Anopheles gambiae*. *Molecular and Cellular Biology, 17*, 5176–5183.

Sasaki, T., & Fujiwara, H. (2000). Detection and distribution patterns of telomerase activity in insects. *European Journal of Biochemistry, 267*, 3025–3031.

Takahashi, H., & Fujiwara, H. (1999). Transcription analysis of the telomeric repeat-specific retrotransposons TRAS1 and SART1 of the silkworm *Bombyx mori*. *Nucleic Acids Research, 27*, 2015–2021. doi:10.1093/nar/27.9.2015.

Takahashi, H., & Fujiwara, H. (2002). Transplantation of target site specificity by swapping the endonuclease domains of two LINEs. *EMBO Journal, 21*, 408–417. doi:10.1093/emboj/21.3.408.

Takahashi, H., Okazaki, S., & Fujiwara, H. (1997). A new family of site-specific retrotransposons, SART1, is inserted into telomeric repeats of the silkworm, *Bombyx mori*. *Nucleic Acids Research, 25*, 1578–1584. doi:10.1093/nar/25.8.1578.

Vitkova, M., Kral, J., Traut, W., Zrzavy, J., & Marec, F. (2005). The evolutionary origin of insect telomeric repeats, (TTAGG)n. *Chromosome Research, 13*, 145–156. doi:10.1007/s10577-007-1910-y.

Zakian, V. A. (1995). Telomeres: beginning to understand the end. *Science, 270*, 1601–1607. doi:10.1126/science.270. 5242.1601.

Xia, Q., Zhou, Z., Lu, C., et al. (2004). A draft sequence for the genome of the domesticated silkworm (*Bombyx mo*ri). *Science, 306*, 1937–1940. doi:10.1126/science.1102210.

# Chapter 14
# Subtelomere Plasticity in the Bacterium *Streptomyces*

**Annabelle Thibessard and Pierre Leblond**

Like *Escherichia coli* or *Bacillus subtilis*, the bacterial models, most bacteria contain one or several circular chromosomes. The first exception identified to this rule is the case of the agent responsible for the Lyme disease, the spirochaete *Borrelia burgdorferi*, whose chromosome linearity has been revealed in 1989 (Ferdows and Barbour 1989). This bacterium carries a single and small chromosome of 0.91 Mb (Fraser et al. 1997). Then came the case of the plant pathogen *Agrobacterium tumefaciens*, a α-proteobacterium inducing crown gall tumours, which carries two chromosomes: one circular (2.84 Mb) and one linear (2.08 Mb) (Allardet-Servent et al. 1993; Wood et al. 2001). Gram-positive bacteria and more precisely actinomycetes (*Streptomyces, Rhodococcus*) also harbour exceptions to the bacterial circular chromosome dogma. The only genus where chromosomal linearity is an exclusive character is *Streptomyces*. Hence, all the wild-type strains so far studied harbour 8–12 Mb unique linear chromosomes (Hopwood 2006).

*Streptomyces* are common bacteria in soil and inhabit various ecological niches such as plant rhizosphere, mycorrhizosphere and mineralosphere. In soils, they interact with many other bacterial genera as well as different symbiotic, pathogenic and saprotrophic fungi. They produce a variety of extracellular enzymes and secondary metabolites, such as the well-known antibiotics by which *Streptomyces* play a key role in biogeochemical cycles and microbial communities' homoeostasis.

The *Streptomyces* chromosome exhibits a remarkable degree of genome plasticity associated with a very specific genetic organization: The genus is characterized

A. Thibessard (✉) · P. Leblond
DynAMic UMR UL-INRA 1128, IFR 110 EFABA, Université de Lorraine,
Faculté des Sciences et Technologies, BP 70239, Vandœuvre-lès-Nancy 54506, France
e-mail: annabelle.thibessard@univ-lorraine.fr

P. Leblond
e-mail: pierre.leblond@univ-lorraine.fr

E. J. Louis and M. M. Becker (eds.), *Subtelomeres*, DOI: 10.1007/978-3-642-41566-1_14, 

by a single large linear chromosome with a conserved central 'core' region and variable chromosomal arms including a large portion of genes assumed to derive from lateral gene transfer. Chromosomal rearrangement events observed in laboratory conditions were shown, using compared genomics, to constitute a driving force shaping the subtelomeres of the chromosome. The strong variability and specificity of the terminal region is assumed to contribute to the adaptation to soil biotic and abiotic environmental changes.

## 14.1 Bacterial Telomeres: Structure and Replication Mechanisms

Linear replicons face progressive loss of genetic information resulting from incomplete replication of the 3′ ends parental strands by the selection of specific strategies to ensure end replication and maintenance.

Hence, *Borrelia* harbour a linear chromosome with covalently closed hairpin ends. This unique end structure requires a telomere resolution mechanism involving the resolvase ResT to form hairpin telomeres from intermediates of replication. This mechanism is assumed to promote plasticity by stabilizing telomere fusions. Hence, telomere resolvases can generate Holliday junctions, which are recombination intermediates (Chaconas and Kobryn 2010). It was noticed that this hairpin is similar to those of some eukaryotic viruses (poxviruses) sharing the same tick vector. This suggests that, in *Borrelia*, linearity resulted from the acquisition of viral sequences by horizontal transfer.

*Streptomyces* linear replicons are typified as 'invertrons', structure consisting of perfect repeats (TIR for terminal inverted repeats) covalently attached by their 5′ end to specific proteins (Sakaguchi 1990). The size of the repeats varies widely from a minimum of about 160 nt constituting the strict telomere (e.g. *Streptomyces avermitilis* chromosome) to several hundreds of kilobases (1 Mb in *Streptomyces coelicolor* A3(2), 550 kb in *Streptomyces rimosus*, 210 kb in *Streptomyces ambofaciens* chromosomes, Hopwood 2006).

In *Streptomyces*, the linear replicons are replicated from a centrally located typical bacterial replication origin. Processing of the replication forks on both replichores leaves a 3′ end single-stranded DNA over 250–300 nt (Chaconas and Chen 2007). This single-stranded DNA region adopts hairpin structures capped with terminal proteins (TP)—two proteins, Tpg and Tap, were characterized—involved in the terminal replication mechanism called 'patching' (Bao and Cohen 2001). The TP complex is absolutely needed to perpetuate the replicons under their linear form. Indeed, when TPs are defective, survival plasmids have circularized. Two telomere maintenance systems have been characterized in *Streptomyces* which differ by the palindrome sequence and by the terminal protein actors. These observations suggest different evolutionary origins. Thus, the 'archetypal' telomere described by Huang et al. (2007) typifies the models *S. coelicolor* A3(2) and *S. lividans* 66 chromosomes; *S. ambofaciens* and *Streptomyces bingchenggensis* whose sequence was recently released share similar telomere motifs (Wang et al. 2010). The 'atypical' telomeres were found on the linear plasmid SCP1 from *S. coelicolor* and on the chromosomes of *Streptomyces griseus* (Goshi et al. 2002). Surprisingly, the telomere sequence of *Streptomyces cattleya* (NRLL 8057, Barbe et al. 2011) did not seem to belong to either group (our analysis). Hence, the chromosome sequence does not show any terminal redundancy, but instead shares one end (100 % identity over 90 nt) with its cognate linear plasmid (1.8 Mb). This is reminiscent of a hybrid chromosome resulting from chromosome–plasmid interaction, but no identity could be found at the other chromosome and plasmid ends ruling out this tempting hypothesis. This genome sequence deserves to be investigated to this regard.

While in rhodococci, linear replicons (plasmids and chromosome) share the 'archetypal' *Streptomyces* telomere and encode classical TP proteins (McLeod et al. 2006), no specific terminal structures were documented in *A. tumefaciens* (Wood et al. 2001).

### *14.1.1 Chromosomal Linearity is an Apomorphic Trait*

The scarcity of linearity and the diversity of the terminal structures argue in favour of the recent acquisition of this character in the some branches of bacterial phyla.

For instance, linearity is an exclusive feature in all wild-type streptomycetes species so far studied, while other actinomycetes including *Mycobacterium tuberculosis* (Cole et al. 1998), *Corynebacterium diphtheriae* (Cerdeno-Tarraga et al. 2003) and *Saccharopolyspora erythraea* (Oliynyk et al. 2007) harbour a circular chromosome. The central region of 8.7 Mb *S. coelicolor* linear chromosome shows a significant synteny with the entire circular chromosome of *M. tuberculosis* (4.4 Mb) (Bentley et al. 2002). This suggests that *Streptomyces* emerged from a bacterial ancestor harbouring a circular genome with a much smaller chromosome similar to that of contemporary mycobacteria. Linear replicons (plasmids, bacteriophages, etc.) widely distributed in gram-positive bacteria might have played a key role in linearization. Subsequently or at the same evolutionary time all or part of the subtelomeric regions might have been acquired. Hence, based on a wide comparison of chromosome structures of Actinomycetales, it was recently suggested that the streptomycetes chromosome expanded in a two-step evolutionary process with the early acquisition of a first chromosomal arm (called the Actinomycetales-specific arm) followed by the late acquisition of the second subtelomeric region (called the *Streptomyces*-specific arm, Kirby 2011).

In rhodococci, the situation is even more interesting to track down linearity acquisition (Letek et al. 2010). Hence, the largest genome sizes are associated with linearity. *R. jostii* and *R. opacus*, which are environmental species harbour large linear chromosomes of 9.7 and 8.17 Mb, respectively, while *R. equi* (an animal pathogen) has a unique circular chromosome of rather smaller size (5 Mb). The smaller genome do not show any trace of genome 'degeneration' (i.e. quasi absence of pseudogene), supporting the fact that the genomes of environmental species experience expansion (Letek et al. 2010). Indeed, linearity appears as an apomorphic trait; the largest chromosomes including the genetic information from the ancestor, which probably possessed a smaller circular chromosome. The acquisition of linearity could have been concomitant to that of a large portion of the contingent (accessory) genome likely to contribute to the occupation of environmental niches. Since linearity is not an exclusive character in rhodococci, the acquisition by horizontal transfer of DNA linearity is a tempting hypothesis.

Whether the chromosomal linearity favoured genome expansion remains an open question. If the latter hypothesis seems to be supported by the rhodococci and streptomycetes genome evolutions, the case of the large circular (8.2 Mb) *S. erythraea* chromosome seems to mitigate this scenario. Hence, while a 4.4-Mb region showed gene-order conservation with both *S. coelicolor* and *S. avermitilis*, the remaining part of its circular of chromosome showed all the features of gene acquisition and genome expansion (Oliynyk et al. 2007).

## *14.1.2 Interactions Between Linear Replicons*

Linearity itself may favour gene exchange between linear replicons since single crossovers are sufficient to promote DNA exchanges. Hence, *Streptomyces* linear plasmids carry traces of multiple recombination events including accretion events revealed by the presence of internal pseudo-telomeres as typified in the 356-kb linear SCP1 plasmid of *S. coelicolor* (Bentley et al. 2004). The recent analysis of the 1.8-Mb megaplasmid of *Streptomyces clavuligerus* also revealed that intimate interactions have shaped the plasmid and chromosome content, presumably by successive terminal exchanges (Medema et al. 2010). Earlier, similar interactions were shown to lead to terminal exchanges in *S. rimosus* transferring the oxytetracycline biosynthetic genes from the chromosome onto the 387-kb pZG101 linear plasmid (Gravius et al. 1994). Similarly, interactions between SCP1 (385 kb) and the chromosome of *S. coelicolor* A3(2) were shown to produce chimerical replicons: SCP1′ (SCP1′-*cysD*) reaching 1.85-Mb and a residual 7.2-Mb linear chromosome (Yamasaki and Kinashi 2004). The recombinant plasmid could not be cured out of the strain, indicating that the plasmid gained at least one essential gene. In *S. ambofaciens*, this kind of event probably promoted the specificity of the terminal inverted repeats (TIR): About 60 and 48 kb of DNA were unique to each of the strains studied (DSM40697 and ATCC23877, Choulet et al. 2006b). These interactions may have also led to genomic expansion as suggested in rhodococci and/or to the strong diversification of the ancestral contingent DNA acquired together with termini. Interactions may also be stimulated by the spatial organization of the chromosomal telomeres. Recently, telomeres (plasmid and chromosome) were found to interact in vivo at both intra- and intermolecular levels through the terminal protein complex (Tsai et al. 2011) confirming the early observation of the co-localization of terminal regions in cells (Yang and Losick 2001). These data are consistent with the circularity of the genetic map (Hopwood and Wright 1979). These interactions may also play a key role in the mobilization of chromosomal DNA via linear plasmids. The model called 'end first' (Chen 1996) was indeed recently supported by the report that linear plasmids mobilize chromosomal DNA from cells harbouring a linear chromosome, while no transfer is observed from cell harbouring a circular one (Lee et al. 2011).

### *14.1.3 Chromosome Compartmentalization and Subtelomere Plasticity in Streptomyces*

In most circular bacterial genomes, accessory genes are concentrated into variable regions, which are prone to DNA loss and gene replacement. These loci were frequently identified in the terminus of the chromosomal replication as in *E. coli* or *B. subtilis*. The high variability of these regions was understood as resulting from (1) the recombinogenic nature of these regions and from (2) the assumed low contribution of their gene content. The replication termini are indeed associated with a high frequency of recombination triggered by the slowdown of the replisome progression at Tus–*ter* sites (antihelicase activity of the Tus–*ter* complex) and by the chromosome dimer resolution (site-specific recombinase XerCD onto the *dif* sites). This region is also deemed to favour acquisition of exogenous sequences. Finally, the dispensability of these regions was evidenced in *E. coli* and *B. subtilis* by deletion in laboratory conditions over several hundreds of kilobases without significant loss of viability (Henson and Kuempel 1985; Iismaa and Wake 1987).

### *14.1.4 Subterminal Plasticity in Laboratory Conditions*

The study of chromosome plasticity associated with genetic instability was the first evidence of the specific genetic organization of the *Streptomyces* genome. Various phenotypic traits are highly unstable in streptomycetes (colony pigmentation, differentiation, antibiotic production, resistance to drugs, etc.) with spontaneous mutants arising at frequencies reaching 0.1–1 %. Spontaneous mutants show overlapping deletions extending over several hundreds of kilobases including the determinants of unstable phenotypes. The mutants also showed DNA amplifications (tandem repeats of amplifiable units of DNA, AUD) at high frequencies which were for long times the landmark of genetic instability. Selective pressure for DNA amplification is mostly unknown although the phenotype conferred by the amplification can sometimes be positively selected (e.g. increased antibiotic). The reiterative structure of the AUDs seems to favour intralocus recombination events (e.g. unequal crossovers) as supported by the decrease in DNA amplification frequencies in *recA* mutants (Volff and Altenbuchner 1997). The AUDs were located, together with the deletable loci, in the same chromosomal region which was called the 'unstable' region. This region which turned out to correspond to the subterminal regions of the chromosome: When the chromosome linearity was proved by Lin et al. (1993) in *S. lividans* and extended to our model *S. ambofaciens* (Leblond et al. 1996), the unstable region overlapped the subterminal regions of both arms by several hundreds of kilobases. In the same way, Windebrant et al. (2007) reported spontaneous terminal duplication exceeding 2,500 kb in *S. coelicolor*.

Mutants having lost both chromosomal telomeres and subtelomeres harboured circular chromosomes, while strains having lost only one arm were shown to keep their chromosome linear. Subterminal deletions internal to a chromosome arm were subsequently characterized. In some other cases, two deleted chromosomes were fused to give a giant duplicated chromosome (reaching 14–15 Mb vs. 8 Mb in the wild type, Wenner et al. 2003; Denapaite 2005). In *S. ambofaciens*, this structure was highly dynamic and produces a strong terminal diversity with TIRs varying in size from 5 to 1,440 kb (Wenner et al. 2003).

The high instability of this structure might result from the impairment of chromosome segregation in daughter cells triggered by the presence of two *cis*-located centromere-like onto the fusion chromosomes and was reminiscent of the break–fusion–bridge described by McClintock (1939) in maize. Interestingly, it was shown that the deficiency in FtsK which is a DNA translocase involved in chromosome segregation during division was associated with an increase in frequency of genome rearrangements in the terminal regions of *Streptomyces* chromosome (Wang et al. 2007). The dynamics of terminal inverted repeats (TIR) associated with the formation of double-strand breaks (DSB) suggests that the formation of TIR may result from recombinational repair of DSB by the break-induced replication mechanism (BIR) as described in yeast (Signon et al. 2001).

These DNA rearrangements resulted from different recombination mechanisms. While homologous recombination was involved in chromosomal arm translocations in *S. ambofaciens* and *S. griseus* (Uchida et al. 2003), transposition was involved in the formation of a large inversion in *S. griseus* (Murata et al. 2011). However, in most cases where several large DNA rearrangements were characterized at the nucleotide level, the involvement of illegitimate recombination was revealed by the finding of either short microhomologies (3–6 bp; Chen et al. 2010; Wenner et al. 2003) or even no nucleotide homology at the break points (Birch et al. 1991; Chen et al. 2010). This suggests the involvement of a single-stranded annealing (SSA) joining mechanism or a non-homologous end joining (NHEJ) repair event (see perspectives).

### *14.1.5 Gene-Order Conservation Along Streptomyces Chromosome*

The genomic comparison of complete genome sequences of Streptomycetes confirmed genome instability studies and showed that the central part is mostly conserved, while terminal regions are variable (Fig. 14.1a). Based on functional annotation of *S. coelicolor*, it was early noticed that the identifiable essential genes were concentrated in the central region ('core') while large terminal areas, the arms (1.6 and 2.3 Mb) did contain either genes with undetermined function or genes assumed to be contingent (Fig. 14.1b; Bentley et al. 2002). Later, the

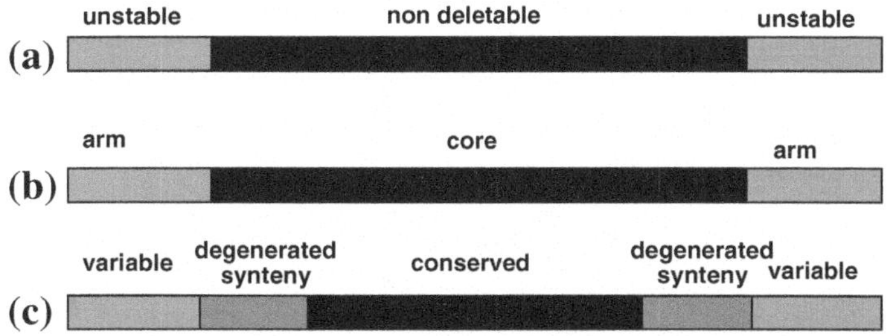


**Fig. 14.1** Schematic organization of the *Streptomyces* chromosome. **a** Definition of the core and arms of the linear chromosome based on functional annotation. This terminology was founded by Bentley et al. (2002) on *S. coelicolor*. **b** Delimitation of the unstable regions at the end of the chromosome through the studies on genetic instability in several species. **c** Definition of the conserved, variable and the degenerated synteny regions by compared genomics of differently related species by Choulet et al. (2006a). The schema does not reflect the respective sizes of the different chromosomal regions

comparison of *S. avermitilis* and *S. coelicolor* (Ikeda et al. 2003) gave a more precise picture with a conserved region constituting the core genome and variable regions overlapping the ends of the chromosome. This picture can be extrapolated to the newly sequenced *Streptomyces* genomes: *S. bingchenggensis* despite its large genome size shows a central conserved region of about 7.2 Mb and two extra large arms of 3.2 and 1.5 Mb for the right and left arms, respectively (Wang et al. 2010) (Fig. 14.2b). Conversely, *S. cattleya* NRRL 8057 which is characterized by the smallest chromosome ever identified up to now in the *Streptomyces* genus (6.3 Mb) shares the common organization with however shorter chromosomal arms (Fig. 14.2c). The organization of its 1.8-Mb linear megaplasmid revealed a complete absence of synteny with the chromosome of *S. coelicolor* (our analysis, Fig. 14.2c). Thus, the megaplasmid does not seem to derive from the interaction between the ancestral chromosome and a smaller linear plasmid.

The comparison of couples of genomes of species sharing more or less close phylogenetic relationships gave a more dynamic view of the evolution of the *Streptomyces* linear chromosome and suggested a specific and original evolutionary scenario. Hence, in a previous study, we compared four complete genome sequences and that, partial, of *S. ambofaciens*, to trace the molecular events, leading to terminal diversity in *Streptomyces* (Fig. 14.2d, e). For that purpose, the level and evolution along the genome of gene-order conservation were assessed by pairwise comparisons (Choulet et al. 2006a). The index used was the GOC, for gene-order conservation, developed by Rocha (2006) to study the architecture of bacterial genomes over evolutionary times. Hence, except for large genomic islands (e.g. 12 islands extending from 26 to 149 kb were specific of *S. coelicolor* compared to the related species *S. ambofaciens*) scattered in the core region, genome variability was found to be mostly confined to the chromosomal arms (Fig. 14.1b). Surprisingly, the variable regions increased in size when the compared species were chosen with more tenuous phylogenetic relationships.

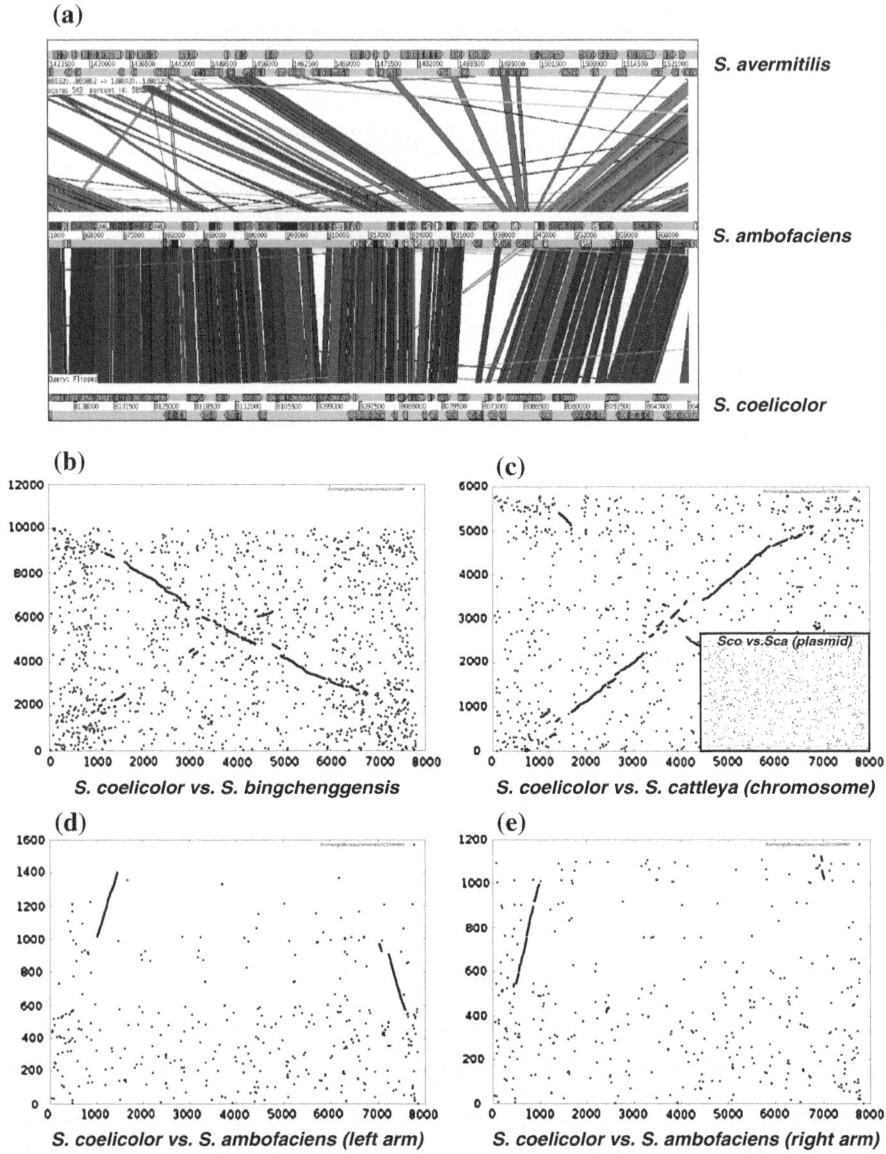


**Fig. 14.2** Subtelomere variability in *Streptomyces*. **a** Degenerated synteny between *S. avermitilis*, *S. ambofaciens* and *S. coelicolor* as revealed by protein-to-protein comparison Artemis Comparison Tools (Rutherford et al. 2000). Putative genes are represented by *arrows* in the *6 frames*, and pairs of homologues are linked by a *grey area*. The window presented here corresponds to a 100-kb region of the *Streptomyces ambofaciens* chromosome included into the degenerated syntenic regions compared with S. *coelicolor* and *S. avermitilis*. **b** Dot-plot comparison of *S. coelicolor* and *S. bingchenggensis*. Each *dot* represents the best reciprocal hit between the two compared genomes. The axes reflect gene numbers according to the annotation of the genomes. **c** Comparison of *S. coelicolor* versus *S. cattleya* (chromosome; main frame, 1.8-Mb linear plasmid, small window, Sco: *S. coelicolor*, Sca: *S. cattleya*). **d**, **e** Illustrate the comparison of *S. coelicolor* chromosome with the *left* and *right* chromosomal arms of *S. ambofaciens*

*S. ambofaciens*-specific region increased from 1.2 Mb (left and right arm sizes cumulated) when compared to the closely related *S. coelicolor* species (1.1 % of divergence between 16S rRNA) to almost 2 Mb when compared to the more distant *S. avermitilis* (2.9 % of divergence between 16S rRNA). Reciprocally, the size of the 'core' region, mostly assessed to be vertically inherited, tended to decrease. Although not all the genes present in the terminal regions are unique to each species, the levels of amino acid homology when existing are quite low, and suggest that these homologues are inherited from horizontal gene transfer (i.e. xenologues). These data suggest (1) that horizontal gene transfer is massive in this bacterial genus and (2) that recombination events promoting the integration of exogenous genetic material either preferentially occur or is less counter-selected (and consequently fixed) in the chromosomal arms; these two hypotheses being non-exclusive.

The central conserved and terminal-specific regions are connected by a region showing a progressive decrease in the gene-order conservation (Fig. 14.1c). This amazing phenomenon was called degenerated synteny. This decrease (intermediate values of GOC) reflected the occurrence of multiple insertions and deletions (indels) of short DNA stretches (including up to several genes) with a frequency that gradually increases until total loss of synteny (Fig. 14.2a). This phenomenon resulted from both the increase in the frequency of fixed DNA rearrangements and the incoming of exogenous information. The degenerated synteny did affect regions which were more internally located on the chromosome when the compared species were more distantly related. The replacement of terminal region by interaction with linear plasmids (as reported in several species) cannot explain by itself this situation. Indeed, such events would have led to the complete loss of synteny at the recombination point and consequently to a sudden fall of GOC.

In contrast, the progressive loss of synteny suggests that (1) integrations of DNA fragments cumulated in this region and (2) more and more integration events were fixed towards the ends of the chromosome. Two hypotheses can explain the contemporary situation: (1) DNA acquisitions and losses occurred all along the genome but are counter-selected according to the contribution of each locus; the corollary of this hypothesis is that the order of the loci along the terminal region reflects their contribution to the bacterium fitness. (2) The insertion and deletion events occur according to a gradient of increasing frequencies towards the ends of the linear chromosome. This would provide a powerful driving force that would have shaped *Streptomyces* chromosome, excluding essential genes from the terminal ends and cumulating accessory genes in the specific regions. The frequent rearrangements would also provide a low cost mean to test new versions of genes (chimeric genes, duplicated and diverged gene sequences) or gene clusters (reassortments of genes). Deleterious alleles or clusters could easily be eliminated through recombination. In both scenarios, the specific regions (over several hundreds of kilobases) would have been saturated with integration events, and the integration flux would have erased the assumed ancestral information.

The gene flux hypothesis is further supported by the composition bias of the subtelomeres. Hence, GC % of the species-specific DNA regions (about 650 kb on each chromosomal arm between *S. ambofaciens* and *S. coelicolor*) declines progressively towards the DNA ends. The lower GC content is a characteristic shared between mobile genetic elements (plasmids, bacteriophages, transposons) and genes inherited from horizontal transfer (Rocha and Danchin 2002). Another noticeable composition bias in favour of the gene flux hypothesis is that of mobile genetic elements: 45 and 79 % of the insertion sequences (IS) are terminally located in *S. avermitilis* and *S. coelicolor*, respectively (Chen et al. 2002).

### *14.1.6 Functions Encoded by Subtelomere Sequences*

As revealed by Konstantinidis and Tiedje (2004), genome expansion in bacterial genomes is correlated with the explosion of functions related to gene regulation and transport. In *S. ambofaciens*, while the terminal regions are assumed to be accessory regions (i.e. variable and dispensable), there is no marked enrichment in regulatory genes with 12.3 % of the CDS (including 27 putative alternative sigma factors) that is equivalent to analysis of the complete genome of *S. coelicolor* (Bentley et al. 2002).

In contrast, the subterminal regions of *S. ambofaciens* genome show a high level of gene redundancy. Hence, a functional categorization of the coding sequences distinguished 33 families with more than 5 paralogues for each in the terminal regions (e.g. the largest gene family corresponds to oxydoreductases). The subtelomeres also encode functions involved in organic polymer degradation (chitin, chitosane, cellulose). Interestingly in some cases, a function may be insured by two redundant gene clusters, one of them being orthologue of a cluster found in streptomyctes, the second one being a clear paralogue acquired by gene transfer (xenologue). Although the presence of paralogues in bacterial genomes is most often explained by gene transfer, gene duplication seems also to occur in the terminal regions. Hence, two sigma factor-encoding genes (*hasR/hasL*) sharing 98 % nucleotide identity were involved in a large chromosomal rearrangement (Fischer et al. 1998). Interestingly, the two chimeric genes formed by the rearrangements were found to be functional (Roth et al. 2004).

The most salient feature of the subtelomeres is probably their richness in gene clusters putatively or actually involved in secondary metabolism. This trend was noticed in *S. coelicolor* (Bentley et al. 2002) and *S. avermitilis* (Ikeda et al. 2003) and was also observed in *S. ambofaciens* with more than half (14 of the 23) of the putative secondary metabolite biosynthetic gene clusters located in the chromosomal arms representing roughly one-third of the total genome size (Aigle 2011). Interestingly in *S. erythraea* which possesses a circular chromosome, most of those gene clusters (21 of 25) were also found located in the 'non-core' region corresponding to half of the total genome size (Oliynyk et al. 2007). This overall striking distribution could reveal frequent transfers of all or part of the gene clusters. Exchange and recombination of such genes could lead to the creation of chimeric gene cluster and could participate to the diversification of secondary metabolites. Such a hybrid locus was found in *S. ambofaciens* (Pang et al. 2004).

Hence, the Type II PKS gene cluster responsible for the biosynthesis of kinamycins (Bunet et al. 2011) is flanked by a block of genes related to the biosynthesis mithramycin. If the synthesis of kinamycins relies on the only kinamycin gene block, the mithramycin block lacks one key function (acyl carrier protein, ACP). Whether this gene block is non-functional or participates to the synthesis of another metabolite is under investigation. A third adjacent gene block is involved in a complex regulation cascade (Bunet et al. 2011) and shows a high level of synteny with the regulatory subcluster controlling tylosin biosynthesis. This is surprising, since tylosin is a macrolide antibiotic produced by a Type I PKS gene cluster (Cundliffe et al. 2001). Horizontal gene transfer is indeed strongly suspected to have shaped this biosynthesis gene cluster.

### 14.1.7 Perspectives

The mechanisms driving the evolution of *Streptomyces* genome since the acquisition of linearity remains to be investigated. We learned from compared genomics that the genetic compartmentalization is extreme with a conserved and constrained central part and highly plastic chromosomal arms.

Our working hypothesis is that plasticity is intensified in the chromosomal arms through a spatiotemporal regulation of the different recombination mechanisms (i.e. homologous recombination, HR, and illegitimate recombination, IR). Beside homologous recombination (HR) which is ubiquitous in prokaryotes, illegitimate recombination pathways were revealed in the genome of about 20 % of the bacterial species whose genome was sequenced including *Streptomyces* (Rocha et al. 2005). While the roles and mechanisms of HR are well-known, illegitimate recombination by non-homologous end joining (NHEJ) has been described only recently in *B. subtilis* or *M. tuberculosis* (Weller et al. 2002; Della et al. 2004). This mechanism mainly involves two proteins (Pitcher et al. 2007): The first one, *ku*, interacts with broken DNA ends and recruits the second one, *ligD*, a multifunctional enzyme carrying at least a DNA polymerase domain, a ligase domain and sometimes a nuclease domain.

While HR is an accurate DSB repair system, NHEJ is mostly mutagenic and triggers endogenous rearrangements as well as exogenous DNA acquisition. In *Streptomyces*, the subterminal variability may well result from a higher tolerance of these regions for DNA rearrangements (error free or prone) rather from a higher frequency of recombination events.

Several hypotheses can be drawn to explain the formation of an 'indel' gradient. The first one, which is not our favourite explanation, is to imagine that the tolerance for DNA rearrangements is increasing towards the ends of the chromosome. The corollary would be that the contribution of the genes to bacterial fitness would decrease progressively in the subtelomere. The second relies on an increasing gradient of DSB occurrence towards the ends of the chromosome. Indeed, DSBs can indeed be triggered by (1) the frequent arrest of the replication fork, chromosomal ends providing natural replication termini, or (2) breaks induced during the partition of the daughter chromatids during sporulation (the sole differentiation phases accompanied by an active partitioning process). The latter hypothesis was supported by the report of increased subtelomeric instability in *ftsK* mutants of *S. coelicolor* (Wang et al. 2007), suggesting that guillotining occurred in the terminal regions in this partitioning defective context. Another origin of frequent recombinogenic ends may also be related to the structure of the incoming exogenous DNA. Hence, it was recently shown that a mechanism of transfer of chromosomal markers would be initiated from the ends of the linear chromosome (Lee et al. 2011). Linear conjugative plasmids indeed mobilize the linear chromosome from its telomeres thanks to interactions between terminal complexes (Tsai et al. 2011) as earlier suggested in the 'end first' model (Chen 1996). The broken ends of the exogenote would then initiate recombination with the recipient chromosome and trigger terminal replacement.

Finally, the gradient may result from a differential efficiency of DSB repair systems along the genome. Such a spatial regulation of repair systems was described in yeast where DSB are more efficiently repaired through HR when occurring in the telomere regions (Ricchetti et al. 2003), while IR keeps the same efficiency along the chromosome. Interestingly, repair of DSB by IR is accompanied by insertion of mitochondrial DNA (Ricchetti et al. 1999) opening the opportunity to insert foreign DNA.

In contrast in *Streptomyces*, we speculate that DNA recombination is favoured in the terminal regions through an increased DSB repair by IR. This could result from a transient (at some growth phase or in response to some stimuli) or constant higher efficiency of IR over HR in these regions. In *Mycobacterium*, a cross-regulation between HR and NHEJ pathways was recently revealed (Gupta et al. 2011). Hence, *ku* appears to suppress HR by binding to the DNA ends. This regulation seems asymmetric since while HR is elevated in *ku* deficient context, NHEJ does not significantly increase in HR mutants. The relative abundance of *ku* during the cell cycle or in response to environmental stimuli may thus influence the choice of DSB repair mechanisms and favour exogenous DNA insertion in specific conditions. In yeast, it is interesting to note that HR shows the highest efficiency to repair DSB when an intact sister chromatid is present, while NHEJ is increased during G1 phase. It was shown that the DNA end resection activities (processing the DNA ends at the DSB) varies along the cell cycle and that *ku* seems to play a key role in that phenomenon by regulating access to the ends by the DNA end resection complexes (Clerici et al. 2008). In *B. subtilis*, *ku* and *ligD* mutants were shown to be sensitive to exposition to damaging agents (Moeller et al. 2007) and the expression of the *ku* and *ligD* genes was found to be under the control of the fore-spore-specific sigma G factor (Wang et al. 2006). In the intracellular symbiont of legume plants *Sinorhizobium meliloti*, 4 *ku* and 5 *ligD* homologues were identified and involved in a NHEJ repair system functioning in free-living cells as well as bacteroid cells located in the host plant (Kobayashi et al. 2008). In both bacterial cases, the role of NHEJ could be to take over HR to carry out DSB repair when a single chromosome is present such as in unigenomic spores (*B. subtilis*) or in stationary phase of growth (*S. meliloti*).

In *Streptomyces*, in silico analyses predict the existence of a NHEJ repair system with a high complexity level. Hence, up to 4 *ku*-like genes are present in the sequenced genomes. No *ligD* gene is present as such, but genes encoding a single functional domain (putative polymerase or putative ligase) could be identified. Our current perspective is to study the relative contributions of the different DNA repair pathways in DSB repair along the chromosome and to understand their role in the mechanisms of subtelomere diversification and in bacterial adaptation.

**Acknowledgments** The UMR UL-INRA 1128 is supported by a grant overseen by the French National Research Agency (ANR) as part of the "Investissements d'Avenir" program (ANR-11-LABX-0002-01, Lab of Excellence ARBRE).

## References

Aigle, B., Bunet, R., Corre, C., Garenaux, A., Hotel, L., Huang, S., et al. (2011). Genome-guided exploration of *Streptomyces ambofaciens* secondary metabolism. In P. Dyson (Ed.), *Streptomyces: Molecular biology and biotechnology*. Swansea: Horizon Scientific Press.

Allardet-Servent, A., Michaux-Charachon, S., Jumas-Bilak, E., Karayan, L., & Ramuz, M. (1993). Presence of one linear and one circular chromosome in the *Agrobacterium tumefaciens* C58 genome. *Journal of Bacteriology, 175*(24), 7869–7874.

Bao, K., & Cohen, S. N. (2001). Terminal proteins essential for the replication of linear plasmids and chromosomes in *Streptomyces*. *Genes and Development, 15*(12), 1518–1527.

Barbe, V., Bouzon, M., Mangenot, S., Badet, B., Poulain, J., Segurens, B., et al. (2011). Complete genome sequence of *Streptomyces cattleya* NRRL 8057, a producer of antibiotics and fluorometabolites. *Journal of Bacteriology, 193*(18), 5055–5056.

Bentley, S. D., Brown, S., Murphy, L. D., Harris, D. E., Quail, M. A., Parkhill, J., et al. (2004). SCP1, a 356,023 bp linear plasmid adapted to the ecology and developmental biology of its host, *Streptomyces coelicolor* A3(2). *Molecular Microbiology, 51*(6), 1615–1628. 3949 [pii].

Bentley, S. D., Chater, K. F., Cerdeno-Tarraga, A. M., Challis, G. L., Thomson, N. R., James, K. D., et al. (2002). Complete genome sequence of the model actinomycete *Streptomyces coelicolor* A3(2). *Nature, 417*(6885), 141–147.

Birch, A., Hausler, A., Ruttener, C., & Hutter, R. (1991). Chromosomal deletion and rearrangement in *Streptomyces glaucescens*. *Journal of Bacteriology, 173*(11), 3531–3538.

Bunet, R., Song, L., Mendes, M. V., Corre, C., Hotel, L., Rouhier, N., et al. (2011). Characterization and manipulation of the pathway-specific late regulator AlpW reveals *Streptomyces ambofaciens* as a new producer of Kinamycins. *Journal of Bacteriology, 193*(5), 1142–1153.

Cerdeno-Tarraga, A. M., Efstratiou, A., Dover, L. G., Holden, M. T., Pallen, M., Bentley, S. D., et al. (2003). The complete genome sequence and analysis of *Corynebacterium diphtheriae* NCTC13129. *Nucleic Acids Research, 31*(22), 6516–6523.

Chaconas, G., & Chen, C. W. (2007). Replication of linear bacterial chromosomes: No longer going around in circles. In N. P. Higgins (Ed.), *The bacterial chromosome* (pp. 525–539). Washington, DC: ASM Press.

Chaconas, G., & Kobryn, K. (2010). Structure, function, and evolution of linear replicons in *Borrelia*. *Annual Review of Microbiology, 64*, 185–202. doi:10.1146/annurev.micro.112408.134037.

Chen, C. W. (1996). Complications and implications of linear bacterial chromosomes. *Trends in Genetics, 12*(5), 192–196.

Chen, C. W., Huang, C. H., Lee, H. H., Tsai, H. H., & Kirby, R. (2002). Once the circle has been broken: Dynamics and evolution of *Streptomyces* chromosomes. *Trends in Genetics, 18*(10), 522–529.

Chen, W., He, F., Zhang, X., Chen, Z., Wen, Y., & Li, J. (2010). Chromosomal instability in *Streptomyces avermitilis*: Major deletion in the central region and stable circularized chromosome. *BMC Microbiology, 10*, 198.

Choulet, F., Aigle, B., Gallois, A., Mangenot, S., Gerbaud, C., Truong, C., et al. (2006a). Evolution of the terminal regions of the *Streptomyces* linear chromosome. *Molecular Biology and Evolution, 23*(12), 2361–2369.

Choulet, F., Gallois, A., Aigle, B., Mangenot, S., Gerbaud, C., Truong, C., et al. (2006b). Intraspecific variability of the terminal inverted repeats of the linear chromosome of *Streptomyces ambofaciens*. *Journal of Bacteriology, 188*(18), 6599–6610.

Clerici, M., Mantiero, D., Guerini, I., Lucchini, G., & Longhese, M. P. (2008). The Yku70-Yku80 complex contributes to regulate double-strand break processing and checkpoint activation during the cell cycle. *EMBO Reports, 9*(8), 810–818.

Cole, S. T., Brosch, R., Parkhill, J., Garnier, T., Churcher, C., Harris, D., et al. (1998). Deciphering the biology of *Mycobacterium tuberculosis* from the complete genome sequence. *Nature, 393*(6685), 537–544.

Cundliffe, E., Bate, N., Butler, A., Fish, S., Gandecha, A., & Merson-Davies, L. (2001). The tylosin-biosynthetic genes of *Streptomyces fradiae*. *Antonie van Leeuwenhoek, 79*(3–4), 229–234.

Della, M., Palmbos, P. L., Tseng, H. M., Tonkin, L. M., Daley, J. M., Topper, L. M., et al. (2004). Mycobacterial *ku* and ligase proteins constitute a two-component NHEJ repair machine. *Science, 306*(5696), 683–685.

Denapaite, D., Paravic Radicevic, B., Hunter, I., Hranueli, D., & Cullum, J. (2005). Persistence of the chromosome end regions at low copy number in mutant strains of *Streptomyces rimosus* and *Streptomyces lividans*. *Food Technology Biotechnology, 43*, 9–17.

Ferdows, M. S., & Barbour, A. G. (1989). Megabase-sized linear DNA in the bacterium *Borrelia burgdorferi*, the Lyme disease agent. *Proceedings of National Academic of Science USA, 86*(15), 5969–5973.

Fischer, G., Wenner, T., Decaris, B., & Leblond, P. (1998). Chromosomal arm replacement generates a high level of intraspecific polymorphism in the terminal inverted repeats of the linear chromosomal DNA of *Streptomyces ambofaciens*. *Proceedings of National Academic of Science USA, 95*(24), 14296–14301.

Fraser, C. M., Casjens, S., Huang, W. M., Sutton, G. G., Clayton, R., Lathigra, R., et al. (1997). Genomic sequence of a Lyme disease spirochaete. *Borrelia burgdorferi. Nature, 390*(6660), 580–586.

Goshi, K., Uchida, T., Lezhava, A., Yamasaki, M., Hiratsu, K., Shinkawa, H., et al. (2002). Cloning and analysis of the telomere and terminal inverted repeat of the linear chromosome of *Streptomyces griseus*. *Journal of Bacteriology, 184*(12), 3411–3415.

Gravius, B., Glocker, D., Pigac, J., Pandza, K., Hranueli, D., & Cullum, J. (1994). The 387 kb linear plasmid pPZG101 of *Streptomyces rimosus* and its interactions with the chromosome. *Microbiology, 140*(Pt 9), 2271–2277.

Gupta, R., Barkan, D., Redelman-Sidi, G., Shuman, S., & Glickman, M. S. (2011). Mycobacteria exploit three genetically distinct DNA double-strand break repair pathways. *Molecular Microbiology, 79*(2), 316–330.

Henson, J. M., & Kuempel, P. L. (1985). Deletion of the terminus region (340 kb pairs of DNA) from the chromosome of *Escherichia coli*. *Proceedings of National Academic of Science USA, 82*(11), 3766–3770.

Hopwood, D. A. (2006). Soil to genomics: the *Streptomyces* chromosome. *Annual Review of Genetics, 40*, 1–23.

Hopwood, D. A., & Wright, H. M. (1979). Factors affecting recombinant frequency in protoplast fusions of *Streptomyces coelicolor*. *Journal of General Microbiology, 111*(1), 137–143.

Huang, C. H., Tsai, H. H., Tsay, Y. G., Chien, Y. N., Wang, S. L., Cheng, M. Y., et al. (2007). The telomere system of the *Streptomyces* linear plasmid SCP1 represents a novel class. *Molecular Microbiology, 63*(6), 1710–1718.

Iismaa, T. P., & Wake, R. G. (1987). The normal replication terminus of the *Bacillus subtilis* chromosome, *terC*, is dispensable for vegetative growth and sporulation. *Journal of Molecular Biology, 195*(2), 299–310.

Ikeda, H., Ishikawa, J., Hanamoto, A., Shinose, M., Kikuchi, H., Shiba, T., et al. (2003). Complete genome sequence and comparative analysis of the industrial microorganism *Streptomyces avermitilis*. *Nature Biotechnology, 21*(5), 526–531.

Kirby, R. (2011). Chromosome diversity and similarity within the Actinomycetales. *FEMS Microbiology Letters, 319*(1), 1–10.

Kobayashi, H., Simmons, L. A., Yuan, D. S., Broughton, W. J., & Walker, G. C. (2008). Multiple *ku* orthologues mediate DNA non-homologous end-joining in the free-living form and during chronic infection of *Sinorhizobium meliloti*. *Molecular Microbiology, 67*(2), 350–363.

Konstantinidis, K. T., & Tiedje, J. M. (2004). Trends between gene content and genome size in prokaryotic species with larger genomes. *Proceedings of National Academic of Science USA, 101*(9), 3160–3165.

Leblond, P., Fischer, G., Francou, F. X., Berger, F., Guerineau, M., & Decaris, B. (1996). The unstable region of *Streptomyces ambofaciens* includes 210 kb terminal inverted repeats flanking the extremities of the linear chromosomal DNA. *Molecular Microbiology, 19*(2), 261–271.

Lee, H. H., Hsu, C. C., Lin, Y. L., & Chen, C. W. (2011). Linear plasmids mobilize linear but not circular chromosomes in *Streptomyces*: Support for the 'end first' model of conjugal transfer. *Microbiology, 157*(Pt 9), 2556–2568.

Letek, M., Gonzalez, P., Macarthur, I., Rodriguez, H., Freeman, T. C., Valero-Rello, A., et al. (2010). The genome of a pathogenic *rhodococcus*: Cooptive virulence underpinned by key gene acquisitions. *PLoS Genet, 6*(9), e1001145.

Lin, Y. S., Kieser, H. M., Hopwood, D. A., & Chen, C. W. (1993). The chromosomal DNA of *Streptomyces lividans* 66 is linear. *Molecular Microbiology, 10*(5), 923–933.

McClintock, B. (1939). The behavior in successive nuclear divisions of a chromosome broken at meiosis. *Proceedings of National Academic of Science, 25*(8), 405–416.

McLeod, M. P., Warren, R. L., Hsiao, W. W., Araki, N., Myhre, M., Fernandes, C., et al. (2006). The complete genome of *Rhodococcus* sp. *RHA1* provides insights into a catabolic powerhouse. *Proceedings of National Academic of Science, 103*(42), 15582–15587.

Medema, M. H., Trefzer, A., Kovalchuk, A., van den Berg, M., Muller, U., Heijne, W., et al. (2010). The sequence of a 1.8-Mb bacterial linear plasmid reveals a rich evolutionary reservoir of secondary metabolic pathways. *Genome Biol Evol, 2*, 212–224.

Moeller, R., Stackebrandt, E., Reitz, G., Berger, T., Rettberg, P., Doherty, A. J., et al. (2007). Role of DNA repair by nonhomologous-end joining in *Bacillus subtilis* spore resistance to extreme dryness, mono- and polychromatic UV, and ionizing radiation. *Journal of Bacteriology, 189*(8), 3306–3311.

Murata, M., Uchida, T., Yang, Y., Lezhava, A., & Kinashi, H. (2011). A large inversion in the linear chromosome of *Streptomyces griseus* caused by replicative transposition of a new Tn3 family transposon. *Archives of Microbiology, 193*(4), 299–306.

Oliynyk, M., Samborskyy, M., Lester, J. B., Mironenko, T., Scott, N., Dickens, S., et al. (2007). Complete genome sequence of the erythromycin-producing bacterium *Saccharopolyspora erythraea* NRRL23338. *Nature Biotechnology, 25*(4), 447–453.

Pang, X., Aigle, B., Girardet, J. M., Mangenot, S., Pernodet, J. L., Decaris, B., et al. (2004). Functional angucycline-like antibiotic gene cluster in the terminal inverted repeats of the *Streptomyces ambofaciens* linear chromosome. *Antimicrobial Agents and Chemotherapy, 48*(2), 575–588.

Pitcher, R. S., Brissett, N. C., Picher, A. J., Andrade, P., Juarez, R., Thompson, D., et al. (2007). Structure and function of a mycobacterial NHEJ DNA repair polymerase. *Journal of Molecular Biology, 366*(2), 391–405.

Ricchetti, M., Dujon, B., & Fairhead, C. (2003). Distance from the chromosome end determines the efficiency of double strand break repair in subtelomeres of haploid yeast. *Journal of Molecular Biology, 328*(4), 847–862.

Ricchetti, M., Fairhead, C., & Dujon, B. (1999). Mitochondrial DNA repairs double-strand breaks in yeast chromosomes. *Nature, 402*(6757), 96–100. doi:10.1038/47076.

Rocha, E. P. (2006). Inference and analysis of the relative stability of bacterial chromosomes. *Molecular Biology and Evolution, 23*(3), 513–522.

Rocha, E. P., Cornet, E., & Michel, B. (2005). Comparative and evolutionary analysis of the bacterial homologous recombination systems. *PLoS Genetics, 1*(2), e15.

Rocha, E. P., & Danchin, A. (2002). Base composition bias might result from competition for metabolic resources. *Trends in Genetics, 18*(6), 291–294.

Roth, V., Aigle, B., Bunet, R., Wenner, T., Fourrier, C., Decaris, B., et al. (2004). Differential and cross-transcriptional control of duplicated genes encoding alternative sigma factors in *Streptomyces ambofaciens*. *Journal of Bacteriology, 186*(16), 5355–5365.

Rutherford, K., Parkhill, J., Crook, J., Horsnell, T., Rice, P., Rajandream, M. A., et al. (2000). Artemis: sequence visualization and annotation. *Bioinformatics, 16*(10), 944–945.

Sakaguchi, K. (1990). Invertrons, a class of structurally and functionally related genetic elements that includes linear DNA plasmids, transposable elements, and genomes of adeno-type viruses. *Microbiological Reviews, 54*(1), 66–74.

Signon, L., Malkova, A., Naylor, M. L., Klein, H., & Haber, J. E. (2001). Genetic requirements for RAD51- and RAD54-independent break-induced replication repair of a chromosomal double-strand break. *Molecular and Cellular Biology, 21*(6), 2048–2056.

Tsai, H. H., Huang, C. H., Tessmer, I., Erie, D. A., & Chen, C. W. (2011). Linear *Streptomyces* plasmids form superhelical circles through interactions between their terminal proteins. *Nucleic Acids Research, 39*(6), 2165–2174.

Uchida, T., Miyawaki, M., & Kinashi, H. (2003). Chromosomal arm replacement in *Streptomyces griseus*. *Journal of Bacteriology, 185*(3), 1120–1124.

Volff, J. N., & Altenbuchner, J. (1997). Influence of disruption of the *recA* gene on genetic instability and genome rearrangement in *Streptomyces lividans*. *Journal of Bacteriology, 179*(7), 2440–2445.

Wang, L., Yu, Y., He, X., Zhou, X., Deng, Z., Chater, K. F., et al. (2007). Role of an FtsK-like protein in genetic stability in *Streptomyces coelicolor* A3(2). *Journal of Bacteriology, 189*(6), 2310–2318.

Wang, S. T., Setlow, B., Conlon, E. M., Lyon, J. L., Imamura, D., Sato, T., et al. (2006). The forespore line of gene expression in *Bacillus subtilis*. *Journal of Molecular Biology, 358*(1), 16–37.

Wang, X. J., Yan, Y. J., Zhang, B., An, J., Wang, J. J., Tian, J., et al. (2010). Genome sequence of the milbemycin-producing bacterium *Streptomyces bingchenggensis*. *Journal of Bacteriology, 192*(17), 4526–4527.

Weller, G. R., Kysela, B., Roy, R., Tonkin, L. M., Scanlan, E., Della, M., et al. (2002). Identification of a DNA nonhomologous end-joining complex in bacteria. *Science, 297*(5587), 1686–1689.

Wenner, T., Roth, V., Fischer, G., Fourrier, C., Aigle, B., Decaris, B., et al. (2003). End-to-end fusion of linear deleted chromosomes initiates a cycle of genome instability in *Streptomyces ambofaciens*. *Molecular Microbiology, 50*(2), 411–425.

Widenbrant, E. M., Tsai, H. H., Chen, C. W., & Kao, C. M. (2007). *Streptomyces coelicolor* undergoes spontaneous chromosomal end replacement. *Journal of Bacteriology, 189*(24), 9117–9121.

Wood, D. W., Setubal, J. C., Kaul, R., Monks, D. E., Kitajima, J. P., Okura, V. K., et al. (2001). The genome of the natural genetic engineer *Agrobacterium tumefaciens* C58. *Science, 294*(5550), 2317–2323.

Yamasaki, M., & Kinashi, H. (2004). Two chimeric chromosomes of *Streptomyces coelicolor* A3(2) generated by single crossover of the wild-type chromosome and linear plasmid scp1. *Journal of Bacteriology, 186*(19), 6553–6559.

Yang, M. C., & Losick, R. (2001). Cytological evidence for association of the ends of the linear chromosome in *Streptomyces coelicolor*. *Journal of Bacteriology, 183*(17), 5180–5186.

# Chapter 15
# Genomics of Subtelomeres: Technical Problems, Solutions and the Future

**Marion M. Becker and Edward J. Louis**

**Abstract** Genome projects invariably are missing the subtelomeric regions due to cloning, sequencing and informatic problems. The repetitive nature of these sequences and the shared homology between different subtelomeres in most organisms preclude assembly of the regions leaving them incomplete. This is clearly an issue when a great deal of interesting biology, as seen in the previous chapters, involves the subtelomeres. In some cases, organism-specific or individual strain-specific approaches have been used to obtain material for sequencing and analysis as well as assembly. However, a more general approach applicable to most organisms has been elusive. Various cloning techniques have been tried and developed with mixed success. The most promising general approach involves individual telomere regions cloned as yeast artificial chromosomes (YACs), isolating them from all the other subtelomeres that have overlapping homology. This is time-consuming, and there is still a bottleneck in sequencing these. New sequencing technologies may solve many of the technical problems, yet there are still informatic problems with assembly because of the repetitive nature of the regions. Although progress has been made, the solution to efficient completion of subtelomeric regions will likely take combined approaches, such as deep sequencing in pedigrees to look for genetic linkage (associations) with known segregating sites near the ends of chromosomes. An efficient and cost-effective approach is needed before individual genome projects can move into population genomics involving the subtelomeres which is crucial for studies of diversity in parasites as well as in quantitative genetic studies in many organisms.

M. M. Becker (✉) · E. J. Louis
Centre for Genetic Architecture of Complex Traits, Department of Genetics, University of Leicester, Leicester LE1 7RH, UK
e-mail: MarionMBecker@gmail.com

E. J. Louis
e-mail: ejl21@le.ac.uk

E. J. Louis and M. M. Becker (eds.), *Subtelomeres*, DOI: 10.1007/978-3-642-41566-1_15, 

## 15.1 Introduction

Most genome projects, including the human genome, are incomplete as they typically are missing the subtelomeric regions. In whole genome shotgun libraries, subtelomeric sequence is frequently missing, rearranged or underrepresented, and despite enormous effort, gaps remain in the subtelomeres. This is true for the *Caenorhabditis elegans* project (Consortium 1998), the *Drosophila genome* project (Adams et al. 2000; Celniker et al. 2002), the human genome project (Riethman et al. 2001), the *Schizosaccharomyces pombe* project (Wood et al. 2002), the *Plasmodium falciparum* project (Gardner et al. 2002), the *Trypanosoma brucei* project (Berriman et al. 2005) and many others. This is not a coincidence, and the following chapter will highlight the problems and some of the solutions that have been used to close the gaps on some of these projects.

The problems historically and currently are as follows:

1. Lack of telomeric and subtelomeric clones
2. Difficulty in cloning large enough fragments to connect with genome contigs
3. Difficulty in sequencing clones
4. Difficulty in assembling sequences

Some of these difficulties have been solved in some cases, but there has been no general approach that solves all of these for a generic genome project, though for some fungi an efficient approach has been developed (Farman 2011; Farman and Leong 1995; Li et al. 2005).

### *15.1.1 Underrepresentation of Subtelomeres in Standard Libraries*

In the early days of genome projects, using first-generation Sanger sequencing and shotgun, cosmid and bacterial artificial chromosome (BAC) libraries, it was clear that telomeres and subtelomeres were underrepresented in these libraries by 10- to 100-fold (Becker et al. 2004). In the case of human subtelomeres, this is a consequence of the proximity to telomeres and the lack of restriction sites used in cloning procedures (Mefford and Trask 2002), the high GC content and the presence of telomeric repeats (Costa et al. 2009). *T. brucei* subtelomeres were more than 10-fold underrepresented in BAC libraries, and all clones isolated and sequenced were incomplete and in some cases rearranged (Berriman et al. 2002, 2005). This is due to several problems, one simply being the structure of the end of the chromosome, which needs to be enzymatically processed before being ligated into a vector. In addition, the telomere repeats are unstable in *Escherichia coli*. Even in BAC libraries, the presence of inverted repeats, AT-rich sequences and Z-DNA sequence structures are extremely unstable in *E. coli* (Kouprina et al. 2003) and cannot be cloned efficiently or at all in bacteria. Interestingly, in fungal genome projects, the telomere sequences were overrepresented in cosmid libraries yet not incorporated into assemblies due to the region being recalcitrant to shearing during shotgun cloning (Schwartz and Farman 2010).

### 15.1.2 Problems with Mapping Subtelomere Clones Onto the Genome

Mapping clones of subtelomeres, if they are obtained, back to the genome is difficult due to the shared homologies between subtelomeres. Inserts in the clones from various libraries are generally smaller than the large regions of shared homology between subtelomeres, precluding a direct mapping onto the core genome. This was true for many genome projects including *C. elegans* (Consortium 1998), *S. pombe* (Wood et al. 2002) and *T. brucei* (Berriman et al. 2005). Gap filling has helped some of these projects with a great deal of effort. One method for direct isolation of chromosomal fragments is PCR, which could help with some of the problems. However, the major limitation is that DNA fragments much larger than ~20 kb cannot be easily amplified due to the shearing of template sequence and the low processivity of thermostable DNA polymerases (Kouprina and Larionov 2006). If shared homology regions are large and there are gaps within several subtelomeres, then it would not be possible to uniquely amplify a given subtelomere region. For most genomes, the only cloning vectors with large enough inserts to cover the long regions of homology are BACs and yeast artificial chromosomes (YACs). BACs are problematic as described above, but YACs have other problems as described next.

### 15.1.3 Problems with Sequencing Large Subtelomere Clones

As will be described below, there are now good methods for obtaining large telomere containing subtelomeric clones that map back to the core of the genome, involving linear YACs. These clones are still difficult to deal with at the sequencing level for different reasons. As already mentioned, even in telomere-enriched libraries of large insert clones, the shotgun approach can result in lack of assembled sequence due to problems with shearing (Schwartz and Farman 2010). Second-generation sequencing that does not have a cloning step in a host organism overcomes much of this problem. However, there is another problem encountered with large-subtelomere-containing clones: subtelomeres cloned as YACs must be isolated away from the yeast host genome before sequencing. This usually involves separation of the YAC from the yeast chromosomes using pulsed-field gels, purification and then sequencing, either through a shotgun library or by second-generation methods without cloning in a vector (Hertz-Fowler et al. 2008). These sequence projects, both first generation and second generation, have large amounts of host yeast genome contamination as seen in a collection of *T. brucei* subtelomere clones (Hertz-Fowler et al. 2008). Better purification methods are required to make the whole process more efficient.

### *15.1.4 Assembly Problems in Repetitive Regions*

The final difficulty is not a subtelomere-specific problem but one involving repeats. Most sequence assemblers have difficulties when reads map to more than one contig location precluding completion of repetitive regions and in particular the subtelomeric regions of most organisms. For fungi where the subtelomeric repetitive regions are relatively small, informatic approaches have solved this problem (Li et al. 2005), but for many projects, including yeast, with smaller genomes than most fungi, the informatic approach has not been sufficient (Liti et al. 2009, 2013). Part of the solution is through isolated individual clones away from the rest of the genome as for example in *T. brucei* (Becker et al. 2004; Hertz-Fowler et al. 2008) and the human subtelomeres (Riethman et al. 2004), but this is time-consuming and still does not help with repeats within a given subtelomere.

## 15.2 Yeast to the Rescue and Other Solutions

Being the first eukaryotic genome project (Goffeau et al. 1996, 1997) and one of the first eukaryotic population genomics projects (Liti et al. 2009), yeast has exposed many of these problems and has lead to the development of various solutions to these problems.

### *15.2.1 Cloning Subtelomeres to Finish Genome Projects*

Approaches have been developed for specific projects, but these are not generally applicable to all genome projects. The yeast genome project is a case in point where these technical difficulties were solved with yeast-specific techniques. Once it was recognized that the standard library approach was not going to complete the ends, as the first eukaryotic chromosome sequenced (Oliver et al. 1992) actually was not complete (Louis 1994), each telomere was marked uniquely by inserting a vector into the telomere repeats (Louis and Borts 1995). Thirty-two strains, one for each telomere, were then used to either clone the sequences adjacent to the inserted vector as a plasmid or generate long-range PCR products using the vector as a unique anchor (Louis 1995). This approach successfully tagged and allowed the sequence and assembly of every telomere despite the shared homologies as the marked telomeres were at specific chromosome ends and the length of the clones or PCR products spanned the large regions of homology. The standard approach for the core of the genome had built contigs for each chromosome that were close enough to overlap the telomere-specific clones (Goffeau et al. 1996, 1997). For other projects, there were small-telomere-containing clones (Consortium 1998; Wood et al. 2002), but these did not map back onto the genomes as the core chromosome contigs did not extend far enough into the subtelomeres. The gaps for some projects such as *C. elegans* and *S. pombe* have been slowly filled with a great deal of effort. For genomes with more chromosomes and therefore more telomeres, such as humans, and those with complex repetitive structures of the subtelomeres, such as *P. falciparum* and *T. brucei*, the screening of libraries and subsequent gap filling would take many person-years of labour (see *T. brucei* for example (Becker et al. 2004; Berriman et al. 2005; Hertz-Fowler et al. 2008) and the human subtelomere project (Riethman 1997, 2008a, b; Riethman et al. 1989, 2001, 2004, 2005). For some genome projects, generalizable techniques have worked well such as in various fungi, where a fosmid library approach for enriching telomere-containing clones resulted in a large insert library that could be mapped back to the core genome (Dean et al. 2005; Farman 2011; Li et al. 2005; Rehmeyer et al. 2006; Wu et al. 2009).

### *15.2.2 Yeast to the Rescue I*

Yeast first came to the rescue by its ability to recognize telomere sequences from other organisms as a telomere (Szostak and Blackburn 1982) and its ability to tolerate large extra chromosomes of foreign genomes as YACs (Burke et al. 1987). In contrast to *E. coli*, AT-rich genomic fragments and long inverted repeats are more stable in yeast (Gardner et al. 2002; Glockner et al. 2002; Hayashi et al. 1993). The original YAC approach involved two telomeres with yeast markers, which were ligated onto the ends of large genomic fragments from another organism. Transformation of this into yeast usually resulted in a YAC with the markers at both ends, though occasionally YACs with only one added telomere came through and these had 'captured' a telomere from the other organism. A half-YAC approach to the human genome projects' telomeres and subtelomeres then ensued (Riethman et al. 1989), though it has taken over 15 years to almost complete a set of human telomere and subtelomere clones as YACs (Riethman 2008b; Riethman et al. 2001, 2004, 2005). This approach has worked for other organisms as well such as *Pneumocystis carinii*, which was unculturable at the time, but was labour intensive and not efficient (Underwood et al. 1996). One problem with this approach was that many of the YACs were chimeras, containing fragments from more than one genomic location (Larionov et al. 1996a, b; Underwood et al. 1996).

### *15.2.3 Yeast to the Rescue II: Transformation-Associated Recombination (TAR) Cloning*

Yeast came to the rescue a second time through the combination of several techniques and approaches into an elegant and generalizible method for cloning genes as well as large chromosomal fragments of up to 300 kb as YACs, called transformation-associated recombination (TAR) cloning (Larionov et al. 1996a). The technique is based on simultaneous transformation of yeast spheroplasts with genomic DNA and a TAR vector containing gene or sequence-specific targeting sequences (hooks) of minimally 60 bp length. Homologous recombination in the yeast cell between targeting sequences in the vector and the complimentary, chromosomal DNA sequence, captures the chromosomal region between the targeting hooks as circular YAC molecules (Kouprina and Larionov 2006; Larionov et al. 1996a; Noskov et al. 2001, 2003). These are faithfully replicated and segregated in the yeast host alongside its usual unaltered set of chromosomes (Kouprina and Larionov 2006). Positive recombinants are selected for further analysis using PCR or hybridization-based screening methods. Yeast has several properties that have made this possible. The high rates of homologous recombination and the use of positive and negative selectable markers (*HIS3, URA3)* produce positive YAC recombinants at high rates (up to 40 %) and suppress negative background caused by vector recircularization from non-homologous end-joining (Kouprina and Larionov 2006; Noskov et al. 2002). The transformation efficiency of yeast is 100 times higher than *E. coli* and some human DNA sequences, including coding DNA, that were instable in *E. coli,* and were therefore entirely missed, are stable in yeast (Kouprina et al. 2003). This combination led to the development of very efficient gap repair of plasmids transformed into yeast by recombination with homologous sequences in the genome (Ma et al. 1987). The amount of homology required could be very small and diverse, less than 60 base pairs and up to 15 % sequence divergence is tolerated. Although chromosomal recombination is greatly reduced in the presence of mismatches in the interacting DNA molecules, the recombination associated with transformation is tolerant of high levels of mismatches (up to 30 % divergence) (Larionov et al. 1994). These were combined to create the TAR cloning method which remarkably could result in large circular YACs using short Alu repeats as homologous targets in their vector (Larionov et al. 1996a). Numerous improvements over the years to increase efficiency have been made, including counter-selectable markers for enriching for recombinants, leaving the yeast origins out of the vector as the short consensus that functions in yeast can be found randomly in foreign genome sequences, development of specific sequence targets, vector improvements for movement into *E. coli* as BACs, etc. (Kouprina and Larionov 2006, 2008). The generation of YACs by TAR has several advantages over the original YAC method including no chimeras and the ability to target specific genomic locations.

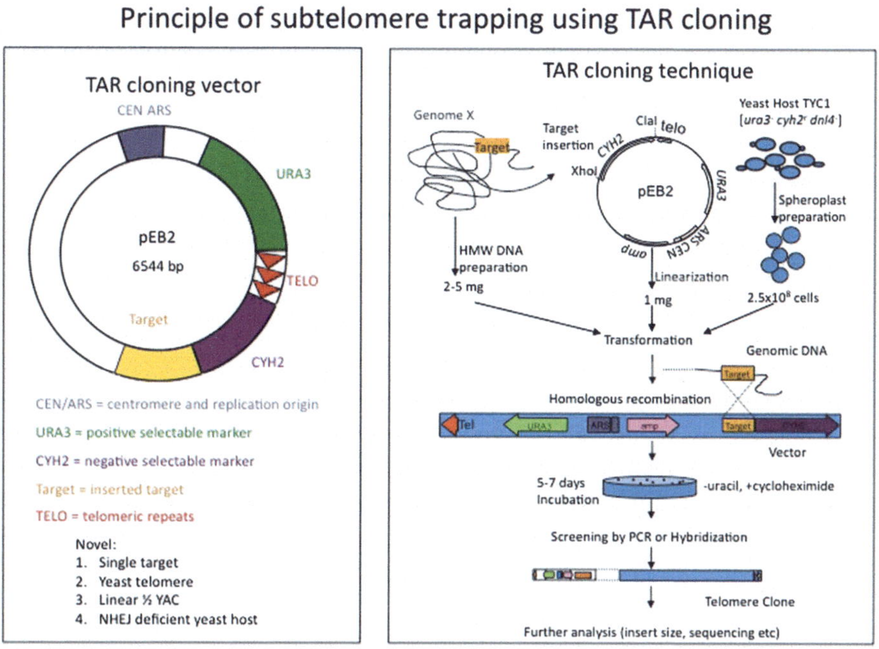


**Fig. 15.1** The principle of TAR cloning of subtelomeres. The trapping of subtelomeres by TAR is based on the use of a target sequence, homologous recombination and the fact that telomere repeats from most organisms function to seed new yeast telomeres, and the requirement of telomeres on both ends of a linear molecule in order to be maintained in yeast. As used to clone the subtelomeres of *T. brucei* (Becker et al. 2004), the vector contains all the necessary elements for replication and maintenance in yeast (centromere, one telomere, origin of replication (ARS element)), as well as a positive selectable marker (URA3), a counter-selectable marker (CYH2 conferring dominant sensitivity to cycloheximide) and the homologous target sequence. The yeast strain is deficient in ura3, has a recessive marker for cycloheximide resistance (cyh2r) and is deficient in the non-homologous end-joining specific ligase (dnl4). Co-transformation of the genomic DNA of interest with the linear vector into appropriately treated yeast cells results in colonies after 1 week on selective media. These are then ready for screening and further analysis

### *15.2.4 TAR Cloning of Subtelomeres*

The existing TAR cloning method was modified to specifically capture telomeric and subtelomeric sequences and was first successfully used to isolate *T. brucei* subtelomeres (Becker et al. 2004). Firstly, a purpose-built basic vector was constructed in which there was a single targeting hook and a yeast telomere. In addition, the vector contains the yeast selectable marker URA3, a counter-selectable marker CYH2, a yeast centromere and an origin of replication (ARS). As shown in Fig. 15.1, a successful targeted recombination event traps a telomere from the hook to the end of the chromosome of interest, as telomeres from virtually any organism function to seed new yeast telomeres. Selection for URA3 and against CYH2 enriches for the desired recombinants. Secondly, the deletion of the non-homologous end-joining specific ligase gene (DNL4) created a highly efficient yeast strain (ura3-, leu2-, dnl4-, cyh2-recessive resistance to cycloheximide), resulting in a threefold increase in the frequency of subtelomere clones over ligase-positive yeast (Becker and Louis unpublished results). The vector acts as a telomere trap whereby the yeast telomere repeats serve as a telomere on one end and the telomere of the genome of interest is captured using subtelomeric DNA sequence as targeting hook, resulting in linear half YACs, with a vector derived telomere on one and captured telomere on the other end of the linear molecule. The entire procedure, as shown in Fig. 15.1, can be conducted for multiple samples simultaneously within 7 days and typically generates thousands of recombinants.

The targeting sequence can be subtelomeric specific, either shared between many subtelomeres or specific to a given chromosome end, or they can be generic repeated elements such as transposable elements, few genomes being without any. Using the shared promoter region for the blood stream form expression sites (BES) of *T. brucei*, most expression sites of several isolates of Trypanosoma have been isolated (Becker et al. 2004; Young et al. 2008) and subsequently sequenced (Hertz-Fowler et al. 2008). The use of more dispersed transposable element repeats has successfully been used on *T. brucei*, *Brugia Malaya* and the planarian *Schmidtea mediterranea* (Becker and Louis, unpublished). Chromosome-end-specific targets have been used to clone individual telomeres in *S. pombe* (Becker and Louis, unpublished) as well as *T. brucei* (see databases (Aslett et al. 2010; Logan-Klumpler et al. 2012)). The use of two hooks flanking subtelomeric genes of the VAR gene family of *P. falciparum* successfully generated a library of the diverse flanking subtelomeric genes virulence factor from a novel isolate and could be used to assess diversity in endemic areas of infection (Gaida et al. 2011).

The frequency of subtelomere-positive clones in these TAR clone libraries was up to 30 % which is up to 100-fold higher than in many standard libraries. This demonstrates that subtelomeres and telomeres can be cloned from any genome even when little information is available using any the following 3 basic strategies: (1) For genomes that contain a known conserved subtelomeric sequence of at least 60 bp, multiple subtelomeres can be cloned simultaneously using a single TAR vector containing this element as targeting hook. This has been used to isolate subtelomeric blood stream form expression sites (BES) from *T. brucei* and *Trypanosoma brucei gambiense* using a conserved promoter element found at all BESs as targeting hook. These subtelomere libraries contained BES-positive clones at a frequency of up to 26 % with clone sizes ranging from 20 to 150 kb. This provided valuable insight into the architecture of BESs and aspects of their use in host adaptation and immune evasion (Becker et al. 2004; Hertz-Fowler et al. 2008; Young et al. 2008). (2) The cloning of specific telomeres using a unique sequence as targeting hook is applicable if telomere-proximal sequences are available that can be used to construct TAR vectors containing chromosome-end-specific targets. This method was successfully used to isolate up to 230 kb of subtelomeric regions of 14 missing chromosome ends of *T. brucei* for the genome project using the end-most unique sequences of chromosome-specific contigs (see genome project databases (Aslett et al. 2010; Logan-Klumpler et al. 2012)). Here, the frequency of positive clones is less, 5.4 %, but still significantly more than standard libraries. (3) The cloning of multiple telomeres using dispersed repeated sequences or mobile genetic elements as targets is a useful strategy to clone subtelomeres in the absence of subtelomeric sequence information. This was successful in cloning subtelomeres from *T. brucei* using the 197-bp-RIME-A element and from *Brugia malayii* using the highly frequent HhaI element (Becker and Louis unpublished).

## 15.3 Bottlenecks

Despite the development of new cloning strategies including TAR cloning to generate subtelomere libraries, there are still bottlenecks in analyzing these clones, which prevent a wider use and affordable high-throughput approaches. Firstly, the purification of subtelomeric clones away from the yeast host genome is a time-consuming process, which involves several rounds of separation and isolation by pulsed-field gel electrophoresis with poor yields of enriched DNA. This has turned out to be very inefficient, for example taking 4 years to sequence a small set of the BES clones (Becker et al. 2004; Hertz-Fowler et al. 2008). The telomere-specific TAR clones for the genome project took longer and are still being analysed (Aslett et al. 2010; Logan-Klumpler et al. 2012). Secondly, the assembly of subtelomeres is still difficult due to their mosaic and repetitive nature. Even with subtelomeric sequences isolated away from others, the assembly of these regions has proven difficult due to the internal repetitive regions. Current assemblers cannot handle this complex repetitive structure. Developing such methodologies is particularly pertinent, considering the increasing reliance on genomic data generated using second-generation sequencing platforms with diminishing resources dedicated to targeted finishing, which is traditionally the only realistic way of tackling assemblies of subtelomeric regions. This is even more of a problem when many individuals are to be sequenced such as with the 1,000 human genome project (Kuehn 2008) or the population genomics of yeast project (Liti et al. 2009) with the desire to map genetic causes of phenotypic variation.

### *15.3.1 Possible Solutions*

Some of the technical problems will be solved soon or could have solutions in existing technologies. Underrepresentation of telomeric and subtelomeric sequences is likely no longer a problem as second-generation sequencing has no cloning steps in *E. coli*. The short reads of most approaches exacerbates the assembly problem; however, the isolation of individual subtelomeres away from the rest of the subtelomeres of an organism helps with some of the assembly issues. There is still the issue of purification of the clones from the yeast host. There are a few potential solutions to this:

1. Sequence the whole yeast genome along with the YAC without any purification. Although the YAC DNA will represent only 2 % of the DNA to be sequenced, current costs and coverage with short second-generation reads make this an affordable and relatively fast solution despite the obvious inefficiency.
2. Oligo-affinity enrichment: Each subtelomere containing YAC has the unique cloning vector at one end. Hybridization with a high-affinity oligo, using custom-made peptide nucleic acids (Chandler et al. 2000) for example, can be used to enrich for the vector and its attached subtelomeric DNA. This may not efficiently enrich the sequences far from the vector on long clones.
3. Use the old standard of CsCl gradients for genome with a different GC content than yeast.
4. Subtract the yeast genome DNA by targeted affinity capture leaving the subtelomeric YAC in solution.

The problem of assembly of repetitive regions is a generic one and in projects where the subtelomeres are not individually cloned remains a big problem, particularly with shorter reads of the current sequencing technologies. A possible solution will come through third-generation sequencing of single long molecules, which may span the shared homology regions of telomeres.

## 15.4 The Future

The study and analysis of subtelomeres has come a long way, and there are now reasonably efficient approaches towards completing individual genomes. One of the remaining big challenges will be population genomics, genome-wide association studies and quantitative genetics involving the subtelomeres. For this, there will have to be more rapid and efficient high-throughput means to obtaining the ends of chromosomes for many individuals. Yeast is a case in point where advances are being made in determining the underlying genetic cause of phenotypic variation. In genetic crosses between 4 different strains of yeast, used to map the genes responsible for a number of phenotypes, 25 % of the genes responsible for any given phenotype mapped beyond the last known segregating marker (Cubillos et al. 2011). This is a significant lack of understanding of quantitative traits and is likely to hold in other organisms such as humans and the study of disease-causing loci through genome-wide association studies. In yeast, the missing subtelomeric sequences represent about 8 % of the genome, indicating an enrichment of genes of interest in this unknown genomic region (Liti and Louis 2012). Even if there is no enrichment for such loci in human studies, there must be a great deal of genetic information on polygenic traits and disease in humans missing as they are in the subtelomeric regions. Not only are the subtelomeres interesting in their own right, having exciting biology as seen in the previous chapters, there is more biology to learn that we are not even aware of yet.

## References


Adams, M. D., Celniker, S. E., Holt, R. A., Evans, C. A., Gocayne, J. D., Amanatides, P. G., et al. (2000). The genome sequence of Drosophila melanogaster. *Science, 287*, 2185–2195.

Aslett, M., Aurrecoechea, C., Berriman, M., Brestelli, J., Brunk, B. P., Carrington, M., et al. (2010). TriTrypDB: a functional genomic resource for the Trypanosomatidae. *Nucleic Acids Research, 38*, D457–D462.

Becker, M., Aitcheson, N., Byles, E., Wickstead, B., Louis, E., & Rudenko, G. (2004). Isolation of the repertoire of VSG expression site containing telomeres of Trypanosoma brucei 427 using transformation-associated recombination in yeast. *Genome Research, 14*, 2319–2329.

Berriman, M., Ghedin, E., Hertz-Fowler, C., Blandin, G., Renauld, H., Bartholomeu, D. C., et al. (2005). The genome of the African trypanosome Trypanosoma brucei. *Science, 309*, 416–422.

Berriman, M., Hall, N., Sheader, K., Bringaud, F., Tiwari, B., Isobe, T., et al. (2002). The architecture of variant surface glycoprotein gene expression sites in Trypanosoma brucei. *Molecular and Biochemical Parasitology, 122*, 131–140.

Burke, D. T., Carle, G. F., & Olson, M. V. (1987). Cloning of large segments of exogenous DNA into yeast by means of artificial chromosome vectors. *Science, 236*, 806–812.

Celniker, S. E., Wheeler, D. A., Kronmiller, B., Carlson, J. W., Halpern, A., Patel, S., et al. (2002). Finishing a whole-genome shotgun: release 3 of the Drosophila melanogaster euchromatic genome sequence. *Genome Biology, 3*, RESEARCH0079.

Chandler, D. P., Stults, J. R., Anderson, K. K., Cebula, S., Schuck, B. L., & Brockman, F. J. (2000). Affinity capture and recovery of DNA at femtomolar concentrations with peptide nucleic acid probes. *Analytical Biochemistry, 283*, 241–249.

Consortium, C. E. G. (1998). Genome sequence of the nematode *C. elegans*: A platform for investigating biology. *Science, 282*, 2012–2018.

Costa, V., Casamassimi, A., Roberto, R., Gianfrancesco, F., Matarazzo, M. R., D'Urso, M., et al. (2009). DDX11L: A novel transcript family emerging from human subtelomeric regions. *BMC Genomics, 10*, 250.

Cubillos, F. A., Billi, E., Zorgo, E., Parts, L., Fargier, P., Omholt, S., et al. (2011). Assessing the complex architecture of polygenic traits in diverged yeast populations. *Molecular Ecology, 20*, 1401–1413.

Dean, R. A., Talbot, N. J., Ebbole, D. J., Farman, M. L., Mitchell, T. K., Orbach, M. J., et al. (2005). The genome sequence of the rice blast fungus Magnaporthe grisea. *Nature, 434*, 980–986.

Farman, M. L. (2011). Targeted cloning of fungal telomeres. *Methods in Molecular Biology, 722*, 11–31.

Farman, M. L., & Leong, S. A. (1995). Genetic and physical mapping of telomeres in the rice blast fungus, Magnaporthe grisea. *Genetics, 140*, 479–492.

Gaida, A., Becker, M. M., Schmid, C. D., Buhlmann, T., Louis, E. J., & Beck, H. P. (2011). Cloning of the repertoire of individual Plasmodium falciparum var genes using transformation associated recombination (TAR). *PLoS ONE, 6*, e17782.

Gardner, M. J., Hall, N., Fung, E., White, O., Berriman, M., Hyman, R. W., et al. (2002). Genome sequence of the human malaria parasite Plasmodium falciparum. *Nature, 419*, 498–511.

Glockner, G., Eichinger, L., Szafranski, K., Pachebat, J. A., Bankier, A. T., Dear, P. H., et al. (2002). Sequence and analysis of chromosome 2 of Dictyostelium discoideum. *Nature, 418*, 79–85.

Goffeau, A., Barrell, B. G., Bussey, H., Davis, R. W., Dujon, B., Feldmann, H., et al. (1996). Life with 6000 genes. *Science, 274*(546), 547–563.

Goffeau, A., et al. (1997). The yeast genome directory. *Nature, 387*, 1–105.

Hayashi, Y., Heard, E., & Fried, M. (1993). A large inverted duplicated DNA region associated with an amplified oncogene is stably maintained in a YAC. *Human Molecular Genetics, 2*, 133–138.

Hertz-Fowler, C., Figueiredo, L. M., Quail, M. A., Becker, M., Jackson, A., Bason, N., et al. (2008). Telomeric expression sites are highly conserved in Trypanosoma brucei. *PLoS ONE, 3*, e3527.

Kouprina, N., & Larionov, V. (2006). TAR cloning: insights into gene function, long-range haplotypes and genome structure and evolution. *Nature Reviews Genetics, 7*, 805–812.

Kouprina, N., & Larionov, V. (2008). Selective isolation of genomic loci from complex genomes by transformation-associated recombination cloning in the yeast Saccharomyces cerevisiae. *Nature Protocols, 3*, 371–377.

Kouprina, N., Leem, S. H., Solomon, G., Ly, A., Koriabine, M., Otstot, J., et al. (2003). Segments missing from the draft human genome sequence can be isolated by transformation-associated recombination cloning in yeast. *EMBO Reports, 4*, 257–262.

Kuehn, B. M. (2008). 1000 Genomes Project promises closer look at variation in human genome. *JAMA, 300*, 2715.

Larionov, V., Kouprina, N., Eldarov, M., Perkins, E., Porter, G., & Resnick, M. A. (1994). Transformation-associated recombination between diverged and homologous DNA repeats is induced by strand breaks. *Yeast, 10*, 93–104.

Larionov, V., Kouprina, N., Graves, J., Chen, X. N., Korenberg, J. R., & Resnick, M. A. (1996a). Specific cloning of human DNA as yeast artificial chromosomes by transformation-associated recombination. *Proceedings of the National Academy of Sciences, 93*, 491–496.

Larionov, V., Kouprina, N., Graves, J., & Resnick, M. A. (1996b). Highly selective isolation of human DNAs from rodent-human hybrid cells as circular yeast artificial chromosomes by transformation-associated recombination cloning. *Proceedings of the National Academy of Sciences, 93*, 13925–13930.

Li, W., Rehmeyer, C. J., Staben, C., & Farman, M. L. (2005). TERMINUS–telomeric end-read mining in unassembled sequences. *Bioinformatics, 21*, 1695–1698.

Liti, G., Carter, D. M., Moses, A. M., Warringer, J., Parts, L., James, S. A., et al. (2009). Population genomics of domestic and wild yeasts. *Nature, 458*, 337–341.

Liti, G., & Louis, E. J. (2012). Advances in quantitative trait analysis in yeast. *PLoS Genetics, 8*, e1002912.

Liti, G., Nguyen Ba, A. N., Blythe, M., Muller, C. A., Bergstrom, A., Cubillos, F. A., et al. (2013). High quality de novo sequencing and assembly of the Saccharomyces arboricolus genome. *BMC Genomics, 14*, 69.

Logan-Klumpler, F. J., De Silva, N., Boehme, U., Rogers, M. B., Velarde, G., McQuillan, J. A., et al. (2012). GeneDB–an annotation database for pathogens. *Nucleic Acids Research, 40*, D98–108.

Louis, E. J. (1994). Corrected sequence for the right telomere of *Saccharomyces cerevisiae* chromosome III. *Yeast, 10*, 271–274.

Louis, E. J. (1995) Use of expand PCR system to complete the telomeres for the yeast genome project. *Biochemica, 3*, 25–26.

Louis, E. J., & Borts, R. H. (1995). A complete set of marked telomeres in *Saccharomyces cerevisiae* for physical mapping and cloning. *Genetics, 139*, 125–136.

Ma, H., Kunes, S., Schatz, P. J., & Botstein, D. (1987). Plasmid construction by homologous recombination in yeast. *Gene, 58*, 201–216.

Mefford, H. C., & Trask, B. J. (2002). The complex structure and dynamic evolution of human subtelomeres. *Nature Reviews Genetics, 3*, 91–102.

Noskov, V., Kouprina, N., Leem, S. H., Koriabine, M., Barrett, J. C., & Larionov, V. (2002). A genetic system for direct selection of gene-positive clones during recombinational cloning in yeast. *Nucleic Acids Research, 30*, E8.

Noskov, V. N., Koriabine, M., Solomon, G., Randolph, M., Barrett, J. C., Leem, S. H., et al. (2001). Defining the minimal length of sequence homology required for selective gene isolation by TAR cloning. *Nucleic Acids Research, 29*, E32.

Noskov, V. N., Kouprina, N., Leem, S. H., Ouspenski, I., Barrett, J. C., & Larionov, V. (2003). A general cloning system to selectively isolate any eukaryotic or prokaryotic genomic region in yeast. *BMC Genomics, 4*, 16.

Oliver, S. G., Van der Aart, Q. J. M., Agostoni-Carbone, M. L., Aigle, M., Alberghina, L., Alexandraki, D., et al. (1992). The complete DNA sequence of yeast chromosome III. *Nature, 357*, 38–46.

Rehmeyer, C., Li, W., Kusaba, M., Kim, Y. S., Brown, D., Staben, C., et al. (2006). Organization of chromosome ends in the rice blast fungus, Magnaporthe oryzae. *Nucleic Acids Research, 34*, 4685–4701.

Riethman, H. (1997). Closing in on telomeric closure. *Genome Research, 7*, 853–855.

Riethman, H. (2008a). Human subtelomeric copy number variations. *Cytogenet Genome Res, 123*, 244–252.

Riethman, H. (2008b). Human telomere structure and biology. *Annual Review of Genomics and Human Genetics, 9*, 1–19.

Riethman, H., Ambrosini, A., Castaneda, C., Finklestein, J., Hu, X. L., Mudunuri, U., et al. (2004). Mapping and initial analysis of human subtelomeric sequence assemblies. *Genome Research, 14*, 18–28.

Riethman, H., Ambrosini, A., & Paul, S. (2005). Human subtelomere structure and variation. *Chromosome Research, 13*, 505–515.

Riethman, H. C., Moyzis, R. K., Meyne, J., Burke, D. T., & Olson, M. V. (1989). Cloning human telomeric DNA fragments into Saccharomyces cerevisiae using a yeast-artificial-chromosome vector. *Proceedings of the National Academy of Sciences, 86*, 6240–6244.

Riethman, H. C., Xiang, Z., Paul, S., Morse, E., Hu, X. L., Flint, J., et al. (2001). Integration of telomere sequences with the draft human genome sequence. *Nature, 409*, 948–951.

Schwartz, S. L., & Farman, M. L. (2010). Systematic overrepresentation of DNA termini and underrepresentation of subterminal regions among sequencing templates prepared from hydrodynamically sheared linear DNA molecules. *BMC Genomics, 11*, 87.

Szostak, J. W., & Blackburn, E. H. (1982). Cloning yeast telomeres on linear plasmid vectors. *Cell, 29*, 245–255.

Underwood, A. P., Louis, E. J., Borts, R. H., Stringer, J. R., & Wakefield, A. E. (1996). Pneumocystis carinii telomere repeats are composed of TTAGGG and the subtelomeric sequence contains a gene encoding the major surface glycoprotein. *Molecular Microbiology, 19*, 273–281.

Wood, V., Gwilliam, R., Rajandream, M. A., Lyne, M., Lyne, R., Stewart, A., et al. (2002). The genome sequence of Schizosaccharomyces pombe. *Nature, 415*, 871–880.

Wu, C., Kim, Y. S., Smith, K. M., Li, W., Hood, H. M., Staben, C., et al. (2009). Characterization of chromosome ends in the filamentous fungus Neurospora crassa. *Genetics, 181*, 1129–1145.

Young, R., Taylor, J. E., Kurioka, A., Becker, M., Louis, E. J., & Rudenko, G. (2008). Isolation and analysis of the genetic diversity of repertoires of VSG expression site containing telomeres from Trypanosoma brucei gambiense, *T. b. brucei* and *T. equiperdum*. *BMC Genomics, 9*, 385.

The manufacturer's authorised representative in the EU is Springer Nature Customer Service Centre GmbH, Europaplatz 3, 69115 Heidelberg, Germany. If you have any concerns regarding our products, please contact ProductSafety@springernature.com

Printed and bound by CPI Group (UK) Ltd, Croydon, CR0 4YY
01/10/2026
02225974-0011